DIGITAL
ELECTRONICS

DIGITAL
ELECTRONICS

THIRD EDITION

ROGER L. TOKHEIM
Henry Sibley High School

McGRAW-HILL PUBLISHING COMPANY

New York   Atlanta   Dallas   St. Louis   San Francisco
Auckland   Bogotá   Caracas   Hamburg   Lisbon
London   Madrid   Mexico   Milan   Montreal   New Delhi
Paris   San Juan   São Paulo   Singapore
Sydney   Tokyo   Toronto

Sponsoring Editor: Gordon Rockmaker
Editing Supervisor: Angela Piliouras
Design and Art Supervisor/
   Cover and Interior Designer: Annette Mastrolia-Tynan
Production Supervisor: Mirabel Flores

Cover Illustration: Noi Viva Design

ACKNOWLEDGMENTS

The *Basic Skills in Electricity and Electronics* series was conceived and developed through the talents and energies of many individuals and organizations.

The original, on-site classroom testing of the texts and manuals in this series was conducted at the Burr D. Coe Vocational Technical High School, East Brunswick, New Jersey; Chantilly Secondary School, Chantilly, Virginia; Nashoba Valley Technical High School, Westford, Massachusetts; Platt Regional Vocational Technical High School, Milford, Connecticut; and the Edgar Thomson, Irvin Works of the United States Steel Corporation, Dravosburg, Pennsylvania. Postpublication testing took place at the Alhambra High School, Phoenix, Arizona; St. Helena High School, St. Helena, California; and Addison Trail High School, Addison, Illinois.

Early in the publication life of this series, the appellation "Rainbow Books" was used. The name stuck and has become a point of identification ever since.

In the years since the first publication of this series, extensive follow-up studies and research have been conducted. Thousands of instructors, students, school administrators, and industrial trainers have shared their experiences and suggestions with the authors and publishers. To each of these people we extend our thanks and appreciation.

Library of Congress Cataloging-in-Publication Data

Tokheim, Roger L.
   Digital electronics / Roger L. Tokheim. — 3rd ed.
      p.   cm. — (Basic skills in electricity and electronics)
   Includes index.
   ISBN 0-07-065035-7 : $36.40
   1. Digital electronics. I. Title. II. Series.
TK7868.D5T65   1989
621.3815—dc19   88-35597
                                                                  CIP

The manuscript for this book was processed electronically.

Digital Electronics, Third Edition

1 2 3 4 5 6 7 8 9 0   DOCDOC   8 9 6 5 4 3 2 1 0 9

ISBN 0-07-065035-7

Contents

Editor's Foreword

The McGraw-Hill *Basic Skills in Electricity and Electronics* series has been designed to provide entry-level competencies in a wide range of occupations in the electrical and electronic fields. The series consists of coordinated instructional materials designed especially for the career-oriented student. Each major subject area covered in the series is supported by a textbook, an activities manual, and a teacher's manual. All the materials focus on the theory, practices, applications, and experiences necessary for those preparing to enter technical careers.

There are two fundamental considerations in the preparation of materials for such a series: the needs of the learner and needs of the employer. The materials in this series meet these needs in an expert fashion. The authors and editors have drawn upon their broad teaching and technical experiences to accurately interpret and meet the needs of the student. The needs of business and industry have been identified through questionnaires, surveys, personal interviews, industry publications, government occupational trend reports, and field studies.

The processes used to produce and refine the series have been ongoing. Technological change is rapid, and the content has been revised to focus on current trends. Refinements in pedagogy have been defined and implemented based on classroom testing and feedback from students and teachers using the series. Every effort has been made to offer the best possible learning materials.

The widespread acceptance of the *Basic Skills in Electricity and Electronics* series and the positive responses from users confirm the basic soundness in content and design of these materials as well as their effectiveness as learning tools. Teachers will find the texts and manuals in each of the subject areas logically structured, well-paced, and developed around a framework of modern objectives. Students will find the materials to be readable, lucidly illustrated, and interesting. They will also find a generous amount of self-study and review materials to help them determine their own progress.

The publisher and editor welcome comments and suggestions from teachers and students using the materials in this series.

Charles A. Schuler
Project Editor

BASIC SKILLS IN ELECTRICITY AND ELECTRONICS

Charles A. Schuler, Project Editor

Books in this series:

Introduction to Television Servicing by Wayne C. Brandenburg
Electricity: Principles and Applications by Richard J. Fowler
Communication Electronics by Louis E. Frenzel, Jr.
Instruments and Measurements by Charles M. Gilmore
Microprocessors: Principles and Applications by Charles M. Gilmore
Small Appliance Repair by Phyllis Palmore and Nevin E. André
Electronics: Principles and Applications by Charles A. Schuler
Digital Electronics by Roger L. Tokheim

Preface

Digital Electronics, Third Edition, is designed to be used as an introductory text for students who are new to the field of electronics. Prerequisites are general mathematics and basic dc circuits. Digital electronics can be studied before or concurrently with a course in basic electronics, since knowledge of active discrete components is not a prerequisite. Binary mathematics and Boolean concepts are introduced and explained in this book as needed.

Digital electronics is not a specialized field in electronics. Digital circuits were first used in computing devices, but they are now commonly found in a broad range of products, for example, automobiles, communications equipment, toys, audio systems, kitchen and laundry appliances, television equipment, test instruments, and, of course, computers and calculators. The advances in microelectronic design and manufacturing have caused a rapid increase in the use of digital circuits.

The text of *Digital Electronics* includes at least 27 topics new to this edition. Many new topics deal with the increased use of power-saving CMOS ICs in digital circuits. An entire chapter details both TTL and CMOS IC specifications and the important topic of interfacing. Both LCD and VF displays receive prominent coverage. To aid motivation, several digital electronic game circuits are used to demonstrate both TTL and CMOS digital technology. Coverage of semiconductor memories has been expanded. The unique and popular Chapter 12, "Digital Systems," has been greatly expanded to emphasize the use of CMOS ICs. Specific A/D converter ICs are detailed, along with applications such as a digital light meter and 3½-digit digital voltmeter. Several useful computer programs written in BASIC are listed in the Appendix.

Student interest and motivation are maintained through the use of:

1. Simple, two-color illustrations emphasizing important ideas and focusing on basic concepts.
2. Frequent, short self-tests (with answers) providing immediate reinforcement and building student confidence.
3. A unique format presenting important terms and concepts in a color-highlighted margin of each page.
4. The systems-subsystems approach making the topics more relevant to the student.
5. Simple analysis techniques building the student's ability to troubleshoot.
6. Electronic game circuits demonstrating the function of digital components.
7. Practical, "hands-on" tasks.

The material in this text is based on carefully selected and formulated performance objectives. Surveys, classroom testing, and feedback from teachers and industry representatives were used in updating these objectives. The objectives are accomplished through the use of digital systems and subsystems. The systems-subsystems approach is fundamental to digital electronics because of the extensive use of medium-, large-, and very-large-scale ICs. Small-scale ICs are used when students learn the fundamentals of subsystems. All circuits in the text can be wired for classroom demonstration using off-the-shelf TTL or CMOS ICs. A companion volume, *Activities Manual for Digital Electronics,* Third Edition, is closely correlated with this textbook. Experiments from the activities manual feature many of the circuits from this textbook. Digital design problems, troubleshooting problems, construction projects, and chapter tests are also available in the companion activities manual.

Appreciation should be given to the many teachers, students, and industry representatives who contributed to this book. I would like to give special thanks to my former student, Darrell Klotzbach, for his outstanding work in verifying the operation of the circuits in this book. Finally, I would like to express my appreciation to members of my family, Daniel, Marshall, and Carrie, for their help and patience.

Roger L. Tokheim

Safety

Electric and electronic circuits can be dangerous. Safe practices are necessary to prevent electrical shock, fires, explosions, mechanical damage, and injuries resulting from the improper use of tools.

Perhaps the greatest hazard is electrical shock. A current through the human body in excess of 10 milliamperes can paralyze the victim and make it impossible to let go of a "live" conductor or component. Ten milliamperes is a rather small amount of electrical flow: It is only *ten one-thousandths* of an ampere. An ordinary flashlight uses more than 100 times that amount of current!

Flashlight cells and batteries are safe to handle because the resistance of human skin is normally high enough to keep the current flow very small. For example, touching an ordinary 1.5-V cell produces a current flow in the microampere range (a microampere is one-millionth of an ampere). This amount of current is too small to be noticed.

High voltage, on the other hand, can force enough current through the skin to produce a shock. If the current approaches 100 milliamperes or more, the shock can be fatal. Thus, the danger of shock increases with voltage. Those who work with high voltage must be properly trained and equipped.

When human skin is moist or cut, its resistance to the flow of electricity can drop drastically. When this happens, even moderate voltages may cause a serious shock. Experienced technicians know this, and they also know that so-called low-voltage equipment may have a high-voltage section or two. In other words, they do not practice two methods of working with circuits: one for high voltage and one for low voltage. They follow safe procedures at all times. They do not assume protective devices are working. They do not assume a circuit is off even though the switch is in the OFF position. They know the switch could be defective.

As your knowledge and experience grows, you will learn many specific safe procedures for dealing with electricity and electronics. In the meantime:

1. Always follow procedures.
2. Use service manuals as often as possible. They often contain specific safety information.
3. Investigate before you act.
4. When in doubt, *do not act*. Ask your instructor or supervisor.

General Safety Rules for Electricity and Electronics

Safe practices will protect you and your fellow workers. Study the following rules. Discuss them with others, and ask your instructor about any you do not understand.

1. Do not work when you are tired or taking medicine that makes you drowsy.
2. Do not work in poor light.
3. Do not work in damp areas or with wet shoes or clothing.
4. Use approved tools, equipment, and protective devices.
5. Avoid wearing rings, bracelets, and similar metal items when working around exposed electric circuits.
6. Never assume that a circuit is off. Double-check it with an instrument that you are sure is operational.
7. Some situations require a "buddy system" to guarantee that power will not be turned on while a technician is still working on a circuit.
8. Never tamper with or try to override safety devices such as an interlock (a type of switch that automatically removes power when a door is opened or a panel removed).
9. Keep tools and test equipment clean and in good working condition. Replace insulated probes and leads at the first sign of deterioration.
10. Some devices, such as capacitors, can store a *lethal* charge. They may store this

charge for long periods of time. You must be certain these devices are discharged before working around them.

11. Do not remove grounds and do not use adaptors that defeat the equipment ground.

12. Use only an approved fire extinguisher for electrical and electronic equipment. Water can conduct electricity and may severely damage equipment. Carbon dioxide (CO_2) or halogenated-type extinguishers are usually preferred. Foam-type extinguishers may also be desired in some cases. Commercial fire extinguishers are rated for the type of fires for which they are effective. Use only those rated for the proper working conditions.

13. Follow directions when using solvents and other chemicals. They may be toxic, flammable, or may damage certain materials such as plastics.

14. A few materials used in electronic equipment are toxic. Examples include tantalum capacitors and beryllium oxide transistor cases. These devices should not be crushed or abraded, and you should wash your hands thoroughly after handling them. Other materials (such as heat shrink tubing) may produce irritating fumes if overheated.

15. Certain circuit components affect the safe performance of equipment and systems. Use only exact or approved replacement parts.

16. Use protective clothing and safety glasses when handling high-vacuum devices such as picture tubes and cathode ray tubes.

17. Don't work on equipment before you know proper procedures and are aware of any potential safety hazards.

18. Many accidents have been caused by people rushing and cutting corners. Take the time required to protect yourself and others. Running, horseplay, and practical jokes are strictly forbidden in shops and laboratories.

Circuits and equipment must be treated with respect. Learn how they work and the proper way of working on them. Always practice safety; your health and life depend on it.

CHAPTER 1

Digital Electronics

This chapter introduces you to digital electronics. You will learn what a digital circuit is and how digital circuits are used in everyday electronic equipment. You will also learn why digital circuits are preferred for some jobs and not others. In addition, you will learn about practical methods of generating a single digital pulse using mechanical switches. Practical free-running clock circuits will be introduced, as well as simple methods of testing for HIGH and LOW digital signals using light-emitting-diode circuits. By the end of this chapter, you will be able to operate and use a logic probe and other digital testing devices.

Digital electronics is the world of the calculator, the computer, the integrated circuit, and the binary numbers 0 and 1. This is an exciting field within electronics because the uses for digital circuits are expanding so rapidly. One small integrated circuit can perform the task of thousands of transistors, diodes, and resistors. You see digital circuits in operation every day. At stores the cash registers read out digital displays. The tiny pocket calculators verge on becoming personal computers. All sizes of computers perform complicated tasks with fantastic speed and accuracy. Factory machines are controlled by digital circuits. Digital clocks and watches flash the time. Some automobiles use microprocessors to control several engine functions. Technicians use digital voltmeters and frequency counters.

All persons working in electronics must now understand digital electronic circuits. The inexpensive integrated circuit has made the subject of digital electronics easy to study. You will use many integrated circuits to construct digital circuits.

1-1 WHAT IS A DIGITAL CIRCUIT?

In your experience with electricity and electronics you have probably used analog circuits. The circuit in Fig. 1-1(*a*) on the next page puts out an *analog* signal or voltage. As the wiper on the potentiometer is moved upward, the voltage from points *A* to *B* *gradually* increases. When the wiper is moved downward, the voltage gradually decreases from 5 to 0 volts (V). The waveform diagram in Fig. 1-1(*b*) is a graph of the analog output. On the left side the voltage from *A* to *B* is gradually increasing to 5 V; on the right side the voltage is gradually decreasing to 0 V. By stopping the potentiometer wiper at any midpoint, we can get an output voltage anywhere between 0 and 5 V. An analog device, then, is one that has a signal which *varies continuously* in step with the input.

A digital device operates with a digital signal. Figure 1-2(*a*) pictures a square-wave generator. The generator produces a square waveform that is displayed on the oscillo-

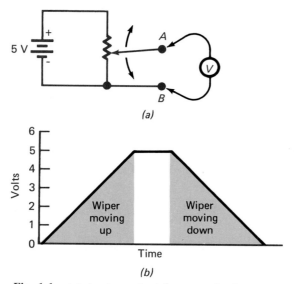

Fig. 1-1 (*a*) **Analog output from a potentiometer.** (*b*) **Analog signal waveform.**

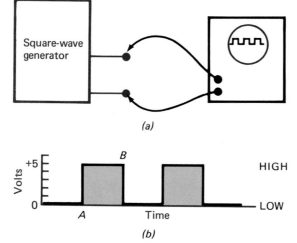

Fig. 1-2 (*a*) **Digital signal displayed on scope.** (*b*) **Digital signal waveform.**

scope. The digital signal is only at +5 V *or* at 0 V, as diagramed in Fig. 1-2(*b*). The voltage at point *A* moves from 0 to 5 V. The voltage then stays at +5 V for a time. At point *B* the voltage drops immediately from +5 to 0 V. The voltage then stays at 0 V for a time. Only two voltages are present in a digital electronic circuit. In the waveform diagram in Fig. 1-2(*b*) these voltages are labeled HIGH and LOW. The HIGH voltage is +5 V; the LOW voltage is 0 V. Later we shall call the HIGH voltage (+5 V) a logical 1 and the LOW voltage (0 V) a logical 0.

Circuits that handle only HIGH and LOW signals are called *digital circuits*. We mentioned that digital electronics is the world of

logical 0s and 1s. The voltages in Fig. 1-2(*b*) are rather typical of the voltages you will be working with in digital electronics.

The digital signal in Fig. 1-2(*b*) could also be generated by a simple on-off switch. A digital signal could also be generated by a transistor turning on and off. In recent years digital electronic signals usually have been generated and processed by integrated circuits (ICs).

The standard volt-ohm-millimmeter (VOM) shown in Fig. 1-3(*a*) is an example of an *analog* measuring device. As the voltage, resistance, or current being measured by the VOM increases, the needle *gradually and continuously moves* up the scale. A *digital multimeter*

(*a*)

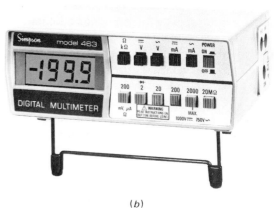

(*b*)

Fig. 1-3 (*a*) **Analog meter (VOM).** (*b*) **Digital multimeter (DMM).** (*Courtesy of Simpson Electric Company.*)

(DMM) is shown in Fig. 1-3(*b*). This is an example of a *digital* measuring device. As the current, resistance, or voltage being measured by the DMM increases, the display *jumps upward in small steps*. The DMM is an example of digital circuitry taking over tasks previously performed only by analog devices. This trend toward digital circuitry is growing. Currently, the modern technician's bench probably has both an analog VOM and a digital DMM.

Self-Test

Supply the missing word in each statement.

1. Refer to Fig. 1-2. The +5-V level of the _____ (analog, digital) signal could also be called a logical 1 or a _____ (HIGH, LOW).
2. A(n) _____ device is one that has a signal which varies continuously in step with the input.

1-2 WHERE ARE DIGITAL CIRCUITS USED?

Digital electronics is a fast-growing field, as witnessed by the widespread use of microcomputers. Although the microcomputer is only about a decade old, there are tens of millions of them used in homes, schools, businesses, and governments. Microcomputers are extremely adaptable. At home a computer might be used for playing video games, managing a household budget, or controlling lights and appliances. At school, students use the same computer to aid them in learning spelling, math, and writing. The staff uses the same computers for word processing, testing, and grading. In business, the same microcomputer might process payrolls, control inventory, and generate mailing lists. In the factories microcomputers are adapted for controlling machines, robots, and processes.

Microcomputers are designed around complex ICs called *microprocessors*. In addition, many IC semiconductor memories make up the microcomputer. Microcomputers with microprocessors and semiconductor memories have started the personal computer (PC) revolution. Small computers that used to cost tens of thousands of dollars now cost only hundreds. Digital circuits housed in ICs are used in both large and small computers.

A hand-held calculator is another example of a digital electronic device used by nearly everyone. Calculators range from the five-dollar models to the sophisticated versions used by engineers and scientists. Only a few decades ago, even simple calculators would have cost thousands of dollars. The more sophisticated programmable calculators can sometimes be connected to *peripheral devices* such as optical wands and printers. Programmable calculators verge on becoming very small computers. Scientists, engineers, and technicians have made great advances in producing digital ICs. As a result of these advances, the field of digital electronics has mushroomed.

The portable or "laptop" computer has come into its own recently with advances in both IC and display technology. A laptop computer is shown in Fig. 1-4. This Zenith Data Systems portable PC has the power of many high-cost desktop microcomputer systems.

Fig. 1-4   Zenith Data systems ZFL-181 laptop computer. *(Courtesy of Heath Company.)*

The digital timepiece is a triumph of electronic technology. Very accurate multifunction digital clocks and wrist watches are available at low cost. The "wrist watch" pictured in Fig. 1-5 on the next page is actually a very small computer terminal marketed by Seiko. This PC Datagraph can be connected to a personal computer for programming. Notice that it has a liquid crystal display (LCD) and contains 2K of memory.

Fig. 1-5 Small "wrist watch" computer terminal, the PC Datagraph by Seiko. *(Courtesy of Hattori Corporation of America.)*

Robots and other computer-controlled machines add to the mystique of electronic technology. Robots have captured the imagination of inventors, science fiction writers, and movie producers. A *robot* may be defined as a machine that can perform humanlike actions and functions. A computer serves as the control center for modern robots. The robot's computer can be reprogrammed to permit the machine to perform a different sequence of operations. The study of the science and technology of the robot is called *robotics*.

Robots are classified according to their use. *Industrial robots* usually take the form of arms and are used in manufacturing and materials handling. *Hobby robots* are used in education, advertising, and entertainment. One popular educational robot is pictured in Fig. 1-6(*a*). The HERO 2000 robot by Heath Company can be assembled from a kit or purchased preassembled. HERO 2000 is controlled by a built-in computer. Besides its mechanical movement, HERO 2000 has an electronically synthesized voice and various sensors. HERO 2000 can be programmed in the BASIC computer language. A floppy disk drive is visible on the left side of the robot. It is possible to control the robot with the remote-control keyboard shown near the base of the machine in Fig. 1-6(*a*).

A second educational robot is featured in Fig. 1-6(*b*). The WAO robot can be programmed directly via the attached keypad on its back or through an interface card connected to your computer. Small robots of this type are favorite projects for students in technology classes.

The technician's bench has a new look. Digital multimeters (DMMs) read out resistance, voltage, and current values. Figure 1-7(*a*) on page 6 illustrates a probe-style DMM that uses a modern low-power LCD display for the 3½-digit readout. The DMM is battery-operated and features autoranging and 0.7 percent direct current voltage (DCV) accuracy. The ohmmeter section features an audible tone for low-resistance continuity checks. It has a data hold switch and automatic direct current (dc) polarity selection. This is all packed into a small unit that weighs less than 3 ounces.

There is also a *digital capacitance meter* on most electronic workbenches. One such unit is pictured in Fig. 1-7(*b*). This hand-held meter accurately measures a wide range of capacitance values.

Frequency counters are also standard test instruments found in most labs. This fantastic digital instrument accurately senses and displays the frequency of alternating current (ac) signals into the hundreds of millions of cycles per second. Figure 1-7(*c*) pictures one such frequency counter. This unit features an eight-digit easy-to-read display with a frequency response from 5 hertz (Hz) to 500 megahertz (MHz). Many modern test instruments make extensive use of digital circuits.

Electronic products for home entertainment (such as TVs and stereos) have traditionally been designed using analog circuits. This is changing, with some digital circuitry used in both television and sound systems. Most modern consumer products used for home entertainment contain both analog and digital circuitry. One popular product is Yamaha's FM Digital Synthesizer, pictured in Fig. 1-8 on page 6. This synthesizer has some characteristics of a computer in that it contains random-access memory (RAM) and a floppy disk drive on the left side. To aid in programming, a 40-character, two-line LCD and two alphanumeric light-emitting diodes (LEDs) grace the front.

Digital electronic circuits are at work in modern automobiles. There is a microprocessor in the ignition circuit of most cars to precisely control several ignition and fuel system vari-

(a)

(b)

Fig. 1-6 *(a)* **HERO 2000 educational robot.** *(Courtesy of Heath Company.)* *(b)* **Small programmable WAO educational robot.** *(Courtesy of OWI, Incorporated.)*

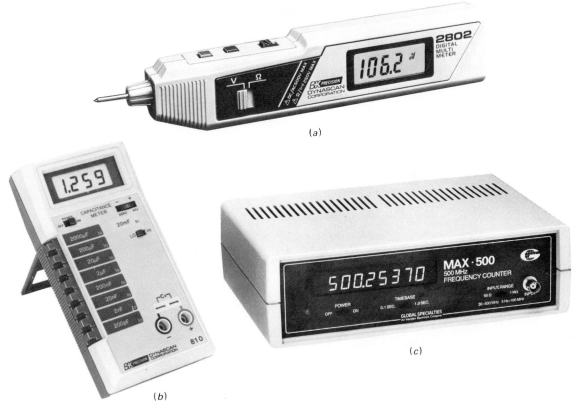

Fig. 1-7 (*a*) **Probe-type digital multimeter.** (*b*) **Capacitance meter.** (*Courtesy of Dynascan Corporation.*) (*c*) **Frequency counter.** (*Courtesy of Interplex Electronics, Inc.*)

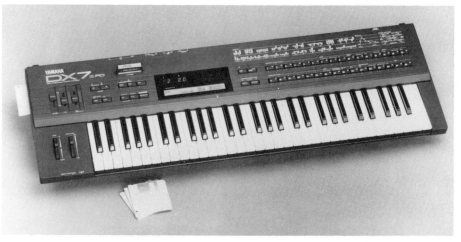

Fig. 1-8 **DX7IIFD FM digital synthesizer with floppy disk drive.** (*Courtesy of Yamaha International Corporation.*)

ables. An inside look at a microprocessor ignition control module is shown in Fig. 1-9. Notice the extensive use of digital ICs.

From the driver's seat of a modern automobile, a digital tachometer stares back at you from the instrument cluster. The glow of a digital display shows the time and the tuning of your radio. The same digital display identifies your selection on the compact disk player. A digital thermometer monitors the interior and exterior temperatures, while a digital compass points the way to your destination.

A digital speedometer blends with the futuristic interior of the Dodge Daytona Concept Sports Car in Fig. 1-10. The display for a satellite map system is at the left on the lower in-

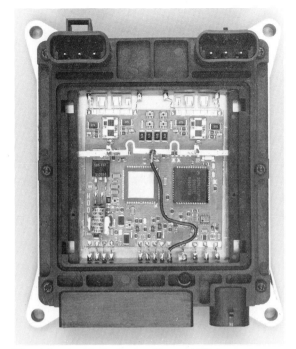

Fig. 1-9 **Inside look at a General Motors micro-processor ignition control module.** *(Courtesy of GM Hughes Electronics.)*

strument cluster, and the diagnostic system display is to the right of the wheel. The steering wheel hub houses the "key card" ignition and steering wheel lock. The driver's identification is encoded on the card along with the driver's preferences as to seating position, climate control, and entertainment. Automobile manufacturers are spending large sums of money on research and development efforts in automotive electronics.

The "cash register" at your local grocery store may be a quite sophisticated digital device. Many of these units are equipped to optically read the *universal product code* (UPC). This is the code found on the packages of store items. It is a rectangular group of parallel bars of various widths with some numbers printed across the bottom. This code identifies the manufacturer as well as the exact product. The point-of-sale register can look up the current price of the product and even perform some inventory control. The data read from the code is processed by digital circuits in the point-of-sale registers.

Diagnostic system display

"Key card" ignition and steering wheel lock

Automotive electronics

Universal product code (UPC)

Fig. 1-10 **Futuristic interior of the Dodge Daytona Concept Sports Car.** *(Courtesy of Chrysler Motors Corporation.)*

In your home there are also pieces of digital equipment. The time, temperature, wind speed and direction, and barometric pressure are available at a glance at the digital weather computer shown in Fig. 1-11(a). This device is a microprocessor-based system that measures and stores weather information. In the past, stereo and radio systems contained only analog circuits, but recently digital circuits have been used in these units. Your home's heating and cooling may be controlled by one of the newer "intelligent" setback thermostats. Electronic and video games all make extensive use of digital electronics. Appliances such as microwave ovens, washers, and dryers may have sophisticated microprocessor-controlled digital circuitry. Your bathroom may even have a digital scale or a digital thermometer like the one pictured in Fig. 1-11(b).

Three large users of very advanced digital devices are the military, medical, and telecommunications fields. Formerly, digital circuits were used mainly in computers. Now these circuits are being used in many other products because of their low cost and great accuracy. Because digital circuits are appearing in almost all electronic equipment, all well-trained technicians need to know how they operate.

Self-Test

Supply the missing word or words in each statement.

3. The DMM, or _____ , in Fig. 1-3(b) uses a modern low-power _____ display.

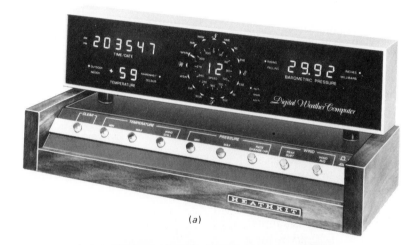

(a)

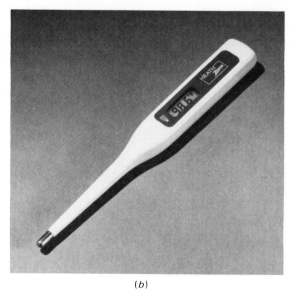

(b)

Fig. 1-11 (a) **Digital weather computer.** (b) **Digital thermometer.** *(Courtesy of Heath Company.)*

4. The field of digital electronics has mushroomed as a result of advances in the making of _____ circuits.

5. A microcomputer, such as the one shown in Fig. 1-4, is designed around a complex IC called a(n) _____ .

1-3 WHY USE DIGITAL CIRCUITS?

Electronic designers and technicians must have a working knowledge of both analog and digital systems. The designer must decide if the system will use analog or digital techniques or a combination of both. The technicians must build a prototype or troubleshoot and repair digital, analog, and combined systems.

Analog electronic systems have been more popular in the past. "Real-world" information dealing with time, speed, weight, pressure, light intensity, and position measurements are all *analog* in nature.

A simple analog electronic system for measuring the amount of liquid in a tank is illustrated in Fig. 1-12. The input to the system is a varying resistance. The processing proceeds according to the Ohm's law formula, $I = V/R$. The output indicator is an ammeter which is calibrated as a water tank gage. In the analog system in Fig. 1-12, as the water rises, the input resistance drops. Decreasing the resistance R causes an increase in current. Increased current causes the ammeter (water tank gage) to read higher.

The analog system in Fig. 1-12 is simple and efficient. The gage in Fig. 1-12 gives an indication of the water level in the tank. If more information is required about the water level, then a digital system such as the one shown in Fig. 1-13 might be used.

Digital systems are required when data must be stored, used for calculations, or displayed as numbers and/or letters. A somewhat more complex arrangement for measuring the amount of liquid in a water tank is the digital system shown in Fig. 1-13 on the next page. The input is still a variable resistance as it was in the analog system. The resistance is converted into numbers by the analog-to-digital (A/D) converter. The central processing unit (CPU) of a computer can manipulate the input data, output the information, store the information, calculate things such as flow rates in and out, calculate the time till the tank is full (or empty) based on flow rates, and so forth. Digital systems are valuable when calculations, data manipulations, and cathode-ray tube (CRT) outputs are required.

Some of the advantages given for using digital circuitry instead of analog are as follows:

1. Inexpensive ICs can be used with few external components.
2. Information can be stored for short periods or indefinitely.
3. Data can be used for precise calculations.
4. Systems can be designed more easily using compatible digital logic families.
5. Systems can be programmed and show some manner of "intelligence."

The limitations of digital circuitry are as follows:

1. Most "real-world" events are analog in nature.
2. Analog processing is usually simpler and faster.

Digital circuits are appearing in more and more products primarily because of low-cost, reliable digital ICs. Other reasons for their growing popularity are accuracy, added stability, computer compatibility, memory, ease of use, and simplicity of design.

Analog electronic system

Analog-to-digital (A/D) converter

Central processing unit (CPU)

Advantages of digital circuitry

Limitations of digital circuitry

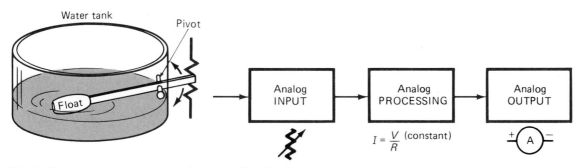

Fig. 1-12 **Analog system used to interpret float level in water tank.**

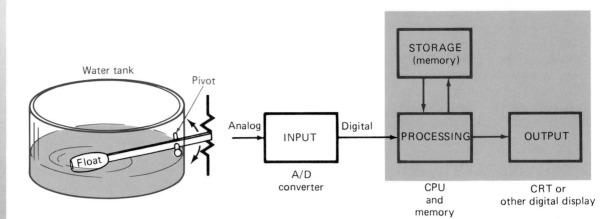

Fig. 1-13 Digital system used to interpret float level in water tank.

Self-Test

Answer the following questions.

6. Generally, electronic circuits are classified as either analog or _____ .
7. Measurements of time, speed, weight, pressure, light intensity, and position are _____ (analog, digital) in nature.
8. Refer to Fig. 1-12. As the water level drops, the input resistance increases. This causes the current I to _____ (decrease, increase) and the water level gage (ammeter) will read _____ (higher, lower).
9. Refer to Figs. 1-12 and 1-13. If this water tank were part of the city water system, where rates of water use are important, the system in Fig. _____ (1-12, 1-13) would be most appropriate.
10. True or false. The most important reason why digital circuitry is becoming more popular is because digital circuits are usually simpler and faster than analog circuits.

1-4 HOW DO YOU MAKE A DIGITAL SIGNAL?

Digital signals are composed of two well-defined voltage levels. For most of the circuits you will use, these voltage levels will be about 0 V (GND) and + 3 to +5 V. These are called *TTL voltage levels* because they are used with the *transistor-transistor logic* family of ICs.

A TTL digital signal could be made manually by using a mechanical switch. Consider the simple circuit shown in Fig. 1-14(*a*). As the common blade of the single-pole, double-throw (SPDT) switch is moved up and down, it pro-duces the *digital waveform* shown at the right. At time period t_1, the voltage is 0 V, or LOW. At t_2 the voltage is +5 V, or HIGH. At t_3, the voltage is again 0 V, or LOW, and at t_4, it is again +5 V, or HIGH.

One problem with a mechanical switch is *contact bounce*. If we could look very carefully at a switch toggling from LOW to HIGH, it might look like the waveform in Fig. 1-14(*b*). The waveform first goes directly from LOW to HIGH (see *A*) but then, because of contact bounce, drops to LOW and then back to HIGH again. Although this happens in a very short time, digital circuits are fast enough to see this as a LOW, HIGH, LOW, HIGH waveform. Note that Fig. 1-14(*b*) shows that there is actually a range of voltages that are defined HIGH and LOW. The *undefined region* between HIGH and LOW causes trouble in digital circuits and should be avoided.

To cure the problem illustrated in Fig. 1-14(*b*), mechanical switches are sometimes *debounced*. A block diagram of a *debounced logic switch* is shown in Fig. 1-14(*c*). Note the use of the debouncing circuit, or latch. Almost all the mechanical logic switches you will use on laboratory equipment will have been debounced with latch circuits. Latches are sometimes called *flip-flops* and will be studied in detail in a later chapter. Notice in Fig. 1-14(*c*) that the output of the latch during time period t_1 is LOW but not quite 0 V. During t_2 the output of the latch is HIGH even though it is something less than a full +5 V. Likewise t_3 is LOW and t_4 is HIGH in Fig. 1-14(*c*).

It might be suggested that a push-button switch be used to make a digital signal. If the button is pressed, a HIGH should be generated. If the push button is released, a LOW

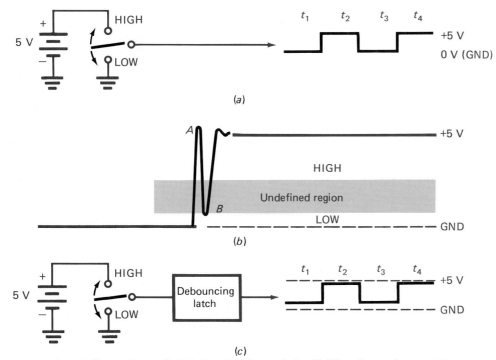

Fig. 1-14 (*a*) **Generating a digital signal with a switch.** (*b*) **Waveform of contact bounce within a mechanical switch.** (*c*) **Adding a debouncing latch to a simple switch to condition the digital signal.**

should be generated. Consider the simple circuit in Fig. 1-15(*a*). When the push button is pressed, a HIGH of about +5 V is generated at the output. When the push button is released, however, the voltage at the output is *undefined*. There is an open circuit between the power supply and the output. This would not work properly as a logic switch.

A normally open push-button switch can be used with a special circuit to generate a digital pulse. Figure 1-15(*b*) shows the push button connected to a *one-shot multivibrator* circuit. Now for each press of the push button, a *single short, positive pulse* is output from the one-shot circuit. The pulse width of the output is determined by the design of the multivibrator and *not* by how long you hold down the push button.

Both the latch circuit and the one-shot circuit were used earlier. Both are *multivibrator* (MV) circuits. The latch is also called a *flip-flop* or a *bistable multivibrator*. The one-shot is also called the *monostable multivibrator*. A third type of MV circuit is the *astable multivibrator*. This is also called a *free-running multivibrator*. In many digital circuits it may be referred to simply as the *clock*.

The free-running MV oscillates by itself without the need for external switching or an external signal. A block diagram of a free-running MV is shown in Fig. 1-16. The free-running MV generates a continuous series of TTL level pulses. The output in Fig. 1-16 on the next page alternately goes LOW and HIGH.

In the laboratory, you will need to generate digital signals. The equipment you will use will

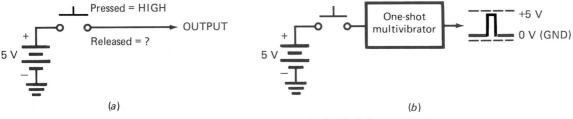

Fig. 1-15 (*a*) **Push button will not generate a digital signal.** (*b*) **Push button used to trigger a one-shot multivibrator for a single-pulse digital signal.**

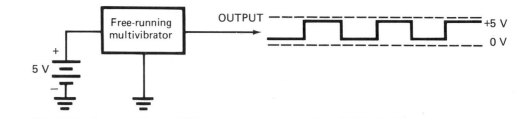

Fig. 1-16   Free-running multivibrator generates a string of digital pulses.

have slide switches, push buttons, and free-running clocks that will generate TTL level signals similar to those shown in Figs. 1-14, 1-15, and 1-16. In the laboratory, you will use *logic switches* which will have been debounced using a latch circuit as in Fig. 1-14(c). You will also use a *single-pulse clock* triggered by a push-button switch. The single-pulse clock push button will be connected to a one-shot multivibrator as shown in Fig. 1-15(b). Finally, your equipment will have a free-running clock. It will generate a continuous series of pulses, as shown in Fig. 1-16.

Astable, monostable, and bistable MVs can all be wired using discrete components (individual resistors, capacitors, and transistors) or purchased in IC form. Because of their superior performance, ease of use, and low cost, the IC forms of these circuits will be used in this course. A schematic diagram for a practical free-running clock circuit is shown in Fig. 1-17(a). This clock circuit produces a low-frequency (1- to 2-Hz) TTL level output. The heart of the free-running clock circuit is a common 555 timer IC. Note that several resistors, a capacitor, and a power supply must also be used in the circuit.

A typical breadboard wiring of this free-running clock is sketched in Fig. 1-17(b). Notice the use of the IC type of mounting board. Also note that pin 1 on the IC is immediately counterclockwise from the notch or dot near the end of the eight-pin IC. The wiring diagram in Fig. 1-17(b) is shown for your convenience. You will normally have to wire circuits on solderless breadboards directly from the schematic diagram.

Self-Test

Supply the missing word in each statement.

11. Refer to Fig. 1-14(c). The digital signal at t_2 is _____ (HIGH, LOW), while it is _____ (HIGH, LOW) at t_3.
12. Refer to Fig. 1-15(a). When the push button is released (open), the output is _____ .

13. Refer to Fig. 1-14(c). The debouncing latch is also called a flip-flop or _____ multivibrator.
14. Refer to Fig. 1-15(b). The one-shot multivibrator used for conditioning the digital signal is also called a(n) _____ multivibrator.
15. Refer to Fig. 1-17. A 555 _____ IC and several discrete components are being used to generate a continuous series of TTL level pulses. This free-running clock is also called a free-running multivibrator or _____ multivibrator.

1-5 HOW DO YOU TEST FOR A DIGITAL SIGNAL?

In the last section you generated digital signals using various MV circuits. These are the methods you will use in the laboratory to generate *input signals* for the digital circuits constructed. In this section, several simple methods of testing the *outputs* of digital circuits will be discussed.

Consider the circuit in Fig. 1-18(a). The *input* is provided by a simple SPDT switch and power supply. The *output indicator* is an LED. The 150-Ω resistor limits the current through the LED to a safe level. When the switch in Fig. 1-18(a) is in the HIGH (up) position, +5 V is applied to the anode end of the LED. The LED is forward-biased, current flows upward, and the LED lights. With the switch in the LOW (down) position, both the anode and cathode ends of the LED are grounded and it does not light. Using this indicator, a light means HIGH and no light generally means LOW.

The simple LED output indicator is shown again in Fig. 1-18(b). This time a simplified diagram of a logic switch forms the input. The logic switch acts like the switch in Fig. 1-18(a) except it is debounced. The output indicator is again the LED with a series-limiting resistor. When the input logic switch in Fig. 1-18(b)

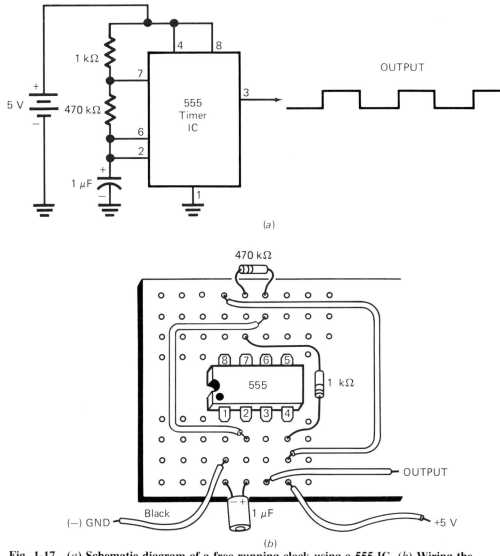

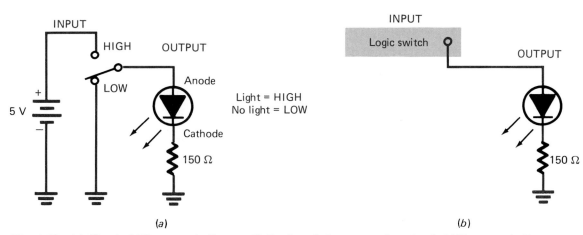

Fig. 1-17 (a) Schematic diagram of a free-running clock using a 555 IC. (b) Wiring the free-running clock circuit on a solderless breadboard.

Fig. 1-18 (a) Simple LED output indicator. (b) Logic switch connected to simple LED output indicator.

generates a LOW, the LED will not light. However, when the logic switch produces a HIGH, the LED will light.

Another LED output indicator is illustrated in Fig. 1-19. The LED acts exactly the same as the one shown previously. It lights to indicate a logical HIGH and does not light to indicate a LOW. The LED in Fig. 1-19 is driven by a transistor instead of directly by the input. The transistorized circuit in Fig. 1-19 holds an advantage over the direct-drive circuit in that it draws less current from the output of the digital circuit under test. Light-emitting diode output indicators wired like those in Figs. 1-18 and 1-19 may be found in your laboratory equipment.

Consider the output indicator circuit using two LEDs shown in Fig. 1-20. When the input is HIGH (+5 V), the bottom LED lights and the top LED does not light. When the input is LOW (GND), only the top LED lights. If point Y in the circuit in Fig. 1-20 enters the undefined region between HIGH and LOW or is not connected to a point in the circuit, both LEDs light.

Output voltages from a digital circuit can be measured with a standard voltmeter. With the TTL family of ICs, a voltage from 0 to 0.8 V is considered a LOW. A voltage from 2 to 5 V is considered a HIGH. Voltages between about 0.8 and 2 V are in the undefined region and signal trouble in TTL circuits.

A handy portable measuring device used to determine output logic levels is the *logic probe*. One such inexpensive, student-constructed logic probe is sketched in Fig. 1-21(a) on page

16. To use this unit for measuring TTL logic levels, follow this procedure:

1. Connect the red power lead to +5 V of the circuit under test.
2. Connect the black (GND) power lead to GND of the circuit under test.
3. Connect the third power lead (TTL) to GND of the circuit under test.
4. Touch the probe tip to the point in the digital circuit to be tested.
5. One or the other LED indicator shown in Fig. 1-21(a) should light. If both light, the tip is disconnected from the circuit or the point is in the undefined region between HIGH and LOW.

The logic probe in Fig 1-21(a) can also be used with the CMOS logic family of ICs; CMOS is short for *complementary metal oxide semiconductor*. If the logic probe is to be used to measure *CMOS logic levels*, the TTL family power lead is left *disconnected*. Then the red power lead is connected to positive (+) of the power supply while the black lead (GND) goes to GND. Touch the probe tip to the test point in the CMOS digital circuit and the LED indicators tell its CMOS logic level, either HIGH or LOW.

A schematic diagram of the logic probe is shown in Fig. 1-21(b). The 555 timer IC is being used in this circuit. The 555 IC uses power supply voltages ranging from about 5 V up to 18 V. TTL circuits always operate on 5 V, whereas some CMOS circuits operate on voltages as high as 15 V. The three power supply connections are shown at the left in Fig.

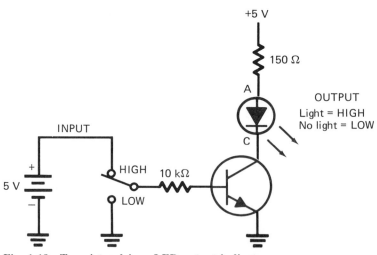

Fig. 1-19   Transistor-driven LED output indicator.

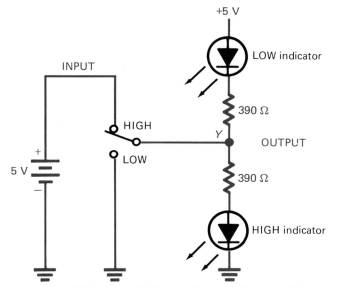

Fig. 1-20 LED output indicators that will show LOW, HIGH, and undefined logic levels.

1-21(*b*). The red lead goes to the positive of the supply while the black lead goes to GND. If the circuit under test is TTL, the TTL (GND) lead is also connected to GND. If a CMOS circuit is being tested, the TTL (GND) power lead is left disconnected. The input to the logic probe is shown on the left, entering pins 2 and 6 of the 555 IC. If this voltage is LOW, the *bottom* LED (D_6) lights. If the input voltage is HIGH, the *top* LED (D_5) lights. With the input probe disconnected, both LEDs light. Note that pin 3 of the 555 IC (output of 555) always goes to the *opposite* logic level of the input. Therefore, if the input (pins 2 and 6) is HIGH, the output of the 555 IC (pin 3) goes LOW. This in turn activates and lights the top LED (the HIGH indicator).

The four silicon diodes (D_1 to D_4) in Fig. 1-21(*b*) protect the IC from reverse polarity. Capacitor C_1 prevents transient voltages from affecting the logic probe when the TTL (GND) lead is not connected. Pin 5 of the 555 IC is grounded through resistor R_1. This resistor "adjusts" for TTL instead of CMOS logic levels.

The logic probe in Fig. 1-21 responds to voltage levels differently in the CMOS and TTL modes. Figure 1-22 on the next page shows the TTL and CMOS logic levels in terms of percentage of total supply voltage. For TTL, which always uses a +5-V supply, this logic probe will indicate a HIGH for 2 V or greater.

For TTL, it will indicate a LOW for 0.8 V or less.

In the laboratory, you may be asked to construct a logic probe like the one in Fig. 1-21 or your instructor might furnish you with a commercial logic probe for checking digital circuits. The operating instructions are different for each logic probe. Read the instruction manual on the unit you will be using.

Self-Test

Supply the missing word or words in each statement.

16. Refer to Fig. 1-18. If the input is HIGH, the LED will _____ (light, not light) because the diode is _____ (forward-, reverse-) biased.
17. Refer to Fig. 1-19. If the input is LOW, the transistor is turned _____ (off, on) and the LED _____ (does, does not) light.
18. Refer to Fig. 1-20. If the input is HIGH, the _____ (bottom, top) LED lights because its _____ (cathode, anode) has +5 V applied, forward biasing the diode.
19. The student-constructed _____ in Fig. 1-21 can be used to test either TTL or _____ digital circuits.

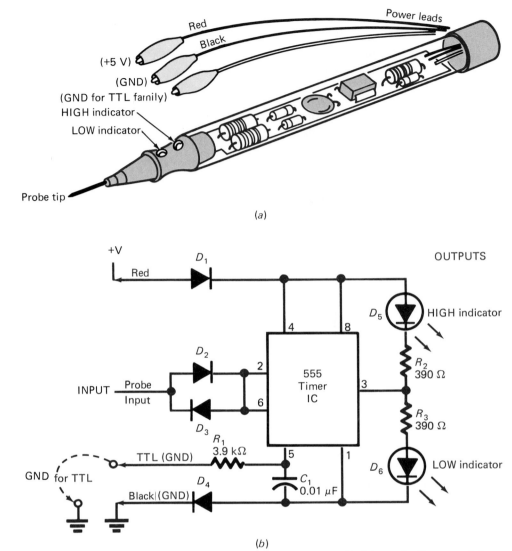

(a)

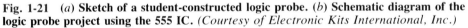

(b)

Fig. 1-21 (a) **Sketch of a student-constructed logic probe.** (b) **Schematic diagram of the logic probe project using the 555 IC.** (*Courtesy of Electronic Kits International, Inc.*)

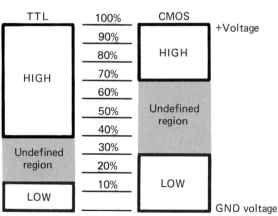

Fig. 1-22 **Defining logic levels for the TTL and CMOS families of digital ICs.**

SUMMARY

1. Analog signals vary gradually and continuously, while digital signals produce discrete voltage levels usually referred to as HIGH and LOW.
2. Computers, including the microcomputer, make extensive use of digital circuits. Calculators are also digital devices.
3. Most modern electronics equipment contains both analog and digital circuitry.
4. Logic levels are different for various digital logic families, such as TTL and CMOS. These logic levels are commonly referred to as HIGH, LOW, and undefined. Fig. 1-22 details these TTL and CMOS logic levels.
5. Digital circuits have become very popular because of the availability of low-cost digital ICs. Other advantages of digital circuitry are computer compatibility, memory, ease of use, simplicity of design, accuracy, and stability.
6. Bistable, monostable, and astable multivibrators are used to generate digital signals. These are sometimes called latches, one-shot, and free-running multivibrators, respectively.
7. Logic level indicators may take the form of simple LED and resistor circuits, voltmeters, or logic probes. Light-emitting diode logic level indicators will probably be found on your laboratory equipment.

CHAPTER REVIEW QUESTIONS

Answer the following questions.

1-1. Define the following:
   a. Analog signal
   b. Digital signal

1-2. Draw a square-wave digital signal. Label the bottom ''0 V'' and the top ''+5 V.'' Label the HIGH and LOW on the waveform. Label the logical 1 and logical 0 on the waveform.

1-3. List two devices that contain digital circuits which do mathematical calculations.

1-4. List three test instruments that contain digital circuits and are used by electronic technicians.

1-5. Refer to Fig. 1-13. The processing, storage of data, and output in this system consist mostly of _____ (analog, digital) circuits.

1-6. List several advantages of digital over analog circuits.

1-7. Traditionally, most consumer electronics devices (TVs, radios) have used _____ (analog, digital) circuitry.

1-8. Refer to Fig. 1-14. When using a SPDT slide switch to produce a digital signal, a _____ latch is used to condition the output.

1-9. Refer to Fig. 1-15. A _____ - _____ multivibrator is commonly used to condition the output of a push-button switch when generating a single digital pulse.

1-10. An astable, or _____ - _____ , multivibrator produces a string of digital pulses.

1-11. The circuit wired in Fig. 1-17(*b*) is classed as a(n) _____ (astable, bistable) multivibrator.

1-12. The LED in Fig. 1-18(*b*) lights when the input logic switch is _____ (HIGH, LOW).

1-13. Refer to Fig. 1-20. The _____ (bottom, top) LED lights when the input switch is LOW.

1-14. The logic probe in Fig. 1-21 can be used with which two digital logic families?

1-15. Refer to Fig. 1-21(*b*). If the input is LOW, pin 3 of the 555 IC will be _____ (HIGH, LOW). This causes the _____ (bottom, top) LED to light.

Answers to Self-Tests

1. digital, HIGH
2. analog
3. digital multimeter, liquid-crystal
4. integrated
5. microprocessor
6. digital

7. analog
8. decrease, lower
9. 1-13
10. F
11. HIGH, LOW
12. undefined
13. bistable

14. monostable
15. timer, astable
16. light, forward-
17. off, does not
18. bottom, anode
19. logic probe, CMOS

CHAPTER 2

Numbers We Use in Digital Electronics

Most people understand us when we say we have nine pennies. The number 9 is part of the decimal *number system we use every day. But digital electronic devices use a "strange" number system called* binary. *Digital computers and microprocessor-based systems use another strange number system called* hexadecimal. *Men and women who work in electronics must know how to convert numbers from the everyday decimal system to the binary or the hexadecimal system. By the end of this chapter you will be able to convert common decimal numbers to binary numbers and binary numbers to common decimal numbers. You will also be able to convert between the binary and hexadecimal and the decimal and hexadecimal number systems. Other computer systems use the* octal *number system. You will learn to make conversions between binary and octal and decimal and octal.*

2-1 COUNTING IN DECIMAL AND BINARY

A number system is a code that uses symbols to refer to a number of items. The decimal number system uses the symbols 0, 1, 2, 3, 4, 5, 6, 7, 8, and 9. The decimal number system contains 10 symbols and is sometimes called the *base 10 system*. The binary number system uses only the two symbols 0 and 1 and is sometimes called the *base 2 system*.

Figure 2-1 compares a number of coins with the symbols we use for counting. The decimal symbols that we commonly use for counting from 0 to 9 are shown in the left column; the right column has the symbols we use to count

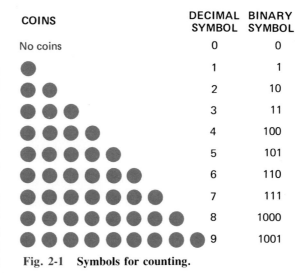

COINS	DECIMAL SYMBOL	BINARY SYMBOL
No coins	0	0
	1	1
	2	10
	3	11
	4	100
	5	101
	6	110
	7	111
	8	1000
	9	1001

Fig. 2-1   Symbols for counting.

nine coins in the binary system. Notice that the 0 and 1 count in binary is the same as in decimal counting. To represent two coins, the binary number 10 (say "one zero") is used. To represent three coins, the binary number 11 (say "one one") is used. To represent nine coins, the binary number 1001 (say "one zero zero one") is used.

For your work in digital electronics you must memorize the binary symbols used to count at least up to 9.

Self-Test

Supply the missing word or number in each statement.

1. The binary number system is sometimes called the _____ system.
2. Decimal 8 equals _____ in binary.
3. The binary number 0110 equals _____ in decimal.
4. The binary number 0111 equals _____ in decimal.

2-2 PLACE VALUE

The clerk at the local restaurant totals your food bill and asks you for $2.43. We all know that this amount equals 243 cents. Instead of paying with 243 pennies, however, you probably would give the clerk the money shown in Fig. 2-2: two $1 dollar bills, four dimes, and three pennies. This money example illustrates the very important idea of *place value*.

Consider the decimal number 648 in Fig. 2-3. The digit 6 represents 600 because of its placement three positions left of the decimal point. The digit 4 represents 40 because of its placement two positions left of the decimal point.

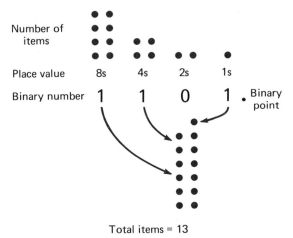

HUNDREDS	TENS	UNITS
648 =	600 +	40 + 8

Fig. 2-3   Place value in the decimal system.

The digit 8 represents eight units because of its placement one position left of the decimal point. The total number 648, then, represents six hundred and forty-eight units. This is an example of place value in the decimal number system.

The binary number system also uses the idea of place value. What does the binary number 1101 (say "one one zero one") mean? Figure 2-4 shows that the digit 1 nearest the binary

Number of items

Place value   8s   4s   2s   1s

Binary number   1   1   0   1   • Binary point

Total items = 13

Fig. 2-4   Place value in the binary number system.

point is the units, or 1s, position, and so we add one item. The digit 0 in the 2s position tells us we have no 2s. The digit 1 in the 4s position tells us to add four items. The digit 1 in the 8s position tells us to add eight more items. When we count all the items, we find that the binary number 1101 represents 13 items.

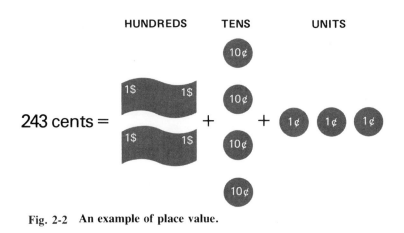

243 cents =

HUNDREDS   TENS   UNITS

1$ 1$ 1$ 1$ +   10¢ 10¢ 10¢ 10¢   +   1¢ 1¢ 1¢

Fig. 2-2   An example of place value.

2⁹	2⁸	2⁷	2⁶	2⁵	2⁴	2³	2²	2¹	2⁰
512s	256s	128s	64s	32s	16s	8s	4s	2s	1s

Binary point

Fig. 2-5 Value of the places left of the binary point in decimal and in powers of 2.

How about the binary number 1100 (say "one one zero zero")? Using the system from Fig. 2-4, we find that we have the following:

8s	4s	2s	1s	place value
yes	yes	no	no	binary
(1)	(1)	(0)	(0)	number
•• •• •• ••	•• ••			number of items

The binary number 1100, then, represents 12 items.

Figure 2-5 shows each place in the value of the binary system. Notice that each place value is determined by multiplying the one to the right by 2. The term "base 2" for binary comes from this idea.

Many times the weight or value of each place in the binary number system is referred to as a power of 2. In Fig. 2-5, place values for a binary number are shown in decimal and also in powers of 2. For instance, the 8s place is the same as the 2^3 position, the 32s place the same as the 2^5 position, etc.

Self-Test

Supply the missing number in each statement.

5. The 1 in the binary number 1000 has a place value of _____ in decimal.
6. The binary number 1010 equals _____ in decimal.
7. The binary number 100000 equals _____ in decimal.
8. The number 2^7 equals _____ in decimal.

2-3 BINARY TO DECIMAL CONVERSION

While working with digital equipment you will have to convert from the binary code to decimal numbers. If you are given the binary number 110011, what would you say it equals in decimal? First write down the binary number as:

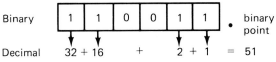

Start at the binary point and work to the left. For each binary 1, place the decimal value of that position (see Fig. 2-5) below the binary digit. Add the four decimal numbers to find the decimal equivalent. You will find that binary 110011 equals the decimal number 51.

Another practical problem is to convert the binary number 101010 to a decimal number. Again write down the binary number as:

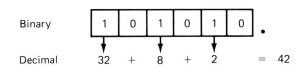

Starting at the binary point, write the place value (see Fig. 2-5) of each digit below the square in decimals. Add the three decimal numbers to get the decimal total. You will find that the binary number 101010 equals the decimal number 42.

Now try a long and difficult binary number: convert the binary number 1111101000 to a decimal number. Write down the binary number as:

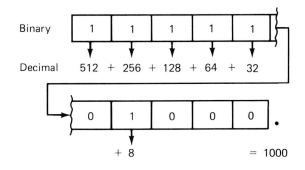

CHAPTER 2   NUMBERS WE USE IN DIGITAL ELECTRONICS   **21**

From Fig. 2-3, convert each binary digit 1 to its correct decimal value. Add the decimal values to get the decimal total. The binary number 1111101000 equals the decimal number 1000.

Self-Test

Supply the missing number in each statement.

9. The binary number 1111 equals _____ in decimal.
10. The binary number 100010 equals _____ in decimal.
11. The binary number 1000001010 equals _____ in decimal.

2-4 DECIMAL TO BINARY CONVERSION

Many times while working with digital electronic equipment you must be able to convert a decimal number to a binary number. We shall teach you a method that will help you to make this conversion.

Suppose you want to convert the decimal number 13 to a binary number. One procedure you can use is:

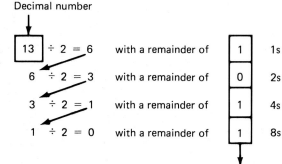

Notice that 13 is first divided by 2, giving a quotient of 6 with a remainder of 1. This remainder becomes the 1s place in the binary number. The 6 is then divided by 2, giving a quotient of 3 with a remainder of 0. This remainder becomes the 2s place in the binary number. The 3 is then divided by 2, giving a quotient of 1 with a remainder of 1. This remainder becomes the 4s place in the binary number. The 1 is then divided by 2, giving a quotient of 0 with a remainder of 1. This re-

mainder becomes the 8s place in the binary number. The decimal number 13 has been converted to the binary number 1101.

Practice this procedure by converting the decimal number 37 to a binary number. Follow the procedure you used before:

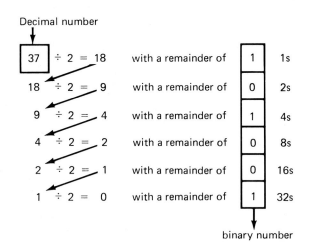

Notice that you stop dividing by 2 when the quotient becomes 0. According to this procedure, the decimal number 37 is equal to the binary number 100101.

Self-Test

Supply the missing number in each statement.

12. The decimal number 39 equals _____ in binary.
13. The decimal number 100 equals _____ in binary.
14. The decimal number 133 equals _____ in binary.

2-5 ELECTRONIC TRANSLATORS

If you were to try to communicate with a French person who did not know the English language, you would need someone to *translate* the English into French and then the French into English. A similar problem exists in digital electronics. Almost all digital circuits (calculators, computers) understand only binary numbers. But most people understand only decimal numbers. Thus we must have electronic devices that can translate from dec-

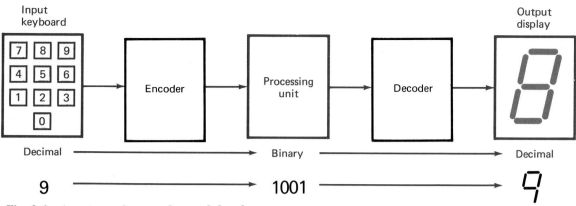

Fig. 2-6   A system using encoders and decoders.

imal to binary numbers and from binary to decimal numbers.

Figure 2-6 diagrams a typical system that might be used to translate from decimal to binary numbers and back to decimals. The device that translates from the keyboard decimal numbers to binary is called an *encoder*; the device labeled *decoder* translates from binary to decimal numbers.

Near the bottom of Fig. 2-6 is a typical conversion. If you press the decimal number 9 on the keyboard, the encoder will convert the 9 into the binary number 1001. The decoder will convert the binary 1001 into the decimal number 9 on the output display.

Encoders and decoders are very common electronic circuits in all digital devices. A pocket calculator, for instance, must have encoders and decoders to electronically translate from decimal to binary numbers and back to decimal. Figure 2-6 is a very basic block diagram of a pocket calculator. When you press the number 9 on the keyboard the number appears on the output display.

You can buy encoders and decoders that translate from any of the commonly used codes in digital electronics. Most of the encoders and decoders you will use will be packaged as single ICs.

Self-Test

Supply the missing word in each statement.

15. A(n) _____ is an electronic device that translates a decimal input number to binary.

16. The processing unit of a calculator outputs binary. This binary is converted to a decimal output display by an electronic device called a(n) _____ .

2-6 HEXADECIMAL NUMBERS

The *hexadecimal number system* uses the 16 symbols 0, 1, 2, 3, 4, 5, 6, 7, 8, 9, A, B, C, D, E, and F and is referred to as the *base 16 system*. The table shown in Fig. 2-7 gives the equivalent binary and hexadecimal representations for the decimal numbers 0 through 17. Note that the letter "A" stands for 10, "B" for 11, and so on. The advantage of the hexadecimal system is its usefulness in converting directly from a 4-bit binary number. For instance, hexadecimal F stands for the 4-bit binary number 1111. Hexadecimal notation is typically used to represent a binary number. For instance, the hexadecimal number A6 would represent the 8-bit binary number 10100110. Hexadecimal notation is widely used in microprocessor-based systems to represent 8-, 16-, or 32-bit binary numbers.

Decimal	Binary	Hexadecimal
0	0000	0
1	0001	1
2	0010	2
3	0011	3
4	0100	4
5	0101	5
6	0110	6
7	0111	7
8	1000	8
9	1001	9
10	1010	A
11	1011	B
12	1100	C
13	1101	D
14	1110	E
15	1111	F
16	10000	10
17	10001	11

Fig. 2-7   Binary and hexadecimal equivalents to decimal numbers 0 through 17.

The number 10 represents how many objects? It can be observed from the table shown in Fig. 2-7 that the number 10 could mean ten objects, two objects, or sixteen objects depending on the base of the number. *Subscripts* are sometimes added to a number to indicate the base of the number. Using subscripts, the number 10_{10} represents ten objects. The subscript (10 in this example) indicates it is a base 10, or decimal number. Using subscripts, the number 10_2 represents two objects since this is in binary (base 2). Again using subscripts, the number 10_{16} represents sixteen objects since this is in hexadecimal (base 16).

Converting hexadecimal numbers to binary and binary numbers to hexadecimal is a common task when working with microprocessors and microcomputers. Consider converting $C3_{16}$ to its binary equivalent. In Fig. 2-8(a), each hexadecimal digit is converted to its 4-bit binary equivalent (see Fig. 2-7). Hexadecimal C equals the 4-bit binary number 1100, while 3_{16} equals 0011. Combining the binary groups yields $C3_{16} = 11000011_2$.

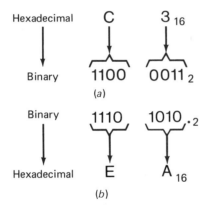

Fig. 2-8 (*a*) **Converting a hexadecimal number to binary.** (*b*) **Converting a binary number to hexadecimal.**

Now reverse the process and convert the binary number 11101010 to its hexadecimal equivalent. The simple process is detailed in Fig. 2-8(*b*). The binary number is divided into 4-bit groups starting at the binary point. Next, each 4-bit group is translated into its equivalent hexadecimal number with the help of the table shown in Fig. 2-7. The example in Fig. 2-8(*b*) shows $11101010_2 = EA_{16}$.

Consider converting hexadecimal $2DB_{16}$ to its decimal equivalent. The place values for the first three places in the hexadecimal number are shown across the top in Fig. 2-9 as

256s, 16s, and 1s. In Fig. 2-9 there are eleven 1s. There are thirteen 16s, which equals 208. There are two 256s, which equals 512. Adding $11 + 208 + 512$ will equal 731_{10}. The example given in Fig. 2-9 shows that $2DB_{16} = 731_{10}$.

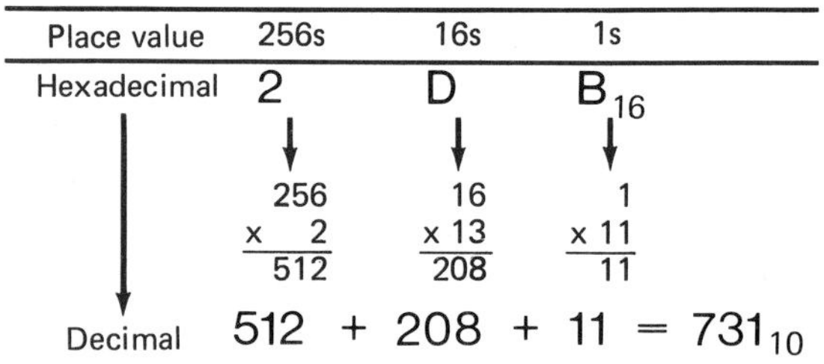

Place value	256s	16s	1s
Hexadecimal	2	D	B₁₆

$$\begin{array}{ccc} 256 & 16 & 1 \\ \times\ 2 & \times\ 13 & \times\ 11 \\ \hline 512 & 208 & 11 \end{array}$$

Decimal $\quad 512 + 208 + 11 = 731_{10}$

Fig. 2-9 Converting a hexadecimal number to a decimal number using the repeated divide-by-16 process.

Now reverse the process and convert the decimal number 47 to its hexadecimal equivalent. Figure 2-10 details the repeated divide-by-16 process. The decimal number 47 is first divided by 16, resulting in a quotient of 2 with a remainder of 15. The remainder of 15 (F in hexadecimal) becomes the least significant digit (LSD) of the hexadecimal number. The quotient (2 in this example) is transferred to the dividend position and divided by 16. This results in a quotient of 0 with a remainder of 2. The 2 becomes the next digit in the hexadecimal number. The process is complete because the integer part of the quotient is 0. The divide-by-16 process shown in Fig. 2-10 converts 47_{10} to its hexadecimal equivalent of $2F_{16}$.

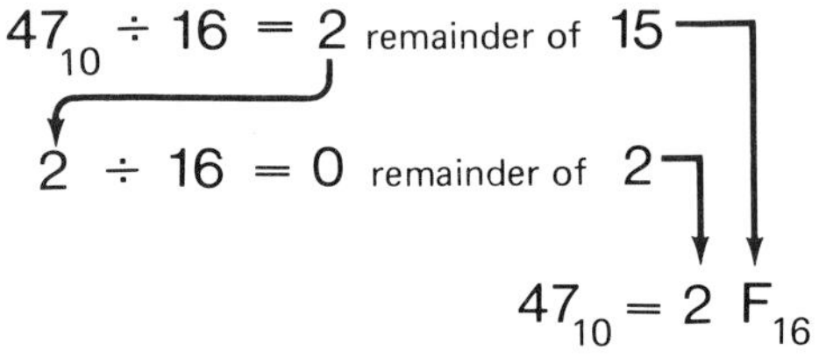

$$47_{10} \div 16 = 2 \text{ remainder of } 15$$
$$2 \div 16 = 0 \text{ remainder of } 2$$
$$47_{10} = 2F_{16}$$

Fig. 2-10 Converting a decimal number to hexadecimal using the repeated divide-by-10 process.

Self-Test

Supply the missing number in each statement.

17. Decimal 15 equals _____ in hexadecimal.
18. Hexadecimal A6 equals _____ in binary.

19. Binary 11110 equals _____ in hexadecimal.
20. Hexadecimal 1F6 equals _____ in decimal.
21. Decimal 63 equals _____ in hexadecimal.

2-7 OCTAL NUMBERS

Some older computer systems use octal numbers to represent binary information. The *octal number system* uses the eight symbols 0, 1, 2, 3, 4, 5, 6, and 7. Octal numbers are also referred to as *base 8* numbers. The table shown in Fig. 2-11 gives the equivalent binary and octal representations for decimal numbers 0 through 17. The advantage of the octal system is its usefulness in converting directly from a 3-bit binary number. Octal notation is used to represent binary numbers.

Decimal	Binary	Octal
0	000	0
1	001	1
2	010	2
3	011	3
4	100	4
5	101	5
6	110	6
7	111	7
8	001 000	10
9	001 001	11
10	001 010	12
11	001 011	13
12	001 100	14
13	001 101	15
14	001 110	16
15	001 111	17
16	010 000	20
17	010 001	21

Fig. 2-11 Binary and octal equivalents to decimal numbers 0 through 17.

Converting octal numbers to binary is a common operation when using some computer systems. Consider converting the octal number 67_8 (say "six seven base eight") to its binary equivalent. In Fig. 2-12(*a*), each octal digit is converted to its 3-bit binary equivalent. Octal 6 equals 110, while 7 equals 111. Combining the binary groups yields $67_8 = 110111_2$.

Now reverse the process and convert binary 100011101 to its octal equivalent. The simple

process is detailed in Fig. 2-12(*b*). The binary number is divided into 3-bit groups (100 011 101) starting at the binary point. Next, each 3-bit group is translated into its equivalent octal number. The example in Fig. 2-12(*b*) shows $100\ 001\ 101_2 = 415_8$.

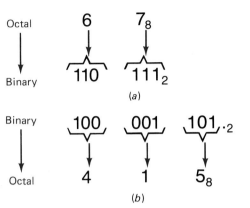

Fig. 2-12 (*a*) Converting an octal number to binary. (*b*) Converting a binary number to octal.

Consider converting octal 415 (say "four one five") to its decimal equivalent. The place values for the first three places in the octal number are shown across the top in Fig. 2-13 as 64s, 8s, and 1s. There are five 1s and one 8. There are four 64s, which equals 256. Adding $5 + 8 + 256 = 269_{10}$. The example in Fig. 2-13 shows that $415_8 = 269_{10}$.

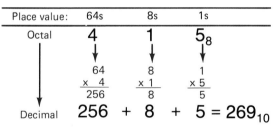

Fig. 2-13 Converting an octal number to decimal.

Now reverse the process and convert the decimal number 498 to its octal equivalent. Figure 2-14 on the next page details the repeated divide-by-8 process. The decimal number 498 is first divided by 8, resulting in a quotient of 62 with a remainder of 2. The remainder (2) becomes the LSD of the octal number. The quotient (62 in this example) is transferred to the dividend position and divided by 8. This results in a quotient of 7 with a remainder of 6. The 6

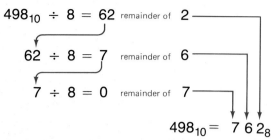

$$498_{10} \div 8 = 62 \text{ remainder of } 2$$

$$62 \div 8 = 7 \text{ remainder of } 6$$

$$7 \div 8 = 0 \text{ remainder of } 7$$

$$498_{10} = 7\,6\,2_8$$

Fig. 2-14 Converting a decimal number to octal using the repeated divide-by-8 process.

becomes the next digit in the octal number. The last quotient (7 in this example) is transferred to the dividend position and divided by 8. The quotient is 0 with a remainder of 7. The 7 is the most significant digit (MSD) in the octal number. The divide-by-8 process shown in Fig. 2-14 converts 498_{10} to its octal equivalent of 762_8. Note that the signal to end the repeated divide-by-8 process is when the quotient becomes 0.

Technicians, engineers, and programmers must be able to convert between number systems. Many commercial calculators can aid in making binary, octal, hexadecimal, and decimal conversions. These calculators also perform arithmetic operations on binary, octal, and hexadecimal as well as decimal numbers.

Self-Test

Supply the missing number in each statement.

22. Octal 73 equals _____ in binary.
23. Binary 100000 equals _____ in octal.
24. Octal 753 equals _____ in decimal.
25. Decimal 63 equals _____ in octal.

SUMMARY

1. The decimal number system contains 10 symbols: 0, 1, 2, 3, 4, 5, 6, 7, 8, and 9.
2. The binary number system contains two symbols: 0 and 1.
3. The place values left of the binary point in binary are 64, 32, 16, 8, 4, 2, and 1.
4. All men and women in the field of digital electronics must be able to convert binary to decimal numbers and decimal to binary numbers.
5. Encoders are electronic circuits that convert decimal numbers to binary numbers.
6. Decoders are electronic circuits that convert binary numbers to decimal numbers.
7. The hexadecimal number system contains 16 symbols: 0, 1, 2, 3, 4, 5, 6, 7, 8, 9, A, B, C, D, E, and F.
8. Hexadecimal digits are widely used to represent binary numbers in the computer field.
9. The octal number system uses eight symbols: 0, 1, 2, 3, 4, 5, 6, and 7. Octal numbers are used to represent binary numbers in a few computer systems.

CHAPTER REVIEW QUESTIONS

Answer the following questions.

2-1. How would you say the decimal number 1001?
2-2. How would you say the binary number 1001?
2-3. Convert the binary numbers in **a** to **j** to decimal numbers:

a. 1	**f.** 10000
b. 100	**g.** 10101
c. 101	**h.** 11111
d. 1011	**i.** 11001100
e. 1000	**j.** 11111111

2-4. Convert the decimal numbers in **a** to **j** to binary numbers:
- **a.** 0
- **b.** 1
- **c.** 18
- **d.** 25
- **e.** 32
- **f.** 64
- **g.** 69
- **h.** 128
- **i.** 145
- **j.** 1001

2-5. Encode the decimal numbers in **a** to **f** to binary numbers:
- **a.** 9
- **b.** 3
- **c.** 15
- **d.** 13
- **e.** 10
- **f.** 2

2-6. Decode the binary numbers in **a** to **f** to decimal numbers:
- **a.** 0010
- **b.** 1011
- **c.** 1110
- **d.** 0111
- **e.** 0110
- **f.** 0000

2-7. What is the job (function) of an encoder?

2-8. What is the job (function) of a decoder?

2-9. Write the decimal numbers from 0 to 15 in binary.

2-10. Convert the hexadecimal numbers in **a** to **d** to binary numbers:
- **a.** 8A
- **b.** B7
- **c.** 6C
- **d.** FF

2-11. Convert the binary numbers in **a** to **d** to hexadecimal numbers:
- **a.** 01011110
- **b.** 00011111
- **c.** 11011011
- **d.** 00110000

2-12. Hexadecimal 3E6 = _____ $_{10}$.

2-13. Decimal 4095 = _____ $_{16}$.

2-14. Octal 156 = _____ $_{10}$.

2-15. Decimal 391 = _____ $_{8}$.

Answers to Self-Tests

1. base 2
2. 1000
3. 6
4. 7
5. 8 or (2^3)
6. 10
7. 32
8. 128
9. 15
10. 34
11. 522
12. 100111
13. 1100100
14. 10000101
15. encoder
16. decoder
17. F
18. 10100110
19. 1E
20. 502
21. 3F
22. 111011
23. 40
24. 491
25. 77

CHAPTER 3

Binary Logic Gates

Computers, calculators, and other digital devices are sometimes looked upon by the general public as being magical. Actually, digital electronic devices are extremely **logical** *in their operation. The basic building block of any digital circuit is a* **logic gate.** *The logic gates you will use operate with binary numbers, hence the term* **binary logic gates.** *Persons working in digital electronics understand and use binary logic gates every day. By the end of this chapter you will have memorized the basic logic gate symbols and characteristics. Remember that logic gates are the building blocks for even the most complex computers. Logic gates can be constructed by using simple switches, relays, vacuum tubes, transistors and diodes, or ICs. Because of their availability, wide use, and low cost, ICs will be used to construct digital circuits. A variety of logic gates are available in all logic families including TTL and CMOS.*

3-1 THE AND GATE

The AND gate is sometimes called the "all or nothing gate." Figure 3-1 shows the basic idea of the AND gate using simple switches.

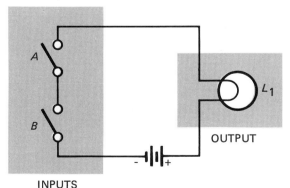

INPUTS

Fig. 3-1   AND circuit using switches.

What must be done in Fig. 3-1 to get the output lamp (L_1) to light? You must close *both* switches *A* and *B* to get the lamp to light. You could say that switch *A and* switch *B* must be closed to get the output to light. The AND gates you will operate most often are constructed of diodes and transistors and packaged inside an IC. To show the AND gate we use the *logic symbol* in Fig. 3-2. This standard AND gate symbol is used whether we are using relays, switches, pneumatic circuits, discrete diodes and transistors, or ICs. This is the

INPUTS $\quad \begin{matrix} A \\ B \end{matrix}$ — Y OUTPUT

Fig. 3-2   AND gate logic symbol.

symbol you will memorize and use from now on for AND gates.

The term "logic" is usually used to refer to a decision-making process. A logic gate, then, is a circuit that can decide to say yes or no at the output based upon the inputs. We already determined that the AND gate circuit in Fig. 3-1 says yes (light on) at the output only when we have a yes (switches closed) at *both* inputs.

Now let us consider an actual circuit similar to one you will set up in the laboratory. The AND gate in Fig. 3-3 is connected to input switches A and B. The output indicator is an LED. If a LOW voltage (GND) appears at inputs A and B, then the output LED is *not lit*. This situation is illustrated in line 1, Fig. 3-4. Notice also in line 1 that the inputs and outputs are given as *binary digits*. Line 1 indicates that if the inputs are binary 0 and 0, then the output will be a binary 0. Carefully look over the four combinations of switches A and B in Fig. 3-4. Notice that only binary 1s at both inputs A and B will produce a binary 1 at the output (see line 4).

It is a +5 V compared to GND appearing at A, B, or Y that is called a binary 1 or a HIGH voltage. A binary 0, or LOW voltage, is defined as a GND voltage (near 0 V compared to GND) appearing at A, B, or Y. We are using *positive logic* because it takes a *positive* 5 V to produce what we call a binary 1. You will use positive logic in most of your work in digital electronics.

The table in Fig. 3-4 is called a *truth table*. The truth table for the AND gate gives all the possible input combinations of A and B and the resulting outputs. Thus the truth table defines very exactly the operation of the AND gate. The truth table in Fig. 3-4 is said to describe the AND *function*. You will memorize the truth table for the AND function. The unique output from the AND gate is a HIGH only when all inputs are HIGH. The output column in Fig. 3-4 shows that *only* line 4 in the AND truth table generates a 1 while the rest of the outputs are 0.

So far you have memorized the logic symbol and the truth table for the AND gate. Now you will learn a shorthand method of writing the statement "input A is ANDed with input B to get output Y." The short method used to represent this statement is called its *Boolean expression* ("Boolean" from Boolean algebra—the algebra of logic). The Boolean expression is a universal language used by engineers and technicians in digital electronics. Figure 3-5 shows the ways to express that input A is ANDed with input B to produce output Y. The top expression in Fig. 3-5 is how you would tell someone in the English language that you are ANDing inputs A and B to get output Y. Next in Fig. 3-5 you see the Boolean expression for ANDing inputs A and B. Note that a multiplication dot ($\cdot$) is used to symbolize the AND function in Boolean expressions. Figure 3-5, then, illustrates the four commonly used ways to express the ANDing of inputs A and B. All these methods are widely used and must be learned by personnel in digital electronics.

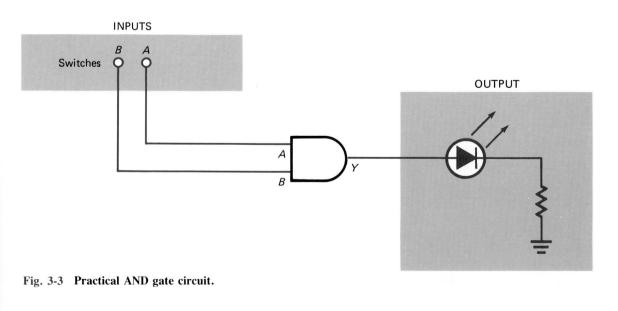

INPUTS

Fig. 3-3 **Practical AND gate circuit.**

INPUTS				OUTPUT	
B		**A**		**Y**	
Switch voltage	Binary	Switch voltage	Binary	Light	Binary
Line 1	LOW	0	LOW	0	No	0
Line 2	LOW	0	HIGH	1	No	0
Line 3	HIGH	1	LOW	0	No	0
Line 4	HIGH	1	HIGH	1	Yes	1

Fig. 3-4   **AND gate truth table.**

OR gate

"Any or all" gate

Inclusive OR function

OR gate truth table

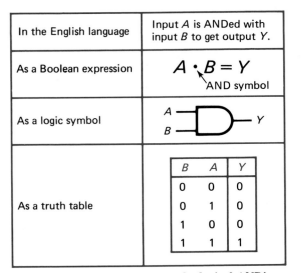

In the English language	Input A is ANDed with input B to get output Y.
As a Boolean expression	$A \cdot B = Y$ AND symbol
As a logic symbol	A, B → Y
As a truth table	B A Y 0 0 0 0 1 0 1 0 0 1 1 1

Fig. 3-5   **Four ways to express the logical ANDing of A and B.**

Self-Test

Answer the following questions.

1. The output side of an AND gate logic symbol is _____ (flat, pointed, round).
2. Write the Boolean expression for a two-input AND gate.
3. Refer to Fig. 3-3. When both inputs are HIGH, output Y will be _____ and the LED will _____ (light, not light).

3-2 THE OR GATE

The OR gate is sometimes called the "any or all gate." Figure 3-6 illustrates the basic idea of the OR gate using simple switches. Looking at the circuit in Fig. 3-6, you can see that the output lamp will light when *either* or *both* of the input switches are closed but not when both are open. A truth table for the OR circuit is shown in Fig. 3-7. The truth table lists the switch and light conditions for the OR gate circuit in Fig. 3-6. The truth table in Fig. 3-7 is said to describe the *inclusive OR function*. The unique output from the OR gate is a LOW only when all inputs are LOW. The output column in Fig. 3-7 shows that *only* line 1 in the OR truth table generates a 0 while all others are 1.

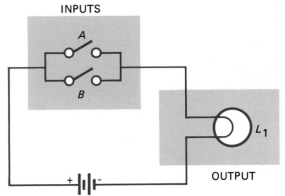

Fig. 3-6   **OR circuit using switches.**

INPUTS				OUTPUT	
B		**A**		**Y**	
Switch	Binary	Switch	Binary	Light	Binary
Open	0	Open	0	No	0
Open	0	Closed	1	Yes	1
Closed	1	Open	0	Yes	1
Closed	1	Closed	1	Yes	1

Fig. 3-7   **OR gate truth table.**

The logic symbol for the OR gate is diagramed in Fig. 3-8 on the next page. Notice in the logic diagram that inputs *A* and *B* are being ORed to produce an output *Y*. The engineer's Boolean expression for the OR function is also illustrated in Fig. 3-8. Note that the plus (+) sign is the Boolean symbol for OR.

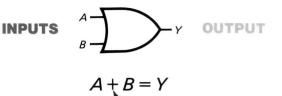

INPUTS ⟶ Y OUTPUT

$$A + B = Y$$

OR symbol

Fig. 3-8 OR gate logic symbol and Boolean expression.

You should now know the logic symbol, Boolean expression, and truth table for the OR gate.

Self-Test

Answer the following questions.

4. The output side of an OR gate logic symbol is _____ (flat, pointed, round).
5. Write the Boolean expression for a two-input OR gate.
6. Refer to Fig. 3-8. If inputs A and B are both LOW, output Y will be _____ .
7. Refer to Fig. 3-7. This truth table describes the _____ (exclusive, inclusive) OR logic function.

3-3 THE INVERTER AND BUFFER

All the gates so far have had at least two inputs and one output. The NOT circuit, however, has only one input and one output. The NOT circuit is often called an *inverter*. The job of the NOT circuit (inverter) is to give an output that is not the same as the input. The logic symbol for the inverter (NOT circuit) is shown in Fig. 3-9.

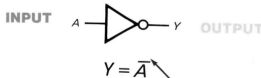

INPUT ⟶ A ⟶ Y OUTPUT

$$Y = \overline{A}$$

NOT symbol

Fig. 3-9 A logic symbol and Boolean expression for an inverter.

If we were to put in a logical 1 at input A in Fig. 3-9, we would get out the opposite, or a logical 0, at output Y. We say that the inverter *complements* or *inverts* the input. Figure 3-9 also shows how we would write a Boolean expression for the NOT, or INVERT, function. Notice the use of the over bar (‾) symbol above the output to show that A has been in-

verted or complemented. We say that the Boolean term "$\overline{A}$" would be "not A."

The truth table for the inverter is shown in Fig. 3-10. If the voltage at the input of the inverter is LOW, then the output voltage is HIGH. However, if the input voltage is HIGH, then the output is LOW. As you learned, the output is always opposite the input. The truth table also gives the characteristics of the inverter in terms of binary 0s and 1s.

INPUT		OUTPUT	
A		*Y*	
Voltages	Binary	Voltages	Binary
LOW	0	HIGH	1
HIGH	1	LOW	0

Fig. 3-10 Truth table for an inverter.

You learned that when a signal passes through an inverter, it can be said that the input is inverted or complemented. We can also say it is *negated*. The terms "negated," "complemented," and "inverted," then, mean the same thing.

The logic diagram in Fig. 3-11 shows an arrangement where input A is run through two inverters. Input A is first inverted to produce a "not A" ($\overline{A}$) and then inverted a second time for a "double not A" ($\overline{\overline{A}}$). In terms of binary digits, we find that when the input 1 is inverted twice, we end up with the original digit. Therefore we find that $\overline{\overline{A}}$ equals A. Thus, a Boolean term with two bars over it is equal to the term under the two bars, as shown at the bottom of Fig. 3-11.

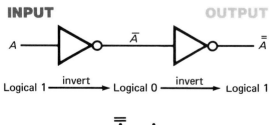

INPUT ⟶ OUTPUT

A ⟶ $\overline{A}$ ⟶ $\overline{\overline{A}}$

Logical 1 ──invert──▶ Logical 0 ──invert──▶ Logical 1

Therefore $\overline{\overline{A}} = A$

Fig. 3-11 Effect of double inverting.

Two symbols found in logic diagrams that look something like an inverter symbol are pictured in Fig. 3-12. The logic symbol in Fig.

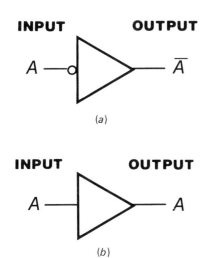

(a)

INPUT   OUTPUT

(b)

Fig. 3-12   (a) **Alternative inverter logic symbol.**
(b) **Noninverting buffer/driver logic symbol.**

3-12(a) is an alternative symbol for an inverter and performs the NOT function.

The symbol in Fig. 3-12(b) is that of a *noninverting buffer/driver*. The noninverting buffer serves no logical purpose but is used to supply greater *drive current* at its output than is normal for a regular gate. Since regular digital ICs have limited drive current capabilities, the noninverting buffer/driver is very important when interfacing ICs with other devices such as LEDs, lamps, etc. Buffer/drivers are available in both noninverting and inverting form.

You now know the logic symbols, Boolean expression, and truth table for an inverter or NOT gate. You also recognize the symbol for a noninverting buffer/driver and know its purpose is to drive LEDs, lamps, and so forth. You will use inverters and buffers every day as you work in the field of digital electronics.

Self-Test

Answer the following questions.

8. Refer to Fig. 3-9. If input *A* is HIGH, output *Y* from the inverter will be _____ .

9. Refer to Fig. 3-11. If input *A* to the left inverter is LOW, the output from the right inverter will be _____ .

10. Write the Boolean expression used to describe the action of an inverter.

11. List two words that are used to mean "inverted."

12. Refer to Fig. 3-12(b). If input *A* is LOW, the output from this buffer will be _____ .

3-4 THE NAND GATE

The AND, OR, and NOT gates are the three basic circuits that make up all digital circuits. The NAND gate is a NOT AND, or an inverted AND function. The standard logic symbol for the NAND gate is diagramed in Fig. 3-13(a). The little *invert bubble* (small circle) on the right end of the symbol means to invert the AND.

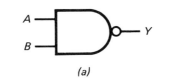

(a)

INPUTS   **OUTPUT**

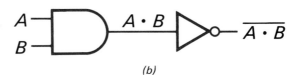

(b)

Fig. 3-13   (a) **NAND gate logic symbol.** (b) **A Boolean expression for the output of a NAND gate.**

Figure 3-13(b) shows a separate AND gate and inverter being used to produce the NAND logic function. Also notice that the Boolean expressions for the AND gate ($A \cdot B$) and the NAND ($\overline{A \cdot B}$) are shown on the logic diagram in Fig. 3-13(b).

The truth table for the NAND gate is shown at the right in Fig. 3-14. Notice that the truth table for the NAND gate is developed by *inverting the outputs* of the AND gate. The AND gate outputs are also given in the table for reference.

INPUTS		OUTPUT	
B	**A**	AND	NAND
0	0	0	1
0	1	0	1
1	0	0	1
1	1	1	0

Fig. 3-14   **Truth tables for AND and NAND gates.**

NAND gates are commonly employed in industrial practice and extensively used in all digital equipment. Do you know the logic symbol, Boolean expression, and truth table for the NAND gate? You must commit these to memory. The unique output from the NAND gate is a LOW only when all inputs are HIGH. The output column in Fig. 3-14 shows that *only* line 4 in the NAND truth table generates a 0 while all other outputs are 1.

Self-Test

Answer the following questions.

13. The output side of a NAND gate logic symbol is _____ (flat with an added invert bubble, pointed with an added invert bubble, round with an added invert bubble).
14. Write the Boolean expression for a two-input NAND gate.
15. Refer to Fig. 3-13(*a*). When both inputs *A* and *B* are HIGH, output *Y* from the NAND gate will be _____ . This _____ (is, is not) the unique output of the NAND gate.

3-5 THE NOR GATE

The NOR gate is actually a NOT OR gate. In other words, the output of an OR gate is inverted to form a NOR gate. The logic symbol for the NOR gate is diagramed in Fig.3-15(*a*). Note that the NOR symbol is an OR symbol with a small invert bubble (small circle) on the right side. The NOR function is being performed by an OR gate and an inverter in Fig. 3-15(*b*). The Boolean expression for the OR function ($A + B$) is shown. The Boolean expression for the final NOR function is $\overline{A + B}$.

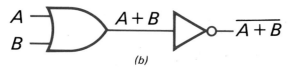

(a)

INPUTS   **OUTPUT**

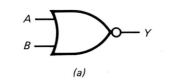
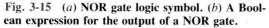

(b)

Fig. 3-15 (*a*) **NOR gate logic symbol.** (*b*) **A Boolean expression for the output of a NOR gate.**

The truth table for the NOR gate is shown at the right in Fig. 3-16. Notice that the NOR gate truth table is just the complement of the output of the OR gate. The output of the OR gate is also included in the truth table in Fig. 3-16 for reference.

INPUTS		OUTPUT	
B	**A**	OR	NOR
0	0	0	1
0	1	1	0
1	0	1	0
1	1	1	0

Fig. 3-16 Truth table for OR and NOR gates.

You now should memorize the symbol, Boolean expression, and truth table for the NOR gate. You will encounter these items often in your work in digital electronics. The unique output from the NOR gate is a HIGH only when all inputs are LOW. The output column in Fig. 3-16 shows that *only* line 1 in the NOR truth table generates a 1 while all other outputs are 0.

Self-Test

Answer the following questions.

16. The output side of a NOR gate logic symbol is _____ (flat with an added invert bubble, pointed with an added invert bubble, round with an added invert bubble).
17. Write the Boolean expression for a two-input NOR logic gate.
18. Refer to Fig. 3-15. If input *A* is LOW and input *B* is HIGH, output *Y* of the NOR gate will be _____ . This _____ (is, is not) the unique output of the NOR gate.
19. Refer to Fig. 3-15. When both inputs are LOW, output *Y* of the NOR gate will be _____ . This _____ (is, is not) the unique output of the NOR gate.

3-6 THE EXCLUSIVE OR GATE

The exclusive OR gate is sometimes referred to as the "any but not all gate." The term "exclusive OR gate" is often shortened to "XOR gate." The logic symbol for the XOR gate is diagramed in Fig. 3-17(*a*); the Boolean expression for the XOR function is illustrated in Fig. 3-17(*b*). The symbol $\oplus$ means the terms are XORed together.

A truth table for the XOR gate is shown at the right in Fig. 3-18. Notice that if any but not all of the inputs are 1, then the output will be a binary, or logical, 1. The OR gate truth table is also given in Fig. 3-18, so that you may compare the OR gate truth table with the XOR gate truth table.

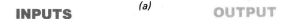

(b)

Fig. 3-17 (*a*) **XOR gate logic symbol.** (*b*) **A Boolean expression for the output of an XOR gate.**

INPUTS		OUTPUT	
B	**A**	OR	XOR
0	0	0	0
0	1	1	1
1	0	1	1
1	1	1	0

Fig. 3-18 Truth table for OR and XOR gates.

An odd number (1, 3, 5, and so on) of HIGH inputs to an XOR gate generates a HIGH output, whereas an even number (0, 2, 4, and so on) of HIGH inputs produces a LOW output from an XOR gate. As an example, consider the XOR truth table in Fig. 3-18. Line 4 shows an *even* number (2) of inputs HIGH, and therefore the output is LOW. Line 3 in Fig. 3-18 shows an *odd* number (1) of inputs HIGH, and therefore the output is HIGH.

Self-Test

Answer the following questions.

20. The exclusive OR gate is sometimes referred to as the _____ (five words).
21. Write the Boolean expression for a two-input XOR gate.
22. Refer to Fig. 3-17(*a*). If both inputs are HIGH, output *Y* from the XOR gate will be _____ .
23. Refer to Fig. 3-17 (*a*). If an odd number of inputs are HIGH, the output from an XOR gate will be _____ .

3-7 THE EXCLUSIVE NOR GATE

The term "exclusive NOR gate" is often shortened to "XNOR gate." The logic symbol for the XNOR gate is shown in Fig. 3-19(*a*). Notice that it is the XOR symbol with the added invert bubble on the output side. Figure 3-19(*b*) illustrates one of the Boolean expressions used for the XNOR function. Observe that the Boolean expression for the XNOR gate is $\overline{A \oplus B}$. The bar over the $A \oplus B$ expression tells us we have inverted the output of the XOR gate. Examine the truth table in Fig. 3-20. Notice that the output of the XNOR gate is the complement of the XNOR truth table. The XOR gate output is also shown in the table in Fig. 3-20 .

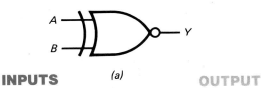

(a)

INPUTS   **OUTPUT**

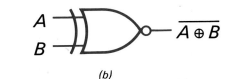

(b)

Fig. 3-19 (*a*) **XNOR gate logic symbol.** (*b*) **A Boolean expression for the output of an XNOR gate.**

INPUTS		OUTPUT	
B	**A**	XOR	XNOR
0	0	0	1
0	1	1	0
1	0	1	0
1	1	0	1

Fig. 3-20 Truth table for XOR and XNOR gates.

You now will have mastered the logic symbol, truth table, and Boolean expression for the XNOR gate.

Self-Test

Answer the following questions.

24. The logic symbol for an XNOR gate is formed by drawing a(n) _____ bubble at the output of the _____ logic symbol.
25. Write the Boolean expression for a two-input XNOR gate.
26. Refer to Fig. 3-19(*a*). If input *A* is LOW and *B* is HIGH, output *Y* of the XNOR gate will be _____ .

27. Refer to Fig. 3-19(a). If an even number of inputs are HIGH, output Y of the XNOR gate will be _____ .

3-8 THE NAND GATE AS A UNIVERSAL GATE

So far in this chapter you have learned the basic building blocks used in all digital circuits. You also have learned about the seven types of gating circuits and now know the characteristics of the AND, OR, NAND, NOR, XOR, and XNOR gates and the inverter. From a manufacturer you can buy ICs that perform any of these seven basic functions.

In looking through manufacturers' literature you will find that NAND gates seem to be more widely available than many other types of gates. Because of the NAND gate's wide use, we shall show how it can be used to make other types of gates. We shall be using the NAND gate as a sort of "universal gate."

The chart in Fig. 3-21 shows how you would wire NAND gates to create any of the other basic logic functions. The logic function to be performed is listed in the left column of the table; the symbol for that function is listed in the center column. In the right column of Fig. 3-21 is a symbol diagram of how NAND gates would be wired to perform the logic function. The chart in Fig. 3-21 need *not* be memorized,

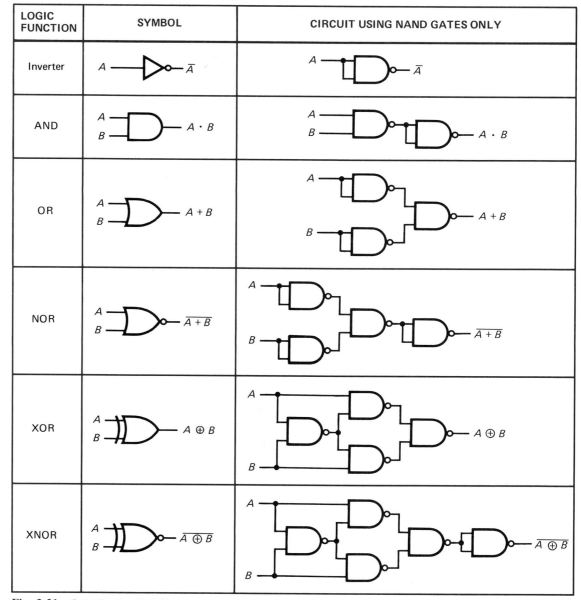

Fig. 3-21   Substituting NAND gates.

but it may be referred to as needed in your future work in digital electronics.

Self-Test

Answer the following questions.

28. The NAND gate can perform the invert function if the inputs are _____ (connected together, left open).
29. How many two-input NAND gates must be used to produce the two-input OR function?

3-9 GATES WITH MORE THAN TWO INPUTS

Until now we have used gates that have had only two or fewer inputs, but often we shall need gates with more than just two inputs. Figure 3-22(a) shows a three-input AND gate. The Boolean expression for the three-input AND gate is $A \cdot B \cdot C = Y$, as illustrated in Fig. 3-22(b). All the possible combinations of inputs A, B, and C are given in the truth table in Fig.

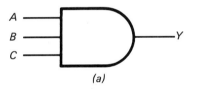

$$A \cdot B \cdot C = Y$$

(b)

INPUTS			OUTPUTS
C	B	A	Y
0	0	0	0
0	0	1	0
0	1	0	0
0	1	1	0
1	0	0	0
1	0	1	0
1	1	0	0
1	1	1	1

(c)

Fig. 3-22   Three-input AND gate. (*a*) Logic symbol. (*b*)Boolean expression. (*c*) Truth table.

3-22(c); the outputs for the three-input AND gate are tabulated in the right column of the truth table. Notice that with three inputs the possible combinations in the truth table have increased to eight.

How could you produce a three-input AND gate as illustrated in Fig. 3-22 if you have only two-input AND gates available? The solution is given in Fig. 3-23(a). Note the wiring of the two-input AND gates on the right side of the diagram to form a three-input AND gate. Figure 3-23(b) illustrates how a four-input AND gate could be wired by using available two-input AND gates.

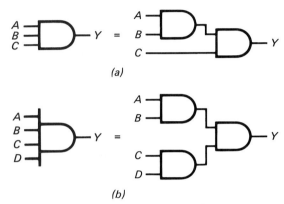

Fig. 3-23   Expanding the number of inputs. (*a*) **using two AND gates to accommodate three inputs.** (*b*) **using three AND gates to accommodate four inputs .**

A four-input OR is illustrated in Fig. 3-24(a) on the next page. The Boolean expression for the four-input OR gate is $A + B + C + D = Y$. This Boolean expression is written in Fig. 3-24(b). Read the Boolean expression $A + B + C + D = Y$ as "input A or input B or input C or input D will equal output Y." Remember that the + symbol means the logic function OR in Boolean expressions. The truth table for the four-input OR gate is shown in Fig. 3-24(c). Notice that because of the four inputs there are 16 possible combinations of A, B, C, and D. To wire the four-input OR gate, you could buy the correct gate from a manufacturer of digital logic circuits or you could use two-input OR gates to wire the four-input OR gate. Figure 3-25(a) diagrams how you could wire a four-input OR gate using two-input OR gates. Figure 3-25(b) shows how to convert two-input OR gates into a three-input OR gate. Notice that the *pattern* of connecting both OR and AND gates to expand the number of inputs is the same (compare Figs. 3-23 and 3-25).

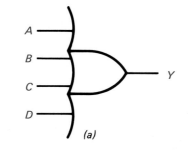

$$A + B + C + D = Y$$

(b)

INPUTS				OUTPUT
D	C	B	A	Y
0	0	0	0	0
0	0	0	1	1
0	0	1	0	1
0	0	1	1	1
0	1	0	0	1
0	1	0	1	1
0	1	1	0	1
0	1	1	1	1
1	0	0	0	1
1	0	0	1	1
1	0	1	0	1
1	0	1	1	1
1	1	0	0	1
1	1	0	1	1
1	1	1	0	1
1	1	1	1	1

(c)

Fig. 3-24 Four-input OR gate. (*a*) **Logic symbol showing the method used to accommodate extra inputs beyond the width of the symbol.** (*b*) **Boolean expression.** (*c*) **Truth table.**

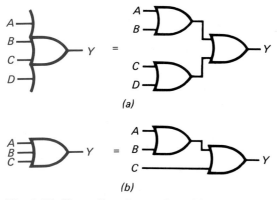

Fig. 3-25 Expanding the number of inputs.

Expanding the number of inputs of a NAND gate is somewhat more difficult than expanding AND and OR gates. Figure 3-26 shows how a 4-input NAND gate can be wired using two 2-input NAND gates and one 2-input OR gate.

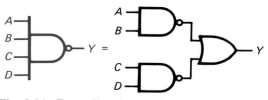

Fig. 3-26 Expanding the number of inputs.

You frequently will run into gates that have from two to as many as eight and more inputs. The basics covered in this section are a handy reference when you need to expand the number of inputs to a gate.

Self-Test

Answer the following questions.

30. Write the Boolean expression for a three-input NAND gate.
31. The truth table for a three-input NAND gate would have _____ lines to cover all the possible input combinations.
32. Write the Boolean expression for a four-input NOR gate.
33. The truth table for a four-input NOR gate would contain _____ lines to cover all the possible input combinations.

3-10 USING INVERTERS TO CONVERT GATES

Frequently it is convenient to convert a basic gate such as an AND, OR, NAND, or NOR to another logic function. This can be done easily with the use of inverters. The chart in Fig. 3-27 is a handy guide for converting any given gate to any other logic function. Look over the chart: notice that in the top section only the outputs are inverted. Inverting the outputs leads to rather predictable results, shown on the right side of the chart.

The center section of the chart shows only the gate inputs being inverted. For instance, if you invert both inputs of an OR gate, the gate becomes a NAND gate. This fact is emphasized in Fig. 3-28(*a*) on page 40. Notice that

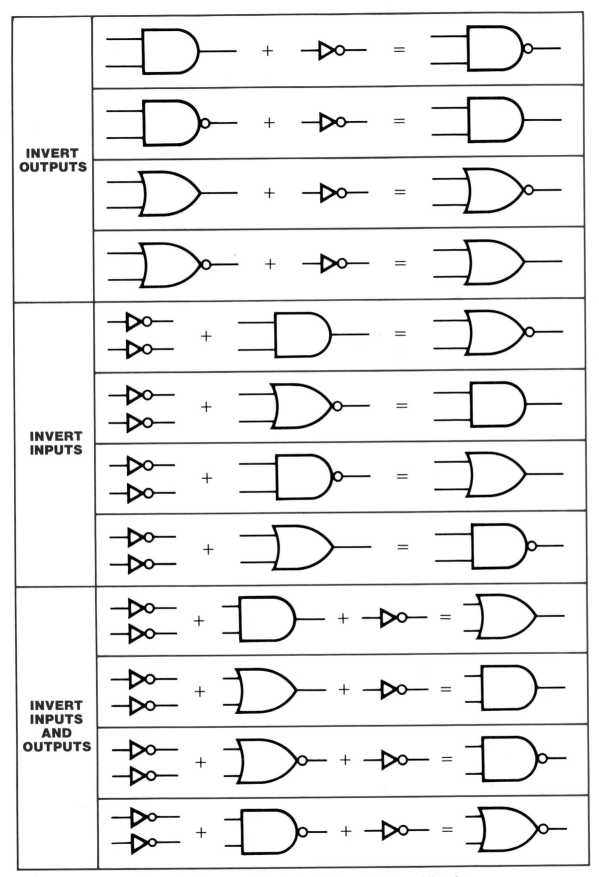

Fig. 3-27 Gate conversions using inverters. The + symbol here indicates adding the functions.

the invert bubbles have been added to the inputs of the OR gate in Fig. 3-28(*a*), which converts the OR gate to a NAND function. Also, in the center section of the chart the inputs of the AND gate are being inverted. This is redrawn in Fig. 3-28(*b*). Notice that the invert bubbles at the input of the AND gate convert it into a NOR gate. The new symbols at the left (with the input invert bubbles) in Fig. 3-28 are used in some logic diagrams in place of the more standard NAND and NOR logic symbols at the right. Be aware of these new symbols because you will run into them in your future work in digital electronics.

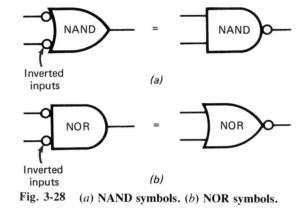

Fig. 3-28 (*a*) **NAND symbols.** (*b*) **NOR symbols.**

The bottom section of the chart in Fig. 3-27 shows both the inputs and the outputs being inverted. Notice that by using inverters at the inputs and outputs you can convert back and forth from AND to OR and from NAND to NOR.

With the 12 conversions shown in the chart in Fig. 3-27 you can convert any basic gate (AND, OR, NAND, and NOR) to any other gate with just the use of inverters. You will *not* need to memorize the chart in Fig. 3-27, but remember it for reference.

Self-Test

Supply the missing word in each sentence.

34. The OR gate can be converted to the NAND function by adding _____ to the inputs of the OR gate.

35. Adding inverters to the inputs of the AND gate produces the _____ logic function.

36. Adding an inverter to the output of an AND gate produces the _____ logic function.

37. Adding inverters to all inputs and outputs of an AND gate produces the _____ logic function.

3-11 PRACTICAL TTL LOGIC GATES

The popularity of digital circuits is due partly to the availability of inexpensive ICs. Manufacturers have developed many *families* of digital ICs. These families are groups of devices that can be used together. The ICs in a family are said to be compatible and can be easily connected to one another.

One group of families is produced using bipolar technology. These ICs contain parts comparable to discrete bipolar transistors, diodes, and resistors. Another group of digital IC families uses metal oxide semiconductor (MOS) technology. In the laboratory, you will probably have an opportunity to use both TTL and CMOS ICs. The CMOS family is a very low power and widely used family from the MOS technology. The CMOS ICs contain parts comparable to insulated-gate field-effect transistors (IGFETs).

A popular type of IC is illustrated in Fig. 3-29(*a*). This case style is referred to as a *dual in-line package* (DIP) by IC manufacturers. This particular IC is called a 14-pin DIP IC.

Just *counterclockwise* from the notch on the IC in Fig. 3-29(*a*) is pin 1. The pins are numbered counterclockwise from 1 to 14 when viewed from the top of the IC. A dot on the top of the DIP IC as in Fig. 3-29(*b*) is another method used to locate pin 1.

Manufacturers of ICs provide pin diagrams similar to the one shown in Fig. 3-30. This IC contains four 2-input AND gates. Thus, it is called a *quadruple two-input AND gate*. This 7408 unit is one of many of the 7400 series of TTL ICs available. The power connections to the IC are the GND (pin 7) and V_{CC} (pin 14) pins. All other pins are the inputs and outputs to the four TTL AND gates.

Given the logic diagram in Fig. 3-31(*a*) on page 42, wire this circuit using the 7408 TTL IC. A wiring diagram for the circuit is shown in Fig. 3-31(*b*). A 5-V dc regulated power supply is typically used with all TTL devices. The positive (V_{CC}) and negative (GND) power connections are made to pins 14 and 7 of the IC. Input switches (*A* and *B*) are wired to pins 1 and 2 of the 7408 IC. Note that if a switch is in the *up position*, a logical 1 (+5 V) is applied to

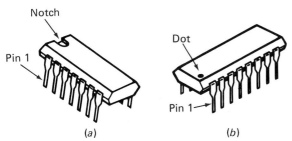

Fig. 3-29 Dual in-line package (DIP) ICs. (a) Locating pin 1 using a notch. (b) Locating pin 1 using a dot.

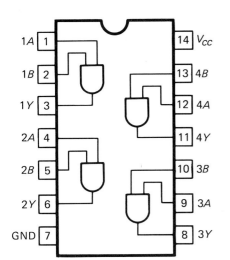

Fig. 3-30 Pin diagram for the 7408 digital IC.

the input of the AND gate. If a switch is in the down position, however, a logical 0 is applied to the input. At the right in Fig. 3-31(b), an LED and 150-Ω limiting resistor are connected to GND. If the output at pin 3 is HIGH (near +5 V), current will flow through the LED. When the LED is lit, it indicates a HIGH output from the AND gate.

The top of a common TTL digital IC is shown in Fig. 3-32(a) on the next page. The block form of the letters "NS" on the top of the IC shows the manufacturer as National Semiconductor. The DM7408N part number can be divided into sections as shown in Fig. 3-32(b). The prefix "DM" is the manufacturer's code (National Semiconductor uses the letters "DM" as a prefix). The core part number is 7408, which is a quadruple two-input AND gate TTL IC. This core part number is the same from manufacturer to manufacturer. The trailing letter "N" (the suffix) is a code used by several manufacturers to designate the DIP.

The top of another digital IC is shown in Fig. 3-33(a). The letter "F" on this IC stands for the manufacturer, Fairchild Camera and Instru-

ment Corporation. On this Fairchild unit, the suffix "P" stands for a plastic DIP packaging and the "C" represents the code for a temperature range of 0 to +70° C. This is typically referred to as the *commercial grade*. The core part number of the IC in Fig. 3-33 is 74LS08. This is about the same as the 7408 quadruple two-input AND gate IC discussed earlier. The letters "LS" in the center of the core number designate the type of TTL circuitry used in the IC. In this case "LS" stands for *low-power Schottky*.

The internal letter(s) in a core part number of a 7400 series IC tell something about the *logic* family or *subfamily*. Typical internal letters used are:

ALS = advanced low-power Schottky
       TTL logic (a subfamily of TTL)
C   = CMOS logic (a separate family)
H   = high-speed TTL logic
       (a subfamily of TTL)
HC = high-speed CMOS logic
       (a separate CMOS family)
HCT = high-speed CMOS logic
       (a separate CMOS family
       with TTL inputs)
L   = low-power TTL logic
       (a subfamily of TTL)
LS = low-power Schottky TTL logic
       (a subfamily of TTL)
S   = Schottky TTL logic
       (a subfamily of TTL)

The internal letters give information about the speed or power consumption of digital ICs. Because of these speed and power consumption differences, manufacturers usually recommend that exact part numbers be used when replacing digital ICs. When the letter "C" is used inside a 7400 series part number, it designates a CMOS and not a TTL digital IC. The internal letters "HC" and "HCT" also designate CMOS ICs.

Data manuals from manufacturers contain much valuable information on digital ICs. They contain pin diagrams and packaging information. Data manuals also contain details on part numbering and other valuable data for the technician, student, or engineer. Most manufacturers bind TTL and CMOS data sheets into separate data manuals.

Pin diagram

TTL digital IC

Commercial grade

Low-power Schottky

Core part number

Subfamily

CMOS ICs

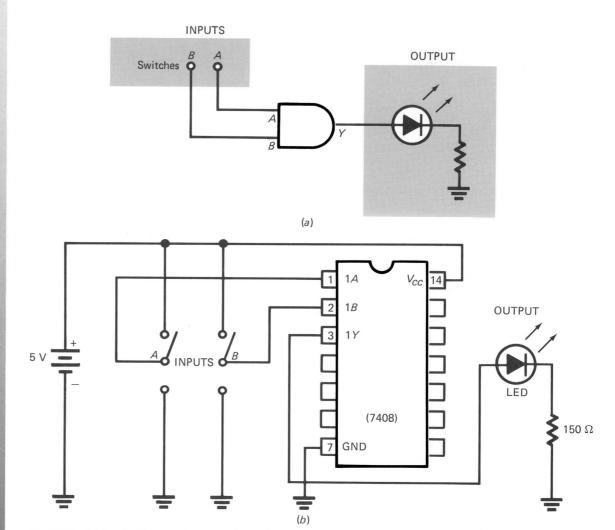

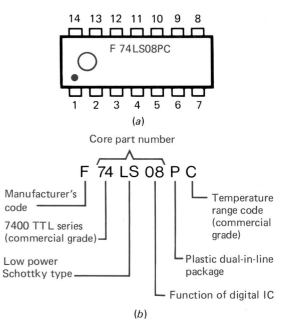

Fig. 3-31 (*a*) **Logic diagram for a two-input AND gate circuit.** (*b*) **Wiring diagram to implement the two-input AND function.**

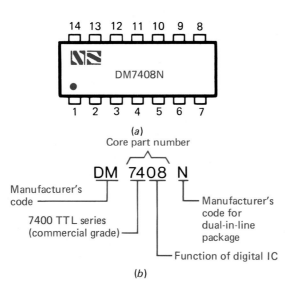

Fig. 3-32 (*a*) **Marking on a typical digital IC.** (*Courtesy of National Semiconductor Corporation.*) (*b*) **Decoding the part number on a typical IC.**

Fig. 3-33 (*a*) **Markings on a Fairchild digital IC.** (*b*) **Decoding the part number on a typical low-power Schottky IC.**

Practical CMOS
logic gates

Complementary
metal oxide

Semiconductor

4000 series

74C00 series

74HC00 serues

Conductive foam

Self-Test

Answer the following questions.

38. List two popular digital IC families.
39. Refer to Fig. 3-29. These ICs use the popular case style called the _____ package.
40. A _____ -V dc power supply is used with TTL ICs. The V_{CC} pin is connected to the _____ ($-$, $+$) while the GND pin goes to the _____ ($-$, $+$) of the supply.
41. Refer to Fig. 3-31(b). How is the 7408 IC described by the manufacturer?
42. What can you tell about a digital IC that has the marking "DM74LS08N" printed on the top?

3-12 PRACTICAL CMOS LOGIC GATES

The older 7400 series of TTL logic devices has been extremely popular for many decades. One of its disadvantages is its higher power consumption. In the late 1960s, manufacturers developed CMOS digital ICs which consume almost no power and were perfect for battery-operated electronic devices. CMOS stands for *complementary metal oxide semiconductor.*

Several families of compatible CMOS ICs have been been developed. The first was the 4000 series. Next came the 74C00 series and more recently the 74HC00 series of CMOS digital ICs. Many large-scale integrated (LSI) circuits such as digital wrist watch and calculator chips are also manufactured using the CMOS technology.

A typical 4000 series CMOS IC is pictured in Fig. 3-34(a). Note that pin 1 is marked as such on the top of the IC immediately counterclockwise from the notch. The manufacturer is RCA. The CD4081BE part number can be divided into sections as shown in Fig. 3-34(b). The prefix CD is RCA's code for CMOS digital ICs. The core part number is 4081B, which stands for a CMOS quadruple two-input AND gate IC. This core part number is almost always the same from manufacturer to manufacturer. The trailing letter "E" is RCA's packaging code for a plastic DIP IC. The letter "B" is a "buffered version" of the original 4000A series. The buffering provides the 4000B series devices with greater output drive and some protection from static electricity.

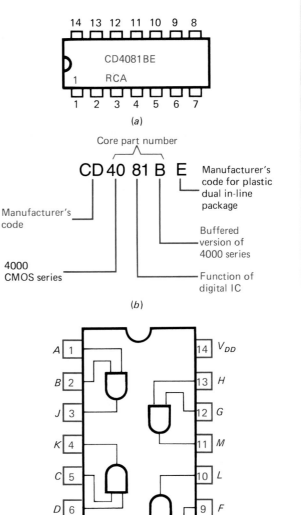

Fig. 3-34 (*a*) **Markings on an RCA CMOS digital IC.** (*b*) **Decoding the part number on a typical 4000B series CMOS IC.** (*c*) **Pin diagram for the 4081B CMOS IC.**

Figure 3-34(c) is a pin diagram for the CD4081BE CMOS quad two-input AND gate IC. The power connections are V_{DD} (positive voltage) and V_{SS} (GND or negative voltage). The labeling of the power connections on TTL and 4000 series CMOS ICs is different. This difference can be observed by comparing Figs. 3-30 and 3-34(c).

Given the schematic diagram in Fig. 3-35(a) on the next page, wire this circuit using the 4081B CMOS IC. A wiring diagram for the circuit is shown in Fig. 3-35(b). A 5-V dc power supply is shown but the 4000 series CMOS ICs can use voltages from 3 to 18 V dc. Great care is taken in removing the 4081 from its conduc-

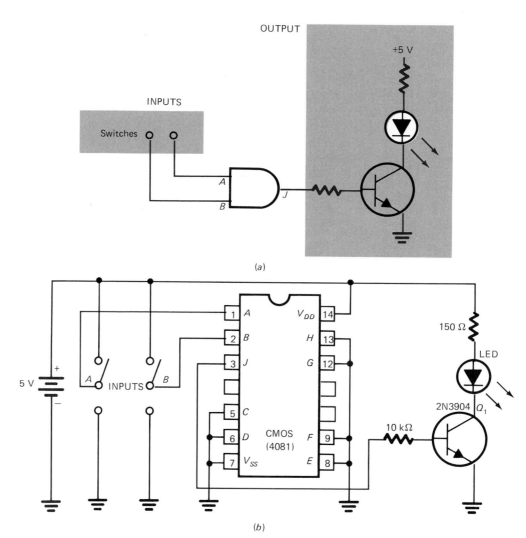

Fig. 3-35 (*a*) Logic diagram for a two-input AND gate circuit. (*b*) Wiring diagram using the 4081 CMOS IC to implement the two-input AND function.

tive foam storage because CMOS ICs can be damaged by static charges. Do not touch the pins when inserting the 4081 CMOS IC in a socket or mounting board. V_{DD} and V_{SS} power connections are made with the power off. *When using CMOS, all unused inputs are tied to GND or* V_{DD}. In this example, unused inputs (*C, D, E, F, H, G*) are grounded. The output of the AND gate (pin 3) is connected to the driver transistor. The transistor turns the LED on when pin 3 is HIGH or off when the output is LOW. Finally, inputs *A* and *B* are connected to the switches.

When the input switches in Fig. 3-35(*b*) are in the up position, they generate a HIGH input. A LOW input is generated when the switches are in the down position. Two LOW inputs to the AND gate produce a LOW output at pin 3 of the IC. The LOW output turns off the transistor and the LED does not light.

Two HIGH inputs to the AND gate produce a HIGH output at pin 3. The HIGH (about +5 V) output at the base of Q1 turns on the transistor and the LED lights. The 4081 CMOS IC will generate a two-input AND truth table like the one in Fig. 3-4 on page 31.

Several families of CMOS digital ICs are available in DIP form. A 4000 series IC was used as an example in this section. The newer 74HC00 series CMOS digital ICs are gaining favor because they are somewhat more compatible with the popular TTL logic. The 74HC00 series of CMOS ICs also has more drive capabilities than the older 4000 and 74C00 series units and operates at high frequencies. The "HC" in the 74HC00 series part number stands for high-speed CMOS.

Caution must be used so that static charges do not damage CMOS ICs. All unused inputs to CMOS gates must be grounded or connected to V_{DD}. Most important, *input voltages must not exceed the GND (V_{SS}) to V_{DD} voltage.*

Self-Test

Answer the following questions.

43. The primary advantage of CMOS digital ICs is their _____ (high, low) power consumption.
44. While TTL must use a 5-V power supply, 4000 series CMOS ICs can operate on dc voltages from _____ to _____ V.
45. Refer to Fig. 3-34. How is this 4081B IC described by the manufacturer?
46. What rule (dealing with unused inputs) must be followed when wiring CMOS ICs?

3-13 TROUBLESHOOTING SIMPLE GATE CIRCUITS

The most basic piece of test equipment used in digital troubleshooting is the *logic probe*. A simple logic probe is pictured in Fig. 3-36. The slide switch on the unit is used to select the type of logic family under test, either TTL or CMOS. The logic probe in Fig. 3-36 is set to test a TTL type of digital circuit in this example. Typically, two leads provide power to the logic probe. The red lead is connected to the positive (+) of the power supply while the black lead of the logic probe is connected to the negative (−) or GND of the power supply. After powering the logic probe, the needlelike probe is touched to a test point or node in the circuit. The logic probe will light either the HIGH or the LOW indicator. If neither indicator lights, it usually means that the voltage is somewhere between the HIGH and LOW.

In practical electronic equipment most digital ICs are mounted on a printed circuit (PC) board. An example is shown in Fig. 3-37(*a*) on the next page. Also available to the student or technician is a circuit wiring diagram or schematic similar to that in Fig. 3-37(*b*). Many times the +5-V (V_{CC}) and GND connections to the ICs are not shown on the wiring diagram. However, they are always understood to be present. Pin numbers are usually given in a wiring diagram. The IC type may not be given on the schematic but is usually listed on a parts list in the equipment manual.

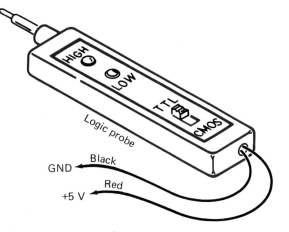

Fig. 3-36 Logic probe.

The first step in troubleshooting is to use your senses. *Feel* the flat top of the ICs to determine if they are hot. Some ICs operate cool, others run slightly warm. CMOS ICs should always be cool. *Look* for broken connections, solder bridges, broken PC board traces, and bent IC pins. *Smell* for possible overheating. *Look* for signs of excessive heat, such as discoloration or charring.

The next step in troubleshooting might involve checking whether each IC has power. With the logic probe connected to power, check at the points labeled *A*, *B* (the V_{CC} pin), *C*, and *D* in Fig. 3-37(*a*). Nodes *A* and *B* should give a bright HIGH light on the logic probe. Nodes *C* and *D* should give a bright LOW light on the logic probe.

The next step might be to trace the path of logic through the circuit. The circuit is equal to a three-input AND gate in this example (Fig. 3-37). Its unique state is a HIGH when all inputs (*A*, *B*, and *C*) are HIGH. Check pins 1, 2, and 5 (inputs *A*, *B*, and *C*) of the IC in Fig. 3-37(*a*) with the logic probe. Manipulate the equipment to get all inputs HIGH. When all inputs are HIGH, the output (pin 6 of the IC) should be HIGH and the circuit LED should light. If the unique state works, try several other input combinations and verify their proper operation.

Refer to Fig. 3-37(*a*). Assume a HIGH reading at node *A* and a LOW reading on the logic probe at node *B* (pin 14 of the IC). This probably means an open circuit in the PC board trace or a faulty solder joint between points *A* and *B*. If DIP IC sockets are used, the thin part of the IC pin can be bent. This common difficulty causes an open between the IC pin and the socket and PC board trace.

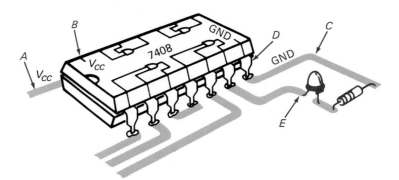

(a)

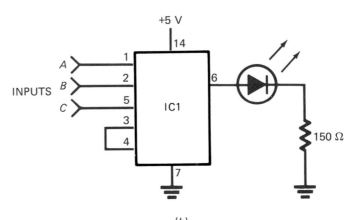

(b)

Fig. 3-37 (a) **Digital IC mounted on a printed circuit (PC) board.**
(b) **Wiring or schematic diagram of a digital gating circuit.**

Refer to Fig. 3-37(a). Assume LOW readings at pins 1, 2, and 3 with no reading (neither LED on the logic probe lit) at pin 4. No reading on most logic probes means a voltage between LOW and HIGH (perhaps 1 to 2 V in TTL). This input (pin 4) is floating (not connected) and is considered to be a HIGH by the TTL circuitry inside the 7408 IC. The output of the first AND gate (pin 3) is supposed to pull the input to the second AND gate (pin 4) LOW. If it does not, the fault could be in the PC board trace, solder connections, or a bent IC pin. Internal opens and shorts also occur in digital ICs.

Troubleshooting a comparable CMOS circuit proceeds in the same way with a few exceptions. The logic probe must be set to CMOS instead of TTL. Floating inputs on CMOS ICs can destroy the IC. A LOW in CMOS is defined as approximately 0 to 20 percent of the power supply voltage. A HIGH in CMOS is defined as approximately 80 to 100 percent of supply voltage.

In summary, troubleshooting first involves using your senses. Second, check with a logic probe to see if each IC has power. Third, determine the exact job of the gating circuit and test for the unique output conditions. Finally, check other input and output conditions. Open and short-circuit conditions can occur inside ICs as well as in the wiring. Digital ICs should be replaced with exact subfamily replacements when possible.

Self-Test

Answer the following questions.

47. Refer to Fig. 3-36. With what two logic families can this logic probe be used?
48. What is the first step in troubleshooting gating circuits using TTL ICs?
49. What is the second step in troubleshooting?
50. Floating inputs to CMOS ICs are _____ (allowed, not allowed).

SUMMARY

1. Binary logic gates are the basic building blocks for all digital circuits.
2. Figure 3-38 shows a summary of the seven basic logic gates. This information should be memorized.
3. NAND gates are very widely employed and can be used to make other logic gates.
4. Logic gates are often needed with 2 to 10 inputs. Several gates may be connected in the proper manner to get more inputs.
5. AND, OR, NAND, and NOR gates can be converted back and forth by using inverters.
6. Logic gates are commonly packaged in DIP ICs. Both TTL and CMOS digital ICs are very widely used in small systems.
7. Very low power consumption is an advantage of CMOS digital ICs.
8. When using CMOS ICs, all unused inputs must go to V_{DD} or GND. Care must be exercised in storing and handling CMOS ICs to avoid static electricity. Input voltages to a CMOS IC must never exceed the power supply voltages.
9. The logic probe, knowledge of the circuit, and your senses of sight, smell, and touch are basic tools used in troubleshooting gating circuits.

LOGIC FUNCTION	LOGIC SYMBOL	BOOLEAN EXPRESSION	TRUTH TABLE		
			INPUTS		OUTPUT
			B	A	Y
AND	$A \cdot B = Y$		0	0	0
			0	1	0
			1	0	0
			1	1	1
OR	$A + B = Y$		0	0	0
			0	1	1
			1	0	1
			1	1	1
Inverter	$A = \overline{A}$			0	1
				1	0
NAND	$\overline{A \cdot B} = Y$		0	0	1
			0	1	1
			1	0	1
			1	1	0
NOR	$\overline{A + B} = Y$		0	0	1
			0	1	0
			1	0	0
			1	1	0
XOR	$A \oplus B = Y$		0	0	0
			0	1	1
			1	0	1
			1	1	0
XNOR	$\overline{A \oplus B} = Y$		0	0	1
			0	1	0
			1	0	0
			1	1	1

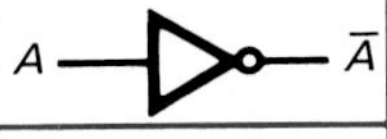

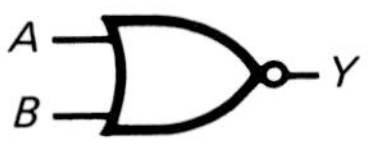

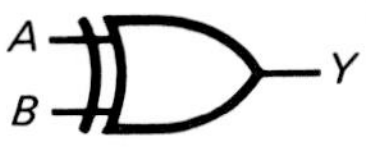

Fig. 3-38   Summary of basic logic gates.

CHAPTER REVIEW QUESTIONS

Answer the following questions.

3-1. Draw the logic symbols for **a** to **j** (label inputs *A*, *B*, *C*, *D* and outputs *Y*):
  a. Two-input AND gate
  b. Three-input OR gate
  c. Inverter (two symbols)
  d. Two-input XOR gate
  e. Four-input NAND gate
  f. Two-input NOR gate
  g. Two-input XNOR gate
  h. Two-input NAND gate (special symbol)
  i. Two-input NOR gate (special symbol)
  j. Buffer (noninverting)

3-2. Write the Boolean expression for each gate in question 1.

3-3. Draw the truth table for each gate in question 1.

3-4. Look at the chart in Fig. 3-38. Which logic gate always responds with an output of logical 1 only with two unlike inputs (0 and 1 or 1 and 0)?

3-5. Which logic gate might be called the "all or nothing gate"?

3-6. Which logic gate might be called the "any or all gate"?

3-7. Which logic circuit complements the input?

3-8. Which logic gate might be called the "any but not all gate"?

3-9. Given an AND gate and inverters, draw how you would produce a NOR function.

3-10. Given a NAND gate and inverters, draw how you would produce an OR function.

3-11. Given a NAND gate and inverters, draw how you would produce an AND function.

3-12. Given four 2-input AND gates, draw how you would produce a 5-input AND gate.

3-13. Given several two-input NAND and OR gates, draw how you would produce a four-input NAND gate.

3-14. Switches arranged in series (see Fig. 3-1) act like what type of logic gate?

3-15. Switches arranged in parallel (see Fig. 3-6) act like what type of logic gate?

3-16. Figure 3-29(*b*) illustrates a(n) _____ (8, 16) pin _____ (three letters) IC.

3-17. Draw a wiring diagram similar to Fig. 3-31(*b*) for a circuit that will perform the three-input AND function. Use a 7408 IC, a 5-V dc power supply, three input switches, and an output indicator.

3-18. The PC board pad labeled _____ (*A*, *C*) is pin 1 of the IC in Fig. 3-39.

3-19. The PC board pad labeled _____ (letter) is the GND pin on the 7408 IC in Fig. 3-39.

3-20. The PC board pad labeled _____ (letter) is the V_{CC} pin on the 7408 IC in Fig. 3-39.

3-21. The IC shown in Fig. 3-40 was manufactured by _____ (Fairchild, Texas Instruments).

3-22. What does the prefix "SN" on the IC in Fig. 3-40 represent?

3-23. The unit shown in Fig. 3-40 is a _____ (low-power, standard) TTL 14-pin DIP IC.

3-24. Pin 1 of the IC shown in Fig. 3-40 is labeled with the letter _____ .

3-25. The pin labeled with a "C" on the IC in Fig. 3-40 is pin number _____ .

3-26. Refer to Fig. 3-37. Draw a logic diagram of this circuit.

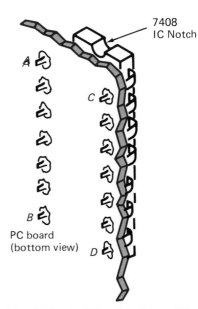

Fig. 3-39 An IC soldered to a PC board.

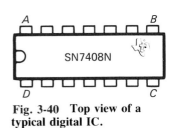

Fig. 3-40 Top view of a typical digital IC.

3-27. Figure 3-37(b) is an example of a _____ (logic, wiring) diagram that might be used by service personnel.

3-28. Refer to Fig. 3-37(a). If all input pins (1, 2, and 5) are HIGH and output pin 6 is HIGH but point E is LOW, the LED _____ (will, will not) light and the circuit _____ (is, is not) working properly.

3-29. Refer to Fig. 3-37(a). List several possible problems if pin 6 is HIGH but point E is LOW.

3-30. Refer to Fig. 3-37(a). An *internal open* between the output of the first AND gate and pin 3 might give neither a HIGH nor a LOW indication on the logic probe. This means that both pins 3 and 4 are floating _____ (HIGH, LOW).

3-31. Refer to Fig. 3-41. The IC shown in Fig. 3-41 was manufactured by _____ (National Semiconductor, RCA).

Fig. 3-41 Top view of a typical digital IC.

3-32. Refer to Fig. 3-41. The core number of this IC is _____ , which means it is a _____ (CMOS, TTL) logic device.

3-33. Pin 1 of the IC shown in Fig. 3-41 is labeled with the letter _____ .

3-34. What precaution should be taken when storing the IC in Fig. 3-41?

3-35. Refer to Fig. 3-35. If input A is HIGH and input B is LOW, output J (pin 3) will be _____ (HIGH, LOW). Transistor Q1 is turned _____ (off, on) and the LED _____ (does, does not) light.

3-36. Refer to Fig. 3-35. Why are pins 5, 6, 8, 9, 12, and 13 grounded in this circuit?

Answers to Self-Tests

1. round
2. $A \cdot B = Y$
3. HIGH, light
4. pointed
5. $A + B = Y$
6. LOW
7. inclusive
8. LOW
9. LOW
10. $Y = \overline{A}$
11. negated, complemented
12. LOW
13. round with an added invert bubble
14. $\overline{A \cdot B} = Y$
15. LOW, is
16. pointed with an added invert bubble
17. $\overline{A + B} = Y$
18. LOW, is not
19. HIGH, is
20. any but not all gate
21. $A \oplus B = Y$
22. LOW
23. HIGH
24. invert, XOR
25. $\overline{A \oplus B} = Y$
26. LOW
27. HIGH
28. connected together
29. three
30. $\overline{A \cdot B \cdot C} = Y$
31. eight
32. $\overline{A + B + C + D} = Y.$
33. 16
34. inverters
35. NOR
36. NAND
37. OR
38. TTL, CMOS
39. dual in-line (DIP)
40. 5, +, −
41. TTL quad two-input AND gate
42. manufacturer—National Semiconductor; package—dual in-line (DIP); family—low-power Schottky; function—two-input AND gate
43. low
44. 3, 18
45. CMOS quad two-input AND gate
46. All unused CMOS inputs must be connected to either GND (−) or positive (+) of the power supply.
47. TTL and CMOS
48. Use your senses to locate possible open or short circuits or possible overheating.
49. Check that each IC has power.
50. not allowed

CHAPTER 4

Using Binary Logic Gates

In Chap. 3 you memorized the symbol, truth table, and Boolean expression for each of the binary logic gates. These gates are the basic building blocks for all digital systems. In this chapter you will see how your knowledge of gate symbols, truth tables, and Boolean expressions can be used to solve real world problems in electronics. You will be connecting gates to form what engineers refer to as combinational logic circuits, and you will be combining gates (ANDs, ORs) and inverters and so on to solve logic problems. There are three "tools of the trade" in solving logic problems: gate symbols, truth tables, and Boolean expressions. Do you have these tools of the trade? Do you know your gate symbols, truth tables, and Boolean expressions? If you need review on logic gates, please refer to the summary of Chap. 3; Fig. 3-38 especially will be helpful. Your understanding of using logic gates is important because to be successful as a technician, troubleshooter, engineer or hobbyist in digital electronics you must master combining gates. It is strongly suggested that you try out your combinational logic circuits in the shop or laboratory. Logic gates are packaged in inexpensive, easy-to-use ICs. Combinational logic circuits can be implemented using either TTL or CMOS ICs.

4-1 CONSTRUCTING CIRCUITS FROM BOOLEAN EXPRESSIONS

We use Boolean expressions to guide us in building logic circuits. Suppose you are given the Boolean expression $A + B + C = Y$ (read as "A or B or C equals output Y") and told to build a logic circuit that will perform this logic.

Looking at the expression, notice that each input must be ORed to get output Y. Figure 4-1 illustrates the *gate* needed to do the job.

Fig. 4-1 **Logic diagram for Boolean expression** $A + B + C = Y$.

From page 51:

Combinational logic circuits

On this page:

Constructing logic circuit from a Boolean expression

Sum-of-products form

Product-of-sums form

Minterm form

Maxterm form

Now suppose you are given the Boolean expression $\overline{A}\cdot B + A\cdot\overline{B} + \overline{B}\cdot C = Y$ (read as "not A and B, or A and not B, or not B and C equals output Y"). How would you construct a circuit that will do the job of this expression? The first step is to look at the Boolean expression and note that you must OR $\overline{A}\cdot B$ with $A\cdot\overline{B}$ with $\overline{B}\cdot C$. Figure 4-2(a) shows that a three-input OR gate will form the output Y. This may be redrawn as in Fig. 4-2(b).

$$\overline{A}\cdot B + A\cdot\overline{B} + \overline{B}\cdot C = Y$$

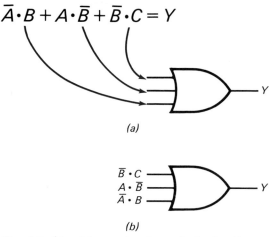

(a)

(b)

Fig. 4-2 Step 1 in constructing a logic circuit.

The second step used in constructing a logic circuit from the given Boolean expression $\overline{A}\cdot B + A\cdot\overline{B} + \overline{B}\cdot C = Y$ is shown in Fig. 4-3. Notice in Fig. 4-3(a) that an AND gate has been added to feed the $\overline{B}\cdot C$ to the OR gate and an inverter has been added to form the $\overline{B}$ for the input to AND gate 2. Figure 4-3(b) adds AND gate 3 to form the $A\cdot\overline{B}$ input to the OR gate. Finally, Fig. 4-3(c) adds AND gate 4 and inverter 6 to form the $\overline{A}\cdot B$ input to the OR gate. Figure 4-3(c) is the circuit that would be constructed to perform the required logic given in the Boolean expression $\overline{A}\cdot B + A\cdot\overline{B} + \overline{B}\cdot C = Y$.

Notice that we started at the output of the logic circuit and worked toward the inputs. You have now experienced how combinational logic circuits are constructed from Boolean expressions.

Boolean expressions come in two forms. The *sum-of-products* form is the type we saw in Fig. 4-2. Another example of this form is $A\cdot B + B\cdot C = Y$. The other Boolean expression form is the *product-of-sums;* an example is $(D + E)\cdot(E + F) = Y$. The sum-of-products form is called the *minterm form* in engineering texts. The product-of-sums form is called the *maxterm form* by engineers, technicians, and scientists.

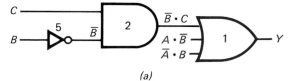

(a)

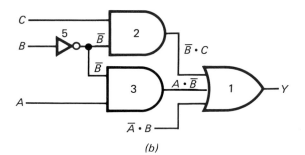

(b)

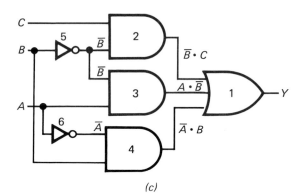

(c)

Fig. 4-3 Step 2 in constructing a logic circuit.

Self-Test

Answer the following questions.

1. Construct logic circuits using AND, OR, and NOT gates for the following Boolean expressions:
   a. $\overline{A}\cdot\overline{B} + A\cdot B = Y$
   b. $\overline{A}\cdot\overline{C} + A\cdot B\cdot C = Y$
2. A minterm Boolean expression is also called the _____ form.
3. A maxterm Boolean expression is also called the _____ form.

4-2 DRAWING A CIRCUIT FROM A MAXTERM BOOLEAN EXPRESSION

Suppose you are given the maxterm Boolean expression $(A + B + C) \cdot (\overline{A} + \overline{B}) = Y$. The first step in constructing a logic circuit for this Boolean expression is shown in Fig. 4-4(a). Notice that the terms $(A + B + C)$ and $(\overline{A} + \overline{B})$ must be ANDed together to form output Y.

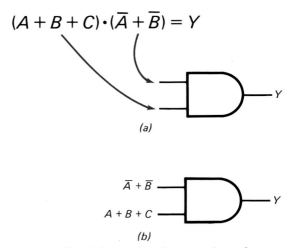

$$(A + B + C) \cdot (\overline{A} + \overline{B}) = Y$$

(a)

(b)

Fig. 4-4 Step 1 in constructing a product-of-sums logic circuit.

Figure 4-4(b) shows the logic circuit redrawn. The second step in drawing the logic circuit is shown in Fig. 4-5. The $(\overline{A} + \overline{B})$ part of the expression is produced by adding OR gate 2 and inverters 3 and 4, as illustrated in Fig. 4-5(a). Then, the expression $(A + B + C)$ is delivered to the AND gate by OR gate 5 in Fig. 4-5(b). The logic circuit shown in Fig. 4-5(b) is the complete logic circuit for the maxterm Boolean expression $(A + B + C) \cdot (\overline{A} + \overline{B}) = Y$.

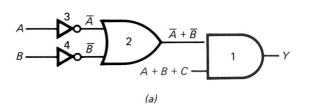

(a)

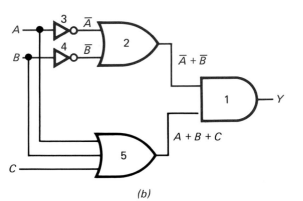

(b)

Fig. 4-5 Step 2 in constructing a product-of-sums logic circuit.

In summary, we work from right to left (from output to input) when converting a Boolean ex-

pression to a logic circuit. Notice that we use only AND, OR, and NOT gates when constructing combinational logic circuits. Maxterm and minterm Boolean expressions both can be converted to logic circuits. Minterm expressions create AND-OR logic circuits similar to that in Fig. 4-3(c), whereas maxterm expressions create OR-AND logic circuits similar to that in Fig. 4-5(b).

You now should be able to identify minterm and maxterm Boolean expressions, and you should be able to convert Boolean expressions to combinational logic circuits by using AND, OR, and NOT gates.

Self-Test

Answer the following questions.

4. Construct a logic circuit using AND, OR, and NOT gates from the following Boolean expressions:
   a. $(A + B) \cdot (\overline{A} + \overline{B}) = Y$
   b. $(A + B) \cdot \overline{C} = Y$
5. Refer to question 4. These Boolean expressions are _____ (maxterm, minterm).
6. Refer to question 4. These Boolean expressions are _____ (product-of-sums, sum-of-products).
7. Maxterm Boolean expressions are used to create _____ (AND-OR, OR-AND) logic circuits.

4-3 TRUTH TABLES AND BOOLEAN EXPRESSIONS

Boolean expressions are a convenient method of describing how a logic circuit operates. The *truth table* is another precise method of describing how a logic circuit works. As you work in digital electronics, you will have to convert information from truth-table form to a Boolean expression.

Look at the truth table in Fig. 4-6(a) on the next page. Notice that only two of the eight possible combinations of inputs A, B, and C generate a logical 1 at the output. The two combinations that generate a 1 output are shown as $\overline{C} \cdot B \cdot A$ (read as "not C and B and A") and $C \cdot \overline{B} \cdot \overline{A}$ (read as "C and not B and not A"). Figure 4-6(b) shows how the combinations are ORed together to form the Boolean expression for the truth table. Both the truth table in Fig. 4-6(a) and the Boolean expression in Fig. 4-6(b) describe how the logic circuit should work.

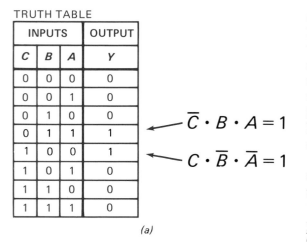

TRUTH TABLE

INPUTS			OUTPUT
C	B	A	Y
0	0	0	0
0	0	1	0
0	1	0	0
0	1	1	1
1	0	0	1
1	0	1	0
1	1	0	0
1	1	1	0

$$\overline{C} \cdot B \cdot A = 1$$
$$C \cdot \overline{B} \cdot \overline{A} = 1$$

(a)

(b) Boolean expression

$$\overline{C} \cdot B \cdot A + C \cdot \overline{B} \cdot \overline{A} = Y$$

Fig. 4-6   Forming a Boolean expression from a truth table.

The truth table is the beginning of most logic circuits. You must be able to convert the truth-table information into a Boolean expression as in this section. Remember to look for combinations of variables that generate a logical 1 in the truth table.

Occasionally you must reverse the procedure you have just learned. That is, you must take a Boolean expression and from it construct a truth table. Consider the Boolean expression in Fig. 4-7(a). It appears that two

(a) Boolean expression

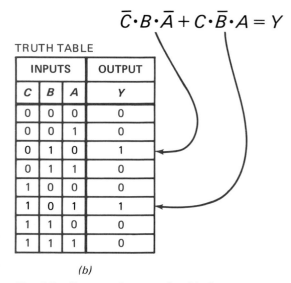

$$\overline{C} \cdot B \cdot \overline{A} + C \cdot \overline{B} \cdot A = Y$$

TRUTH TABLE

INPUTS			OUTPUT
C	B	A	Y
0	0	0	0
0	0	1	0
0	1	0	1
0	1	1	0
1	0	0	0
1	0	1	1
1	1	0	0
1	1	1	0

(b)

Fig. 4-7   Constructing a truth table from a Boolean expression.

combinations of inputs *A*, *B*, and *C* generate a logical 1 at the output. In Fig. 4-7(*b*) we find the correct combinations of *A*, *B*, and *C* that are given in the Boolean expression and mark a 1 in the output column. All other outputs in the truth table are 0. The Boolean expression in Fig. 4-7(*a*) and the truth table in Fig. 4-7(*b*) both accurately describe the operation of the same logic circuit.

Suppose you are given the Boolean expression in Fig. 4-8(*a*). At first glance it seems that this would produce two outputs with a logical 1. However, if you look closely at Fig. 4-8(*b*) you will see that the Boolean expression $\overline{C} \cdot \overline{A} + C \cdot B \cdot A = Y$ actually generates three logical 1s in the output column. The "trick" illustrated in Fig. 4-8 should make you very cautious. Make sure you have all the combinations that generate a logical 1 in the truth table. The Boolean expression in Fig. 4-8(*a*) and the truth table in Fig. 4-8(*b*) both describe the same logic circuit.

(a) Boolean expression

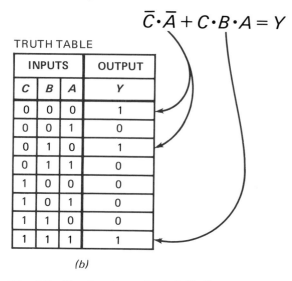

$$\overline{C} \cdot \overline{A} + C \cdot B \cdot A = Y$$

TRUTH TABLE

INPUTS			OUTPUT
C	B	A	Y
0	0	0	1
0	0	1	0
0	1	0	1
0	1	1	0
1	0	0	0
1	0	1	0
1	1	0	0
1	1	1	1

(b)

Fig. 4-8   Constructing a truth table from a Boolean expression.

You have now converted truth tables to Boolean expressions and Boolean expressions to truth tables. You were reminded that the Boolean expressions you worked with were minterm Boolean expressions. The procedure for producing maxterm Boolean expressions from a truth table is quite different.

Self-Test

Answer the following questions.

8. Refer to Fig. 4-6(*a*). Assume that only the bottom two lines of the truth table produce an output of 1 (all other outputs = 0). Write the sum-of-products Boolean expression for this situation.

9. Refer to Fig. 4-6(*a*). The Boolean expression $\overline{C}\cdot\overline{B}\cdot\overline{A} + \overline{C}\cdot\overline{B}\cdot A = Y$ produces a truth table that has HIGH outputs in what two lines?

10. Construct a truth table for the Boolean expression $C\cdot B\cdot\overline{A} + C\cdot\overline{B}\cdot A = Y$.

4-4 SAMPLE PROBLEM

The procedures in Secs. 4-1 to 4-3 are needed skills as you work in digital electronics. To assist you we shall take an everyday logic problem and work from truth table to Boolean expression to logic circuit as shown in Fig. 4-9.

TRUTH TABLE

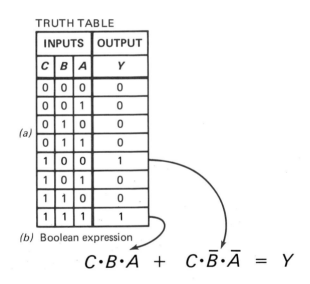

	INPUTS		OUTPUT
	C	B	A	Y
	0	0	0	0
	0	0	1	0
	0	1	0	0
(*a*)	0	1	1	0
	1	0	0	1
	1	0	1	0
	1	1	0	0
	1	1	1	1

(*b*) Boolean expression

$$C\cdot B\cdot A \ + \ C\cdot\overline{B}\cdot\overline{A} \ = \ Y$$

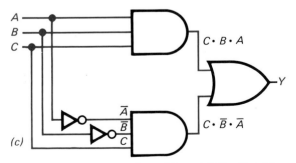

(*c*)

Fig. 4-9 Electronic lock problem. (*a*) **Truth table.** (*b*) **Boolean expression.** (*c*) **Logic circuit.**

Let us assume that we are designing a simple *electronic lock*. The lock will open only when certain switches are pressed. Figure 4-9(*a*) is the truth table for the electronic lock. Notice that the two combinations of input switches A, B, and C generate a 1 at the output. A 1 output will open the lock. Figure 4-9(*b*) shows how we form the minterm Boolean expression for the electronic lock circuit. The logic circuit in Fig. 4-9(*c*) is then drawn from the Boolean expression. Look over the sample problem in Fig. 4-9 and be sure you can follow how we converted from the truth table to the Boolean expression and then to the logic circuit.

You now should be able to solve a problem such as the one in Fig. 4-9. The following self-test will give you some practice in solving problems dealing with truth tables, Boolean expressions, and combinational logic circuits.

Self-Test

Answer the following questions.

11. Using the truth table below for an electronic lock, write the minterm Boolean expression for this truth table.

Truth Table for Question 11—Lock Problem

Input Switches			Output
C	B	A	Y
0	0	0	0
0	0	1	0
0	1	0	1
0	1	1	0
1	0	0	0
1	0	1	1
1	1	0	0
1	1	1	0

12. From the Boolean expression developed in question 11, draw a logic symbol diagram for the electronic lock problem.

4-5 SIMPLIFYING BOOLEAN EXPRESSIONS

Consider the Boolean expression $\overline{A}\cdot B + A\cdot\overline{B} + A\cdot B = Y$ in Fig. 4-10(*a*). In constructing a logic circuit for this Boolean expression, we find that we need three AND gates, two inverters, and one 3-input OR gate. Figure 4-10(b) is a logic circuit that would perform the logic of the Boolean expression $\overline{A} \cdot B + A \cdot \overline{B} + A \cdot B = Y$. Figure 4-10(c) details the truth table for the Boolean expression and logic circuit in Fig. 4-10(a) and (b). Immediately you recognize the truth table in Fig. 4-10(c) as the truth table for a two-input OR gate. The simple Boolean expression for a two-input OR gate is $A + B = Y$, as shown in Fig. 4-10(d). The logic circuit for a two-input OR gate in its simplest form is diagrammed in Fig. 4-10(e).

(a) Original Boolean expression

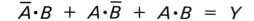
$$\overline{A} \cdot B + A \cdot \overline{B} + A \cdot B = Y$$

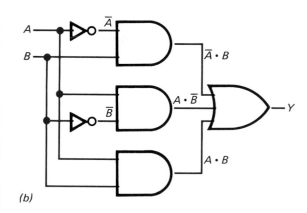

(b)

TRUTH TABLE

INPUTS		OUTPUT
B	**A**	**Y**
0	0	0
0	1	1
1	0	1
1	1	1

(c)

(d) Simplified Boolean expression

$$A + B = Y$$

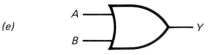

(e)

Fig. 4-10 Simplifying Boolean expressions.

The example summarized in Fig. 4-10 shows how we must try to simplify our original Boolean expression to get a simple, inexpensive logic circuit. In this case we were lucky enough to notice that the truth table belonged to an OR gate. However, usually we must use more systematic methods of simplifying our Boolean expression. Such methods include applying Boolean algebra and *Karnaugh mapping*.

Boolean algebra was originated by George Boole (1815–1864). Boole's algebra was adapted in the 1930s for use in digital logic circuits; it is the basis for the tricks we shall use to simplify Boolean expressions. We shall not deal directly with Boolean algebra in this text. Many of you who continue on in digital electronics and engineering will study Boolean algebra in detail.

Karnaugh mapping, an easy-to-use method of simplifying Boolean expressions, is covered in detail in Secs. 4-6 to 4-9. Several other simplification methods are available, including Veitch diagrams, Venn diagrams, and the tabular method of simplification.

Self-Test

Supply the missing word or words in each statement.

13. The logic circuits in Fig. 4-10(b) and (e) produce _____ (different, identical) truth tables.
14. Boolean expressions can many times be simplified by inspection or by using methods which include _____ algebra or _____ mapping.

4-6 KARNAUGH MAPS

In 1953 Maurice Karnaugh published an article about his system of mapping and thus simplifying Boolean expressions. Figure 4-11 illustrates a Karnaugh map. The four squares (1, 2, 3, 4) represent the four possible combinations of A and B in a two-variable truth table. Square 1 in the Karnaugh map, then, stands for $\overline{A} \cdot \overline{B}$, square 2 for $\overline{A} \cdot B$, and so forth.

Let us map the familiar problem from Fig. 4-10. The original Boolean expression $\overline{A} \cdot B + A \cdot \overline{B} + A \cdot B = Y$ is rewritten in Fig. 4-12(a) for your convenience. Next, a 1 is placed in each square of the Karnaugh map represented in the original Boolean expression, as shown in Fig. 4-12(a). The filled-in

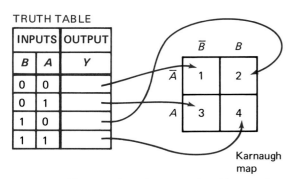

Fig. 4-11 The meaning of squares in a Karnaugh map.

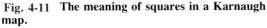

(a)

(b)

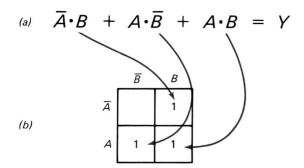

Fig. 4-12 Marking 1s on a Karnaugh map.

Karnaugh map is now ready for *looping*. The looping technique is shown in Fig. 4-13. *Adjacent 1s* are *looped together* in groups of two, four, or eight. Looping continues until all 1s are included inside a loop. Each loop is a new term in the simplified Boolean expression. Notice that we have two loops in Fig. 4-13. These two loops mean that we shall have two terms ORed together in our new simplified Boolean expression.

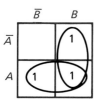

Fig. 4-13 Looping 1s together on a Karnaugh map.

Now let us simplify the Boolean expression based upon the two loops that are redrawn in Fig. 4-14. First the bottom loop:

Fig. 4-14 Simplifying a Boolean expression from a Karnaugh map.

notice that an *A* is included along with a *B* and a $\overline{B}$. The *B* and $\overline{B}$ terms can be *eliminated* according to the rules of Boolean algebra. This leaves the *A* term in the bottom loop. Likewise, the vertical loop contains an *A* and a $\overline{A}$, which are eliminated, leaving only a *B* term. The leftover *A* and *B* terms are then ORed together, giving the simplified Boolean expression $A + B = Y$.

The procedure for simplifying a Boolean expression sounds complicated. Actually, this procedure is quite easy after some practice. Here is a summary of the six steps:

1. Start with a minterm Boolean expression.
2. Record 1s on a Karnaugh map.
3. Loop adjacent 1s (loops of two, four, or eight squares).
4. Simplify by dropping terms that contain a term and its complement within a loop.
5. OR the remaining terms (one term per loop).
6. Write the simplified minterm Boolean expression.

Self-Test

Answer the following questions.

15. The map shown in Fig. 4-12 was developed by _____ .
16. List the six steps used in simplifying a Boolean expression using a Karnaugh map.

4-7 KARNAUGH MAPS WITH THREE VARIABLES

Consider the unsimplified Boolean expression $A \cdot \overline{B} \cdot \overline{C} + \overline{A} \cdot \overline{B} \cdot \overline{C} + \overline{A} \cdot \overline{B} \cdot C + A \cdot B \cdot \overline{C} = Y$, as given in Fig. 4-15(*a*). A three-variable Karnaugh map is illustrated in Fig. 4-15(*b*). Notice the eight possible combinations of *A*, *B*, and *C*, which are represented by the eight squares in the map. Tabulated on the map are four 1s, which represent each of the four terms in the original Boolean expression. The filled-in Karnaugh map is redrawn in Fig. 4-15(*c*). Adjacent groups of two 1s are looped. The bottom loop contains both a *B* and a $\overline{B}$. The *B* and $\overline{B}$ terms are eliminated. The bottom loop still contains the *A* and $\overline{C}$, giving the $A \cdot \overline{C}$ term. The upper loop contains both a *C* and a $\overline{C}$. The *C* and $\overline{C}$ terms are eliminated, leaving the $\overline{A} \cdot \overline{B}$ term. A minterm Boolean expression is formed by adding the OR symbol. The simplified Boolean expression is written in Fig. 4-15(*d*) as $A \cdot \overline{C} + \overline{A} \cdot \overline{B} = Y$.

You can see that the simplified Boolean expression in Fig. 4-15 would take fewer electronic parts than the original expression. Remember that the much different looking simplified Boolean expression produces the same truth table as the original Boolean expression.

It is critical that the Karnaugh map be prepared just as the one shown in Fig. 4-15 was. Note that as you progress downward on the left side of the map, only one variable changes for each step. At the top left $\overline{A}\ \overline{B}$ is listed, while directly below is $\overline{A}\ B$ ($\overline{B}$ changed to *B*). Then, progressing downward from $\overline{A}\ A$ to *AB* the $\overline{A}$ term is changed to *A*. Finally, moving downward from *AB* to $A\overline{B}$ the *B* term is changed to $\overline{B}$. The Karnaugh map will not work properly if it is not laid out correctly.

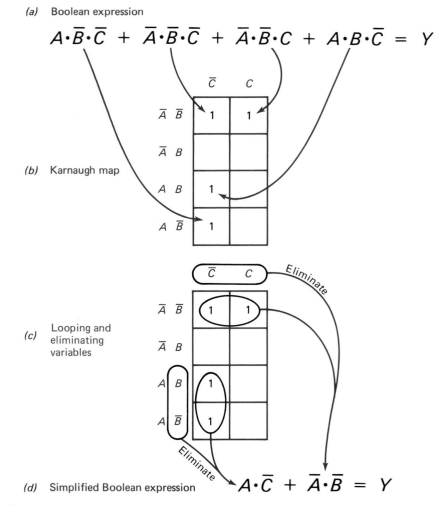

(*a*) Boolean expression

$$A \cdot \overline{B} \cdot \overline{C} + \overline{A} \cdot \overline{B} \cdot \overline{C} + \overline{A} \cdot \overline{B} \cdot C + A \cdot B \cdot \overline{C} = Y$$

(*b*) Karnaugh map

(*c*) Looping and eliminating variables

(*d*) Simplified Boolean expression

$$A \cdot \overline{C} + \overline{A} \cdot \overline{B} = Y$$

Fig. 4-15   Simplifying a Boolean expression using a Karnaugh map. (*a*) Unsimplified expression. (*b*) Mapping 1s. (*c*) Looping 1s and eliminating variables. (*d*) Forming simplified minterm expression.

Self-Test

Answer the following question.

17. Simplify the Boolean expression $\overline{A} \cdot B \cdot \overline{C}$ $+ \overline{A} \cdot B \cdot C + A \cdot B \cdot \overline{C} + A \cdot B \cdot C = Y$ by:
    a. Plotting 1s on a three-variable Karnaugh map
    b. Looping groups of two or four 1s
    c. Eliminating variables whose complement appears within the loop(s)
    d. Writing the simplified Boolean expression

4-8 KARNAUGH MAPS WITH FOUR VARIABLES

The truth table for four variables has 16 possible combinations. Simplifying a Boolean expression that has four variables sounds complicated, but a Karnaugh map makes the job of simplifying a Boolean expression easy.

Consider the Boolean expression $A \cdot \overline{B} \cdot \overline{C} \cdot \overline{D} + \overline{A} \cdot B \cdot \overline{C} \cdot D + \overline{A} \cdot \overline{B} \cdot \overline{C} \cdot D + \overline{A} \cdot B \cdot C \cdot D + \overline{A} \cdot B \cdot C \cdot D + A \cdot \overline{B} \cdot \overline{C} \cdot D = Y$, as in Fig. 4-16(a). The four-variable Karnaugh map in Fig. 4-16(b) gives the 16 possible combinations of A, B, C, and D. These are represented in the 16 squares of the map. Tabulated on the map are six 1s, which represent the six terms in the original Boolean expression. The Karnaugh map is redrawn in Fig. 4-16(c). Adjacent groups of two 1s and four 1s are looped. The bottom loop of two 1s eliminates the D and $\overline{D}$ terms. The bottom loop then produces the $A \cdot \overline{B} \cdot \overline{C}$ term. The upper loop of four 1s eliminates the C and $\overline{C}$ and B and $\overline{B}$ terms. The upper loop then produces the $\overline{A} \cdot D$ term. The $A \cdot \overline{B} \cdot \overline{C}$ and $\overline{A} \cdot D$ terms are then ORed together. The simplified minterm Boolean expression is written in Fig. 4-16(d) as $A \cdot \overline{B} \cdot \overline{C} + \overline{A} \cdot D = Y$.

Observe that the same procedure and rules are used for simplifying Boolean expressions with two, three, or four variables and that larger loops in a Karnaugh map eliminate more variables. You must take care to make sure that the maps look just like the ones in Figs. 4-14 to 4-16 (on next page).

Self-Test

Answer the following question.

18. Simplify the Boolean expression $\overline{A} \cdot B \cdot \overline{C} \cdot \overline{D} + A \cdot B \cdot \overline{C} \cdot \overline{D} + \overline{A} \cdot B \cdot \overline{C} \cdot D + A \cdot B \cdot \overline{C} \cdot D + A \cdot \overline{B} \cdot C \cdot D + A \cdot \overline{B} \cdot C \cdot \overline{D} = Y$ by:

    a. Plotting 1s on a four-variable Karnaugh map
    b. Looping groups of two or four 1s
    c. Eliminating variables whose complements appear within loops
    d. Writing the simplified Boolean expression

4-9 MORE KARNAUGH MAPS

This section presents some sample Karnaugh maps. Notice the unusual looping procedures used on most maps in this section.

Consider the Boolean expression in Fig. 4-17(a) on page 61. The four terms are shown as four 1s on the Karnaugh map in Fig. 4-17(b). The correct looping procedure is shown. Notice that the Karnaugh map is considered to be wrapped in a cylinder, with the left side adjacent to the right side. Also notice the elimination of the A and $\overline{A}$ and C and $\overline{C}$ terms. The simplified Boolean expression of $B \cdot \overline{D} = Y$ is shown in Fig. 4-17(c).

Another unusual looping variation is illustrated in Fig. 4-18(a). Notice that the top and bottom of the map are adjacent to one another, as if rolled into a cylinder while looping. The simplified Boolean expression for this map is given as $\overline{B} \cdot \overline{C} = Y$ in Fig. 4-18(b). The A and $\overline{A}$ as well as the D and $\overline{D}$ terms have been eliminated in Fig. 4-18.

Figure 4-19(a) shows still another unusual looping pattern. The four corners of the Karnaugh map are considered connected, as if the map were formed into a ball. The four corners are then adjacent and may be formed into one loop as shown. The simplified Boolean expression is $\overline{B} \cdot \overline{D} = Y$, given in Fig. 4-19(b). In this example, the A and $\overline{A}$ as well as the C and $\overline{C}$ terms have been eliminated.

Self-Test

Answer the following questions.

19. Simplify the following Boolean expression $\overline{A} \cdot B \cdot \overline{C} \cdot \overline{D} + \overline{A} \cdot \overline{B} \cdot \overline{C} \cdot D + \overline{A} \cdot \overline{B} \cdot C \cdot D + \overline{A} \cdot B \cdot C \cdot \overline{D} + A \cdot \overline{B} \cdot \overline{C} \cdot D + A \cdot \overline{B} \cdot C \cdot D = Y$ by
    a. Plotting 1s on a four-variable Karnaugh map
    b. Looping groups of two or four 1s
    c. Eliminating variables whose complements appear within loops
    d. Writing the simplified Boolean expression

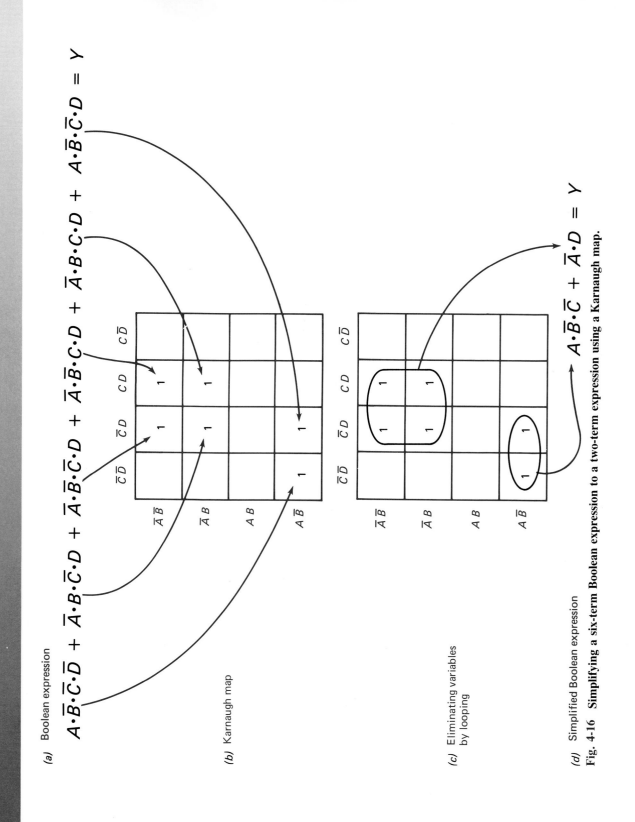

(a) Boolean expression

$$A \cdot \overline{B} \cdot \overline{C} \cdot \overline{D} + \overline{A} \cdot \overline{B} \cdot \overline{C} \cdot D + \overline{A} \cdot \overline{B} \cdot \overline{C} \cdot D + \overline{A} \cdot \overline{B} \cdot C \cdot D + A \cdot \overline{B} \cdot \overline{C} \cdot D = Y$$

(b) Karnaugh map

(c) Eliminating variables by looping

(d) Simplified Boolean expression

$$A \cdot \overline{B} \cdot \overline{C} + \overline{A} \cdot D = Y$$

Fig. 4-16 **Simplifying a six-term Boolean expression to a two-term expression using a Karnaugh map.**

K map looping
variations

Five variable
Karnaugh map

Three-dimensional
Karnaugh map

Looping cylinder

(a) Boolean expression

$$A \cdot B \cdot \overline{C} \cdot \overline{D} + \overline{A} \cdot B \cdot \overline{C} \cdot \overline{D} + \overline{A} \cdot B \cdot C \cdot \overline{D} + A \cdot B \cdot C \cdot \overline{D} = Y$$

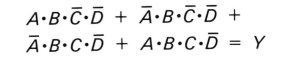

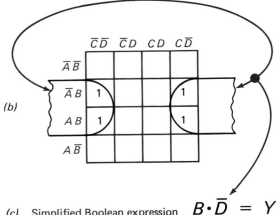

(b)

(c) Simplified Boolean expression $B \cdot \overline{D} = Y$

Fig. 4-17 Simplifying a Boolean expression using a Karnaugh map. By considering the map as a vertical cylinder, the four 1s can be looped as shown.

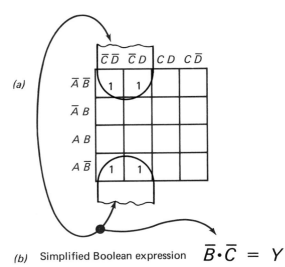

(a)

(b) Simplified Boolean expression $\overline{B} \cdot \overline{C} = Y$

Fig. 4-18 Simplifying a Boolean expression by considering the Karnaugh map as a horizontal cylinder. In this way, the four 1s can be looped.

20. Simplify the following Boolean expression $\overline{A} \cdot \overline{B} \cdot \overline{C} + \overline{A} \cdot \overline{B} \cdot C + A \cdot \overline{B} \cdot \overline{C} + A \cdot \overline{B} \cdot C + A \cdot B \cdot C = Y$ by
   a. Plotting 1s on a three-variable Karnaugh map
   b. Looping groups of two or four 1s
   c. Eliminating variables whose complements appear within loops
   d. Writing the simplified Boolean expression

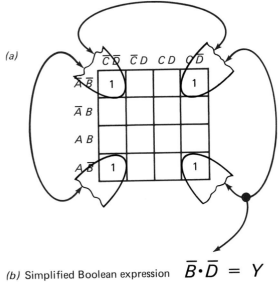

(a)

(b) Simplified Boolean expression $\overline{B} \cdot \overline{D} = Y$

Fig. 4-19 Simplifying a Boolean expression by thinking of the Karnaugh map as a ball. In this way, the 1s in the four corners can be looped.

4-10 A FIVE-VARIABLE KARNAUGH MAP

The Karnaugh map becomes three-dimensional when solving logic problems with more than four variables. A three-dimensional Karnaugh map will be used in this section.

Consider the truth table shown in Fig. 4-20(a) on the next page. The truth table must have 32 (2^5) rows to show all the combinations for five variables. The unsimplified Boolean expression for the truth table is detailed in Fig. 4-20(b).

A five-variable Karnaugh map is drawn in Fig. 4-20(c). Notice that it has two 4-variable Karnaugh maps stacked to make it three dimensional. The top map is the E plane while the bottom is the $\overline{E}$ plane.

Each of the nine terms in the unsimplified Boolean expression is plotted as a 1 on the Karnaugh map in Fig. 4-20(c). Adjacent groups of two, four, and eight are looped. The four 1s on the E and $\overline{E}$ planes are also adjacent so that the entire group is enclosed in a cylinder and is considered a single group of eight 1s.

The next step is the conversion of the looped 1s on the Karnaugh map to a simplified Boolean expression. The lone 1 on the $\overline{E}$ plane of the Karnaugh map in Fig. 4-20(c) cannot be simplified and is written as $\overline{E} \cdot \overline{D} \cdot \overline{C} \cdot \overline{B} \cdot A$ in Fig. 4-20(d). The eight 1s enclosed in the looping cylinder can be simplified. The E and $\overline{E}$, the C and $\overline{C}$, and the B and $\overline{B}$ variables are eliminated leaving the term $D \cdot \overline{A}$. The terms

Simplifying
five-variable
Boolean
expression

Five-variable
Karnaugh map

TRUTH TABLE

INPUTS					OUTPUT
E	D	C	B	A	Y
0	0	0	0	0	0
0	0	0	0	1	1
0	0	0	1	0	0
0	0	0	1	1	0
0	0	1	0	0	0
0	0	1	0	1	0
0	0	1	1	0	0
0	0	1	1	1	0
0	1	0	0	0	1
0	1	0	0	1	0
0	1	0	1	0	1
0	1	0	1	1	0
0	1	1	0	0	1
0	1	1	0	1	0
0	1	1	1	0	1
0	1	1	1	1	0
1	0	0	0	0	0
1	0	0	0	1	0
1	0	0	1	0	0
1	0	0	1	1	0
1	0	1	0	0	0
1	0	1	0	1	0
1	0	1	1	0	0
1	0	1	1	1	0
1	1	0	0	0	1
1	1	0	0	1	0
1	1	0	1	0	1
1	1	0	1	1	0
1	1	1	0	0	1
1	1	1	0	1	0
1	1	1	1	0	1
1	1	1	1	1	0

(a)

$$\overline{E} \cdot \overline{D} \cdot \overline{C} \cdot \overline{B} \cdot A + \overline{E} \cdot D \cdot \overline{C} \cdot \overline{B} \cdot \overline{A} +$$
$$\overline{E} \cdot D \cdot \overline{C} \cdot B \cdot \overline{A} + \overline{E} \cdot D \cdot C \cdot \overline{B} \cdot \overline{A} +$$
$$\overline{E} \cdot D \cdot C \cdot B \cdot \overline{A} + E \cdot D \cdot \overline{C} \cdot \overline{B} \cdot \overline{A} +$$
$$E \cdot D \cdot \overline{C} \cdot B \cdot \overline{A} + E \cdot D \cdot C \cdot \overline{B} \cdot \overline{A} +$$
$$E \cdot D \cdot C \cdot B \cdot \overline{A} = Y$$

(b) Unsimplified Boolean expression

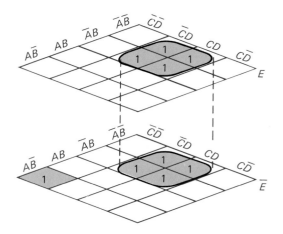

(c) Karnaugh map—plotting 1s and looping

$$\overline{E} \cdot \overline{D} \cdot \overline{C} \cdot \overline{B} \cdot A + D \cdot \overline{A} = Y$$

(d) Simplified Boolean expression

Fig. 4-20 **A logic problem with five variables.** (a) **Truth table.** (b) **Unsimplified Boolean expression.** (c) **Plotting 1s and looping on a five-variable Karnaugh map.** (d) **Simplified Boolean expression.**

$\overline{E} \cdot \overline{D} \cdot \overline{C} \cdot \overline{B} \cdot A$ are ORed with $D \cdot \overline{A}$, yielding the simplified Boolean expression shown in Fig. 4-20(d).

Self-Test

Answer the following question.

21. Simplify the Boolean expression $A \cdot \overline{B} \cdot \overline{C} \cdot \overline{D} \cdot \overline{E} + A \cdot \overline{B} \cdot \overline{C} D \cdot \overline{E} + A \cdot \overline{B} \cdot \overline{C} \cdot \overline{D} \cdot E + A \cdot \overline{B} \cdot \overline{C} \cdot D \cdot E + \overline{A} \cdot \overline{B} \cdot C \cdot D \cdot E + \overline{A} \cdot \overline{B} \cdot C \cdot DE = Y$ by:
    a. Plotting 1s on a five-variable Karnaugh map
    b. Looping groups of two, four, or eight adjacent 1s

c. Eliminating variables whose complements appear within loops or cylinders
d. Writing the simplified Boolean expression in minterm form

4-11 USING NAND LOGIC

Section 3-8 explained how NAND gates could be wired to form other gates or inverters (see Fig. 3-21). We mentioned that the NAND gate can be used as a universal gate. In this section, you shall see how NAND gates are used in wiring combinational logic circuits. NAND gates are widely employed in industry because they are easy to use and readily available.

Suppose your supervisor gives you the Boolean expression $A \cdot B + A \cdot \overline{C} = Y$, as shown in Fig. 4-21(a). You are told to solve this logic problem at the least cost. You first draw the logic circuit for the Boolean expression shown in Fig. 4-21(b), using AND gates, an OR gate, and an inverter. Checking a manufacturer's catalog, you determine that you must use three different ICs to do the job.

Your supervisor suggests that you try using NAND logic. You redraw your logic circuit to look like the NAND–NAND logic circuit in Fig. 4-21(c). Upon checking a catalog, you find you need only one IC that contains the four NAND gates to do the job. Recall from Chap. 3 that the OR symbol with invert bubbles at the inputs is another symbol for a NAND gate. You finally test the circuit in Fig. 4-21(c) and find that it performs the logic $A \cdot B + A \cdot \overline{C} = Y$. Your supervisor is pleased you have found a circuit that requires only one IC, as compared to the circuit in Fig. 4-21(b), which uses three ICs.

Remembering this trick will help you appreciate *why* NAND gates are used in many logic circuits. If your future job is in digital circuit design, this can be a useful tool for making your final circuit the best for the least cost.

You may have questioned why the NAND gates in Fig. 4-21(c) could be substituted for the AND and OR gates in Fig. 4-21(b). If you look carefully at Fig. 4-21(c), you will see two AND symbols feeding into an OR symbol. From previous experience we know that if we invert twice we have the original logic state. Hence the two invert bubbles in Fig. 4-21(c) between the AND and OR symbols cancel one another. Because the two invert bubbles cancel one another, we end up with two AND gates feeding an OR gate.

In summary, using NAND gates involves these steps:

1. Start with a minterm (sum-of-products) Boolean expression.
2. Draw the AND-OR logic diagram using AND, OR, and NOT symbols.
3. Substitute NAND symbols for each AND and OR symbol, keeping all connections the same.

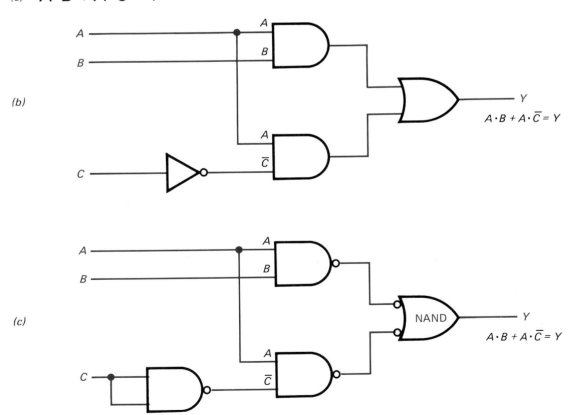

(a) $A \cdot B + A \cdot \overline{C} = Y$

(b)

(c)

Fig. 4-21 Using NAND gates in logic circuits. (*a*) Boolean expression. (*b*) AND-OR logic circuit. (*c*) Equivalent NAND-NAND logic circuit.

4. Substitute NAND symbols with all inputs tied together for each inverter.
5. Test the logic circuit containing all NAND gates to determine if it generates the proper truth table.

Self-Test

Answer the following questions.

22. The logic circuit in Fig. 4-21(*b*) is called a(n) _____ (AND-OR, NAND-NAND) circuit.
23. The logic circuits in Fig. 4-21(*b*) and (*c*) generate _____ (different, identical) truth tables.
24. List five steps in converting a sum-of-products Boolean expression to a NAND-NAND logic circuit.

4-12 SOLVING LOGIC PROBLEMS THE EASY WAY

Manufacturers of ICs have simplified the job of solving many combinational logic problems by producing *data selectors*. A data selector is often a *one-package solution* to a complicated logic problem. The data selector actually contains a rather large number of gates packaged inside a single IC. In this chapter the data selector will be used as a "universal package" for solving combinational logic problems.

A *1-of-8 data selector* is illustrated in Fig. 4-22. Notice the eight *data inputs* numbered from 0 to 7 on the left. Also notice the three *data selector inputs* labeled *A*, *B*, and *C* at the bottom of the data selector. The output of the data selector is labeled *W*.

The basic job the data selector performs is transferring data from a *given* data input (0 to 7) to the output (*W*). Which data input is selected is determined by which binary number you place on the data selector inputs at the bottom (see Fig. 4-22). The data selector in Fig. 4-22 functions in the same manner as a rotary switch. Figure 4-23 shows the data at input 3 being transferred to the output by the rotary switch contacts. In like manner the data from data input 3 in Fig. 4-22 is being transferred to output *W* of the data selector. In the rotary switch you must mechanically change the switch position to transfer data from another input. In the 1-of-8 data selector in Fig. 4-22 you need only change the binary input at the data selector inputs to transfer data from another data input to the output. Remember that the data selector operates somewhat as a rotary switch in transferring logical 0s or 1s from a given input to the single output.

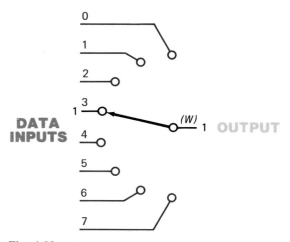

Fig. 4-23 **Single-pole 8-position rotary switch works as a data selector.**

Now you will learn how data selectors can be used to solve logic problems. Consider the *simplified* Boolean expression shown in Fig. 4-24(*a*). For your convenience a logic circuit for this complicated Boolean expression is drawn in Fig. 4-24(*b*). Using standard ICs, we probably would have to use from six to nine ICs to solve this problem. This would be quite expensive because of the cost of the ICs and PC board space.

A less costly solution to the logic problem is to use a data selector. The Boolean expression from Fig. 4-24(*a*) is repeated in truth-table form in Fig. 4-25(*a*). A *1-of-16 data selector* is added in Fig. 4-25(*b*). Notice that logical 0s and 1s are placed at the 16 data inputs of the data selector corresponding to the truth-table output column

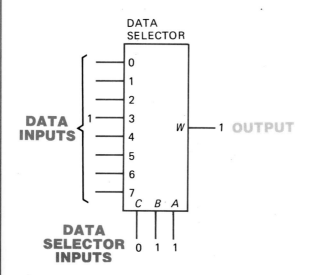

Fig. 4-22 **Logic symbol for a 1-of-8 data selector.**

Y. These are *permanently* connected for this truth table. Data selector inputs (*D*, *C*, *B*, and *A*) are switched to the binary numbers on the input side of the truth table. If the data selector inputs *D*, *C*, *B*, and *A* are at binary 0000, then a logical 1 is transferred to output *W* of the data selector. The first line of the truth table requires that a logical 1 appear at output *W* when *D*, *C*, *B*, and *A* are all 0s. If data selector inputs *D*, *C*, *B*, and *A* are at binary 0001, a logical 0 appears at output *W*, as required by the truth table. Any combination of *D*, *C*, *B*, and *A* generates the proper output according to the truth table.

We used the data selector to solve a complicated logic problem. In Fig. 4-24 we found we needed at least six ICs to solve this logic problem. Using the data selector in Fig. 4-25, we solved this problem by using only one IC.

The data selector seems to be an easy-to-use and efficient way to solve combinational logic problems. Commonly available data selectors can solve logic problems with three, four, or five variables. When using manufacturers' data manuals, you will notice that data selectors are also called *multiplexers*.

Self-Test

Supply the missing word, letter, or number in each statement.

25. Figure 4-22 illustrates the logic symbol for a 1-of-8 _____ .
26. Refer to Fig. 4-22. If all data select inputs are HIGH, data at input _____ (number) is selected and transferred to output _____ (letter) of the data selector.

(a) Simplified Boolean expression

$$A \cdot B \cdot C \cdot D + \overline{A} \cdot \overline{B} \cdot \overline{C} \cdot \overline{D} + A \cdot \overline{B} \cdot \overline{C} \cdot D + A \cdot B \cdot \overline{C} \cdot \overline{D} +$$
$$\overline{A} \cdot B \cdot C \cdot \overline{D} + \overline{A} \cdot B \cdot \overline{C} \cdot D + \overline{A} \cdot \overline{B} \cdot C \cdot D = Y$$

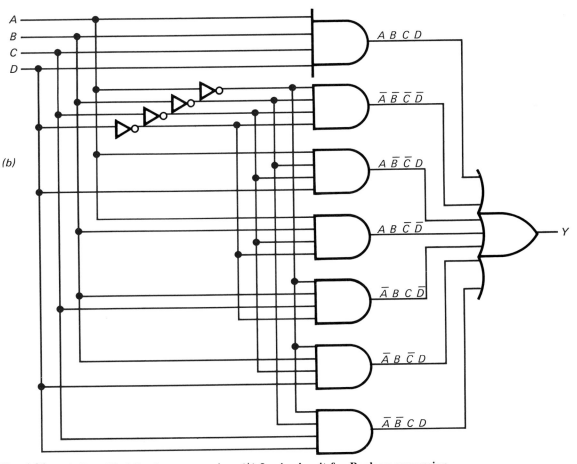

(b)

Fig. 4-24 *(a)* **Simplified Boolean expression.** *(b)* **Logic circuit for Boolean expression.**

TRUTH TABLE

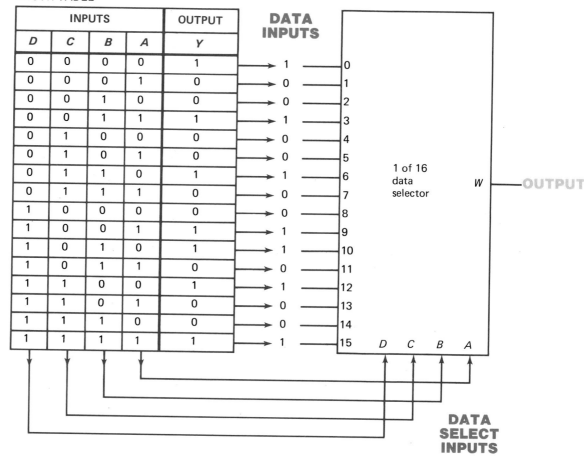

INPUTS				OUTPUT
D	C	B	A	Y
0	0	0	0	1
0	0	0	1	0
0	0	1	0	0
0	0	1	1	1
0	1	0	0	0
0	1	0	1	0
0	1	1	0	1
0	1	1	1	0
1	0	0	0	0
1	0	0	1	1
1	0	1	0	1
1	0	1	1	0
1	1	0	0	1
1	1	0	1	0
1	1	1	0	0
1	1	1	1	1

Fig. 4-25 Solving logic problem with a data selector.

27. The action of a data selector is many times compared to that of a mechanical _____ switch.
28. Refer to Fig. 4-25. If all data select inputs are HIGH, data from input _____ (number) will be transferred to output W. Under these conditions, output W will be _____ (HIGH, LOW).
29. Logic problems can many times be solved using a single _____ IC.

4-13 MORE DATA SELECTOR PROBLEMS

The previous section used a 1-of-16 data selector to solve a four-variable logic problem. A similar logic problem can be solved using a less expensive, 1-of-8 data selector. This is done by using what is sometimes called the "folding" technique.

Consider the four-variable truth table shown in Fig. 4-26. Note that the pattern of inputs C, B, and A is the same in lines 0 through 7 as in lines 8 through 15. These areas are circled on the truth table in Fig. 4-26. To solve this logic problem using a 1-of-8 data selector, inputs C, B, and A are connected to the data select inputs of the unit. This is shown in the lower part of Fig. 4-26.

The eight data inputs (D_0 to D_7) shown in Fig. 4-27(i), on page 68, must now be determined one by one. The input to D_0 on the 74151 IC is determined in Fig. 4-27(a). The truth table from Fig. 4-26 is folded over to compare lines 0 and 8. Inputs C, B, and A (which connect to the 74151 IC's data select or inputs) are each 000. Whether input D is 0 or 1, output Y is always 0 according to Fig. 4-27(a). Therefore, a logical 0 (GND) is applied to the D_0 input to the 74151 1-of-8 data selector IC. This is shown in Fig. 4-27(i).

The input to D_1 of the 74151 IC is determined in Fig. 4-27(b). The folding technique is

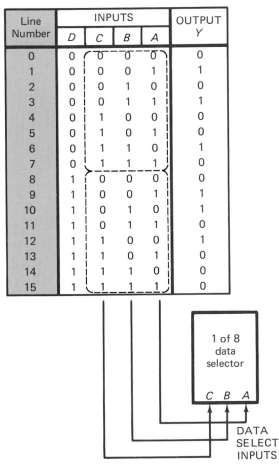

Line Number	INPUTS				OUTPUT Y
	D	C	B	A	
0	0	0	0	0	0
1	0	0	0	1	1
2	0	0	1	0	0
3	0	0	1	1	1
4	0	1	0	0	0
5	0	1	0	1	0
6	0	1	1	0	1
7	0	1	1	1	0
8	1	0	0	0	0
9	1	0	0	1	1
10	1	0	1	0	1
11	1	0	1	1	0
12	1	1	0	0	1
13	1	1	0	1	0
14	1	1	1	0	0
15	1	1	1	1	0

Fig. 4-26 First step in using a 1-of-8 data selector to solve a four-variable logic problem.

used to compare lines 1 and 9 of the truth table. Inputs C, B, and A must be the same. Whether input D is 0 or 1, output Y is always 1. Therefore, a logical 1 ($+5$ V) is applied to the D_1 input to the 74151 data selector IC. This is shown in Fig. 4-27(i).

The input to D_2 of the 74151 IC is determined in Fig. 4-27(c). Folding the truth table compares lines 2 and 10. Inputs C, B, and A are the same. The outputs are different. In each case, the output is the same as the D input. Therefore, data input D_2 to the 74151 IC is equal to input D from the truth table. The logic symbol for the 74151 IC shows a D written to the left of input D_2.

The input to D_3 of the 74151 data selector IC is determined in Fig. 4-27(d). Folding the truth table compares lines 3 and 11. Inputs C, B, and A are the same. The outputs are different. In each case, the output is the complement of the D input. Therefore, data input D_3 to the 74151 IC is equal to a not D ($\overline{D}$). The logic symbol for the 74151 IC in Fig. 4-27 (i) shows a $\overline{D}$ written to the left of input D_3.

In like manner, the input to D_4 is determined in Fig. 4-27(e). Data input D_4 to the 1-of-8 data selector is equal to D.

The input to D_5 is determined in Fig. 4-27(f). Data input D_5 to the 74151 IC is equal to 0 (GND).

The input to D_6 is determined in Fig. 4-27(g). Data input D_6 to the 1-of-8 data selector is equal to $\overline{D}$ (not D).

Finally, the input to D_7 is determined in Fig. 4-27(h). Data input D_7 of the 74151 data selector IC is equal to 0 (GND).

Note in Fig. 4-27(i) that data inputs D_0, D_5, and D_7 of the 74151 IC are permanently grounded. Data input D_1 is permanently connected to $+5$ V. Data inputs D_2 and D_4 are connected directly to input D from the truth table. Data inputs D_3 and D_6 are connected through an inverter to the complement of input D. The enable, or strobe, input to the 74151 1-of-8 data selector must be held LOW (at logical 0) for the unit to operate. The small bubble on the logic symbol in Fig. 4-27(i) means that the enable input is an active LOW input.

The data selector (multiplexer) has been used as a universal logic element in the last two sections. It is a simple, low-cost solution to many logic problems with from three to five input variables.

Self-Test

Select the correct word in each statement.

30. Refer to Fig. 4-27(i). With inputs of $D = 1$, $C = 0$, $B = 0$, $A = 0$, and enable $= 0$, output Y of the data selector will be _____ (HIGH, LOW).

31. Refer to Fig. 4-27(i). With inputs of $D = 1$, $C = 1$, $B = 0$, $A = 0$, and enable $= 0$, output Y of the data selector will be _____ (HIGH, LOW).

32. Refer to Fig. 4-27(i). With inputs of $D = 1$, $C = 0$, $B = 1$, $A = 1$, and enable $= 0$, output Y of the data selector will be _____ (HIGH, LOW).

4-14 DATA SELECTORS: A FIVE-VARIABLE PROBLEM

Logic problems with five variables could be solved using Karnaugh maps and logic gates. These complicated logic problems can also be solved using several data selectors.

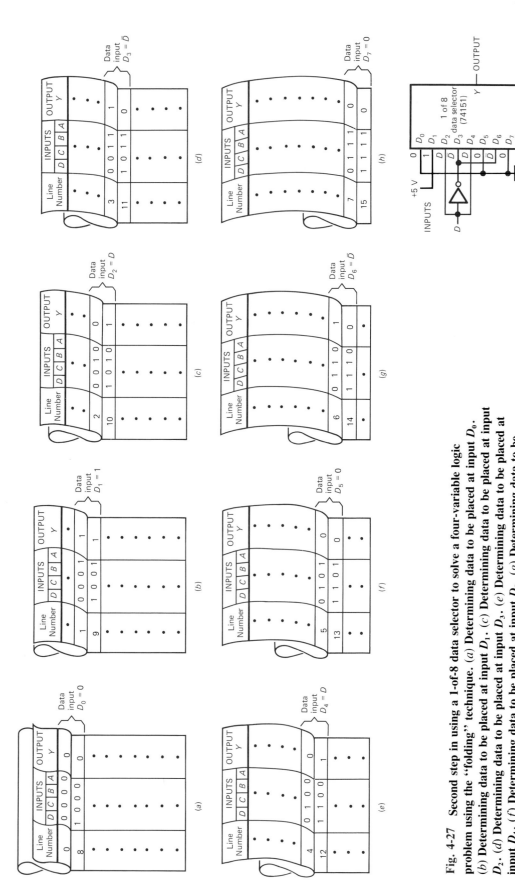

Fig. 4-27 Second step in using a 1-of-8 data selector to solve a four-variable logic problem using the "folding" technique. (a) Determining data to be placed at input D_0. (b) Determining data to be placed at input D_1. (c) Determining data to be placed at input D_2. (d) Determining data to be placed at input D_3. (e) Determining data to be placed at input D_4. (f) Determining data to be placed at input D_5. (g) Determining data to be placed at input D_6. (h) Determining data to be placed at input D_7. (i) Solution to four-variable logic problem posed in truth table.

Consider the truth table in Fig. 4-28(*a*). The problem is plotted on a Karnaugh map in Fig. 4-28(*b*). Observe that there are no 1s adjacent to one another either in rows, columns, or from plane to plane on the Karnaugh map. This results in *no simplification*. The very complex Boolean expression that describes the truth table is written in Fig. 4-28(*c*). It would probably take 10 to 20 ICs to implement this Boolean expression using logic gates.

A simple, less expensive solution might be to use a data selector network to solve the logic problem posed in the truth table in Fig. 4-28(*a*). The solution shown in Fig. 4-29 should solve this complicated logic problem. The truth table for the logic problem is reproduced at the left in Fig.

4-29. The data inputs to two 1-of-16 data selectors are permanently connected to either *H* or *L* of the power supply corresponding to their data input from column *Y* of the truth table. For instance, input 0 of the top 1-of-16 data selector would be grounded. However, input 0 of the bottom 1-of-16 data selector would be tied to +5 V of the power supply. The two 1-of-16 data selector outputs feed a smaller 1-of-2 data selector which is connected to the output indicator.

The data selector inputs (*D, C, B, A*) and select (*E*) on the three data selectors are switches to the binary numbers on the input side of the truth table. If the data selector inputs *E, D, C, B,* and *A* are at binary 00000, then logical 0 is transferred to output *W* of the *top* 1-of-16 data selector,

TRUTH TABLE

INPUTS					OUTPUT
E	*D*	*C*	*B*	*A*	*Y*
0	0	0	0	0	0
0	0	0	0	1	1
0	0	0	1	0	1
0	0	0	1	1	0
0	0	1	0	0	0
0	0	1	0	1	0
0	0	1	1	0	0
0	0	1	1	1	1
0	1	0	0	0	0
0	1	0	0	1	0
0	1	0	1	0	0
0	1	0	1	1	0
0	1	1	0	0	0
0	1	1	0	1	1
0	1	1	1	0	1
0	1	1	1	1	0
1	0	0	0	0	1
1	0	0	0	1	0
1	0	0	1	0	0
1	0	0	1	1	1
1	0	1	0	0	0
1	0	1	0	1	0
1	0	1	1	0	0
1	0	1	1	1	0
1	1	0	0	0	0
1	1	0	0	1	1
1	1	0	1	0	1
1	1	0	1	1	0
1	1	1	0	0	1
1	1	1	0	1	0
1	1	1	1	0	0
1	1	1	1	1	0

(*a*)

(*b*) Karnaugh map

$$\overline{E} \cdot \overline{D} \cdot \overline{C} \cdot B \cdot A + \overline{E} \cdot \overline{D} \cdot \overline{C} \cdot B \cdot \overline{A} +$$
$$\overline{E} \cdot \overline{D} \cdot C \cdot B \cdot A + \overline{E} \cdot D \cdot C \cdot \overline{B} \cdot A +$$
$$\overline{E} \cdot D \cdot C \cdot B \cdot \overline{A} + E \cdot \overline{D} \cdot \overline{C} \cdot \overline{B} \cdot \overline{A} +$$
$$E \cdot \overline{D} \cdot \overline{C} \cdot B \cdot A + E \cdot D \cdot \overline{C} \cdot \overline{B} \cdot A +$$
$$E \cdot D \cdot \overline{C} \cdot B \cdot \overline{A} + E \cdot D \cdot C \cdot \overline{B} \cdot \overline{A} = Y$$

(*c*) Boolean expression

Fig. 4-28 **Five-variable logic problem.** (*a*) **Truth table.** (*b*) **1s plotted on Karnaugh map.** (*c*) **Boolean expression—no simplification possible in this problem.**

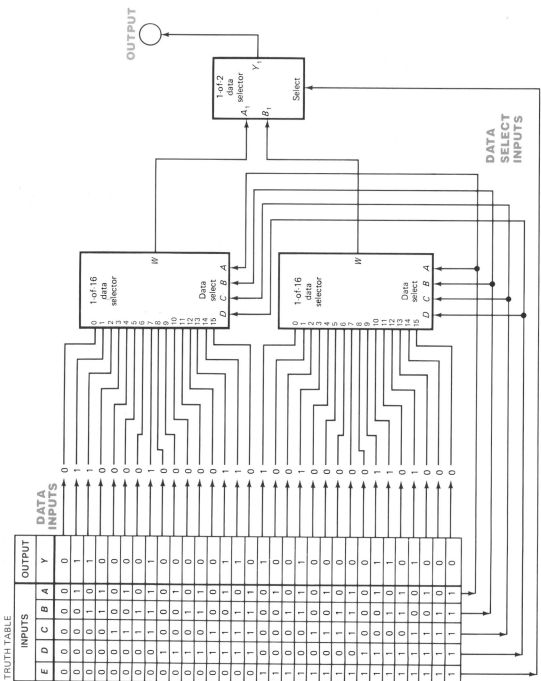

Fig. 4-29 Using a data selector network to solve a five-variable logic problem.

through the 1-of-2 data selector, on to the output indicator. However, if the data selector inputs E, D, C, B, and A are at binary 10000, then logical 1 is transferred to output W of the *bottom* 1-of-16 data selector, through the 1-of-2 data selector and on to the output indicator. Any combination of inputs (E, D, C, B, and A) generates the proper output as given in the truth table.

Data selectors reduced the IC package count when used to solve this five-variable logic problem. Only three ICs were needed using the solution shown in Fig. 4-29. This is compared to 10 to 20 ICs using logic gates.

Simplified gate circuits and data selector ICs have been used to implement logic problems. More complex logic problems are created when there are more variables or when the logic circuit has several outputs. For these problems, designers can use a programmable array of logic gates within a single IC. This device is called *programmable array logic* or *PAL*. The PAL is based on programmable AND/OR architecture. These programmable logic devices are available in both TTL and CMOS. These devices are user-programmable. A PAL may have 16 inputs and 8 outputs.

Very complicated logic problems can be solved using either factory-programmed *gate arrays* or *read-only memories* (ROMs). User-programmable gate arrays and ROMs are also available in the form of *PROMs* (*programmable ROMs*), *EPROMs* (*erasable PROMs*), and *programmable gate arrays*.

Self-Test

Answer the following questions.

33. Refer to Fig. 4-28(*b*). Is simplification possible using the Karnaugh map in this example?
34. Refer to Fig. 4-29. A data input of 0 means that input to the 1-of-16 data selector should be permanently connected to _____ of the power supply.
35. Refer to Fig. 4-29. If the data select inputs (E, D, C, B, A) are 10111, the output indicator would read _____ (HIGH, LOW).
36. List several devices that are used to solve very complicated logic problems.

Programmable array logic (PAL)

ROM

Programmable ROM (PROM)

Erasable PROM (EPROM)

Programmable gate arrays

SUMMARY

1. Combining gates in combinational logic circuits from Boolean expressions is a necessary skill for most competent technicians and engineers.
2. Workers in digital electronics must have an excellent knowledge of gate symbols, truth tables, and Boolean expressions and know how to convert from one form to another.
3. The minterm Boolean expression (sum-of-products form) might look like the expression in Fig. 4-30(*a*). The Boolean expression $A \cdot B + \overline{A} \cdot \overline{C} = Y$ would be wired as shown in Fig. 4-30(*b*).
4. The pattern of gates shown in Fig. 4-30(*b*) is called an AND-OR logic circuit.
5. The maxterm Boolean expression (product-of-sums form) might look like the expression in Fig. 4-30(*c*). The Boolean expression $(A + \overline{C}) \cdot (\overline{A} + B) = Y$ would be wired as shown in Fig. 4-30(*d*). This is an OR-AND logic circuit.
6. A Karnaugh map is a convenient method of simplifying Boolean expressions.

(a) Minterm Boolean expression

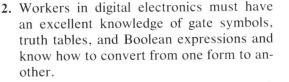

$$A \cdot B + \overline{A} \cdot \overline{C} = Y$$

(b)

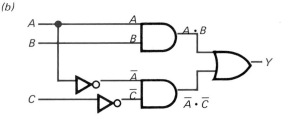

(c) Maxterm Boolean expression

$$(A + \overline{C}) \cdot (\overline{A} + B) = Y$$

(d)

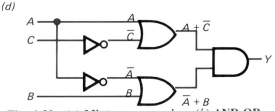

Fig. 4-30 (*a*) **Minterm expression.** (*b*) **AND-OR logic circuit.** (*c*) **Maxterm expression.** (*d*) **OR-AND logic circuit.**

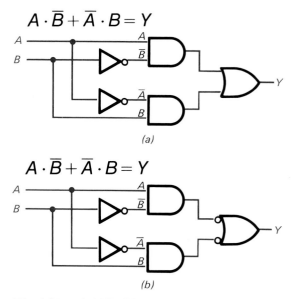

$$A \cdot \overline{B} + \overline{A} \cdot B = Y$$

(a)

$$A \cdot \overline{B} + \overline{A} \cdot B = Y$$

(b)

Fig. 4-31 (*a*) AND-OR logic circuit. (*b*) Equivalent NAND-NAND logic circuit.

7. AND-OR logic circuits can be wired easily by using only NAND gates, as shown in Fig. 4-31.
8. Data selectors are a simple, one-package method of solving many gating problems. Less expensive data selectors can be used when the folding design technique is utilized.
9. Very complex logic problems can be solved using ROMs, PROMs, gate arrays, or user-programmable gate arrays.

CHAPTER REVIEW QUESTIONS

Answer the following questions.

4-1. Engineers and technicians refer to circuits that are a combination of different gates as _____ logic circuits.

4-2. Draw a logic diagram for the Boolean expression $\overline{A} \cdot \overline{B} + B \cdot C = Y$. Use one OR gate, two AND gates, and two inverters.

4-3. The Boolean expression $\overline{A} \cdot \overline{B} + B \cdot C = Y$ is in _____ (product-of-sums, sum-of-products) form.

4-4. The Boolean expression $(A + B) \cdot (C + D) = Y$ is in _____ (product-of-sums, sum-of-products) form.

4-5. A Boolean expression in product-of-sums form is also called a _____ expression.

4-6. A Boolean expression in sum-of-products form is also called a _____ expression.

4-7. Write the minterm Boolean expression that would describe the truth table in Fig. 4-32. Do not simplify the Boolean expression.

4-8. Draw a truth table (three-variable) that represents the Boolean expression $\overline{C} \cdot \overline{B} + C \cdot \overline{B} \cdot A = Y$.

4-9. The truth table in Fig. 4-33 is for an electronic lock. The lock will open only when a logical 1 appears at the output. First, write the minterm Boolean expression for the lock. Second, draw the logic circuit for the lock (use AND, OR, and NOT gates).

4-10. List the six steps for simplifying a Boolean expression as discussed in Sec. 4-6.

INPUTS			OUTPUT
C	**B**	**A**	**Y**
0	0	0	1
0	0	1	0
0	1	0	1
0	1	1	0
1	0	0	0
1	0	1	1
1	1	0	0
1	1	1	1

Fig. 4-32   **Truth table.**

TRUTH TABLE

INPUTS			OUTPUT
C	**B**	**A**	**Y**
0	0	0	0
0	0	1	0
0	1	0	0
0	1	1	1
1	0	0	1
1	0	1	0
1	1	0	0
1	1	1	0

Fig. 4-33   **Truth table for electronic lock.**

4-11. Use a Karnaugh map to simplify the Boolean expression $\overline{A}\cdot\overline{B}\cdot\overline{C} + \overline{A}\cdot\overline{B}\cdot C + A\cdot B\cdot\overline{C} + A\cdot\overline{B}\cdot\overline{C} = Y$. Write the simplified Boolean expression in minterm form.

4-12. Use a Karnaugh map to simplify the Boolean expression $A\cdot\overline{B}\cdot\overline{C}\cdot\overline{D} + A\cdot\overline{B}\cdot\overline{C}\cdot D + A\cdot\overline{B}\cdot C\cdot D + A\cdot\overline{B}\cdot C\cdot\overline{D} = Y$.

4-13. From the truth table in Fig. 4-32 do the following:
   a. Write the unsimplified Boolean expression.
   b. Use a Karnaugh map to simplify the Boolean expression from **a**.
   c. Write the simplified minterm Boolean expression for the truth table.
   d. Draw a logic circuit from the simplified Boolean expression (use AND, OR, and NOT gates).
   e. Redraw the logic circuit from **d** using only NAND gates.

4-14. Use a Karnaugh map to simplify the Boolean expression $\overline{A}\cdot\overline{B}\cdot C\cdot D + A\cdot B\cdot\overline{C}\cdot\overline{D} + A\cdot B\cdot C\cdot\overline{D} + A\,\overline{B}\cdot C\cdot D = Y$. Write the answer as a minterm Boolean expression.

4-15. From the Boolean expression $\overline{A}\cdot\overline{B}\cdot\overline{C}\cdot\overline{D} + \overline{A}\cdot\overline{B}\cdot C\cdot D + \overline{A}\cdot B\cdot\overline{C}\cdot D + A\cdot B\cdot C\cdot D + A\cdot B\cdot C\cdot\overline{D} + A\cdot\overline{B}\cdot\overline{C}\cdot\overline{D} = Y$ do the following:
   a. Draw a truth table for the expression.
   b. Use a Karnaugh map to simplify.
   c. Draw a logic circuit of the simplified Boolean expression (use AND, OR, and NOT gates).
   d. Draw a circuit to solve this problem using a 1-of-16 data selector.
   e. Draw a circuit to solve this problem using the folding technique and a 1-of-8 data selector.

4-16. From the Boolean expression $\overline{A}\cdot\overline{B}\cdot\overline{C}\cdot DE + \overline{A}\cdot B\cdot\overline{C}\cdot D\cdot E + A\cdot B\cdot\overline{C}\cdot D\cdot E + A\cdot\overline{B}\cdot\overline{C}\cdot D\cdot E + A\cdot B\cdot\overline{C}\cdot D\cdot\overline{E} + \overline{A}\cdot\overline{B}\cdot C\cdot D\cdot E + \overline{A}\cdot B\cdot C\cdot\overline{D}\cdot\overline{E} = Y$ do the following:

a. Use a Karnaugh map to simplify.
b. Write the simplified minterm Boolean expression.
c. Draw a logic circuit from the simplified Boolean expression (use AND, OR, and NOT gates).

4-17. Refer to Fig. 4-34. Draw a logic circuit to solve the above problem using two 1-of-16 and one 1-of-2 data selector ICs.

4-18. From the five-variable truth table in Fig. 4-34 do the following:
a. Write an unsimplified Boolean expression.
b. Draw a logic symbol circuit to solve this problem using the folding technique and a 1-of-16 data selector.

| INPUTS | | | | | OUTPUT | INPUTS | | | | | OUTPUT |
E	D	C	B	A	Y	E	D	C	B	A	Y
0	0	0	0	0	0	1	0	0	0	0	0
0	0	0	0	1	1	1	0	0	0	1	1
0	0	0	1	0	0	1	0	0	1	0	0
0	0	0	1	1	0	1	0	0	1	1	0
0	0	1	0	0	0	1	0	1	0	0	1
0	0	1	0	1	1	1	0	1	0	1	0
0	0	1	1	0	0	1	0	1	1	0	0
0	0	1	1	1	0	1	0	1	1	1	0
0	1	0	0	0	1	1	1	0	0	0	0
0	1	0	0	1	0	1	1	0	0	1	0
0	1	0	1	0	0	1	1	0	1	0	0
0	1	0	1	1	0	1	1	0	1	1	0
0	1	1	0	0	1	1	1	1	0	0	1
0	1	1	0	1	0	1	1	1	0	1	1
0	1	1	1	0	0	1	1	1	1	0	0
0	1	1	1	1	0	1	1	1	1	1	0

Fig. 4-34 Five-variable truth table.

Answers to Self-Tests

1. See a and b below

4. See a and b below

a.

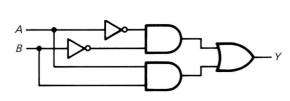

a.

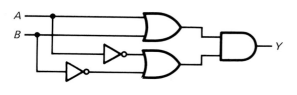

b.

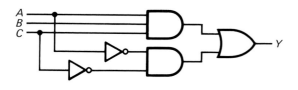

b.

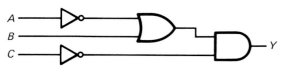

2. sum-of-products
3. product-of-sums

5. maxterm
6. product-of-sums
7. OR-AND

8. $C \cdot B \cdot \overline{A} + C \cdot B \cdot A = Y$
9. lines 1 and 2
10. See table below.

Truth Table

Input Switches			Output
C	B	A	Y
0	0	0	0
0	0	1	0
0	1	0	0
0	1	1	0
1	0	0	0
1	0	1	1
1	1	0	1
1	1	1	0

11. $\overline{C} \cdot B \cdot \overline{A} + C \cdot \overline{B} \cdot A = Y$
12. See figure below.

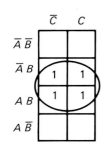

13. identical
14. Boolean, Karnaugh
15. Maurice Karnaugh
16. 1. Start with a minterm Boolean expression.
    2. Record 1s on a Karnaugh map.
    3. Loop adjacent 1s (loops of two, four, or eight squares).
    4. Simplify by dropping terms that contain a term and its complement within a loop.
    5. OR the remaining terms (one term per loop).
    6. Write the simplified minterm Boolean expression.
17. See a–c below

d. $B = Y$

18. See a–c below.

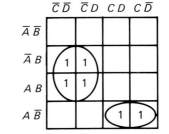

d. $B \cdot \overline{C} + A \cdot \overline{B} \cdot C = Y$

19. See a–c below.

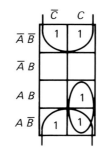

d. $\overline{A} \cdot B \cdot \overline{D} + \overline{B} \cdot D = Y$

20. See a–c below.

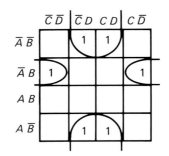

d. $\overline{B} + A \cdot C = Y$

21. See a–c below.

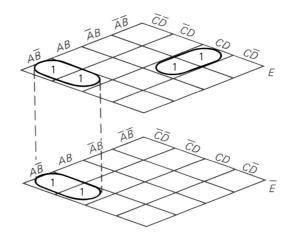

d. $A \cdot \overline{B} \cdot \overline{C} + \overline{A} \cdot C \cdot D \cdot E = Y$

22. AND-OR
23. identical
24. 1. Start with a minterm Boolean expression.
    2. Draw the AND-OR logic diagram using AND, OR, and NOT symbols.
    3. Substitute NAND symbols for each AND and OR symbol, keeping all connections the same.
    4. Substitute NAND symbols with all inputs tied together for each inverter.
    5. Test the logic circuit containing all NAND gates to determine that it generates the proper truth table.

25. data selector
26. 7, *W*.
27. rotary
28. 15, HIGH
29. data selector
30. LOW
31. HIGH
32. LOW
33. No
34. GND or ground
35. LOW
36. ROMs, PROMs, PALs, gate arrays, programmable gate arrays

CHAPTER 5

IC Specifications and Simple Interfacing

The driving force behind the increased use of digital circuits has been the availability of a variety of logic families. *Integrated circuits within a logic family are designed to* interface *easily with one another. For instance, in the TTL logic family you may connect an output directly into the input of several other TTL inputs with no extra parts. The designer can have confidence that ICs from the same logic family will interface properly.* Interfacing between *logic families and between digital ICs and the outside world is a bit more complicated. In this chapter you will study a few of the most important IC specifications. You will also learn some very simple interfacing techniques. You will interface switches with both TTL and CMOS ICs. You will drive LEDs and buzzers with TTL and CMOS logic gates. You will use relays to isolate high-voltage and/or high-current devices such as electric motors and solenoids from digital logic circuits. Also, you will interface TTL to CMOS and CMOS to TTL.* Interfacing *can be defined as the design of the interconnections between circuits that shift the levels of voltage and current to make them compatible. A fundamental knowledge of simple interfacing techniques is required of technicians and engineers who work with digital circuits. Most logic circuits are of no value if they are not interfaced with* "real world" *devices.*

5-1 LOGIC LEVELS AND NOISE MARGIN

In any field of electronics most technicians and engineers start investigating a new device in terms of voltage, current, and resistance or impedance. In this section just the *voltage char-*

acteristics of both TTL and CMOS ICs will be studied.

How is a logical 0 (LOW) or logical 1 (HIGH) defined? Figure 5-1 on the next page shows an inverter (such as the 7404 IC) from the bipolar TTL logic family. Manufacturers specify that for correct operation, a LOW *input* must range

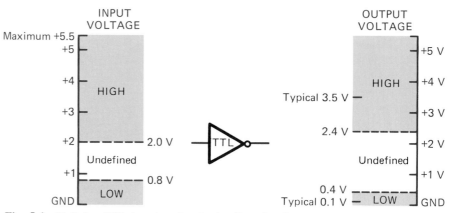

Fig. 5-1 Defining TTL input and output voltage levels.

from GND to 0.8 V. Also, a HIGH *input* must be in the range from 2.0 to 5.5 V. The unshaded section from 0.8 to 2.0 V on the input side is the undefined area, or indeterminate region. Therefore, an input of 3.2 is a HIGH input. An input of 0.5 V is considered a LOW input. An input of 1.6 V is in the undefined region and should be avoided. Inputs in the undefined region yield unpredictable results at the output. Expected outputs from the TTL inverter are shown on the right in Fig. 5-1. A typical LOW output is about 0.1 V. A typical HIGH output might be about 3.5 V. However, a HIGH output could be as low as 2.4 V according to the voltage profile diagram in Fig. 5-1. The HIGH output depends on the resistance value of the load placed at the output. The greater the load current, the lower the HIGH output voltage. The unshaded section on the output voltage side in Fig. 5-1 is the undefined region. Suspect trouble if the output voltage is in the undefined region (0.4 to 2.4 V).

The voltages given for LOW and HIGH logic levels in Fig. 5-1 are for a TTL device. These voltages are different for other logic families.

The popular 4000 and 74C00 series CMOS logic families of ICs operate on a wide range of power supply voltages (from +3 to +15 V). The definition of a HIGH and LOW logic level for a typical CMOS inverter is illustrated in Fig. 5-2(a). A 10-V power supply is being used in this voltage profile diagram.

The CMOS inverter shown in Fig. 5-2(a) will respond to any input voltage within 70–100 percent of V_{DD} (+10 V in this example) as a HIGH. Likewise, any voltage within 0 to 30 percent of V_{DD} is regarded as a LOW input to ICs in the 4000 and 74C00 series.

Typical output voltages for CMOS ICs are

shown in Fig. 5-2(a). Output voltages are normally almost at the *voltage rails* of the power supply. In this example, a HIGH output would be about +10 V while a LOW output would be about 0 V or GND.

The modern 74HC00 series of CMOS ICs operates on a lower-voltage power supply (from +2 to +6 V). The input and output voltage characteristics are summarized in the voltage profile diagram in Fig. 5-2(b). The definition of HIGH and LOW for both input and output on the 74HC00 series is approximately the same as for other CMOS ICs. This can be seen by comparing the two voltage profiles in Figs. 5-1(a) and (b).

The specialized 74HCT00 series of CMOS ICs is designed to operate on a 5-V power supply. The function of the 74HCT00 series of ICs is to interface between TTL and CMOS devices.

A voltage profile diagram for the 74HCT00 series of CMOS ICs is drawn in Fig. 5-2(c). Notice that the definition of LOW and HIGH is the same for the 74HCT00 series inputs as they are for TTL inputs. This can be seen by comparing the input side of the voltage profiles of TTL and the 74HCT00 series [see Figs. 5-1 and 5-2(c)]. The output voltage profile for the 74HCT04 is similar to that of other CMOS ICs. This can be observed by comparing output voltage profiles in Figs. 5-2(a), (b), and (c). In summary, the 74HCT00 series has a typical TTL input voltage profile with a CMOS output.

The most often cited advantages of CMOS are its low power requirements and good noise immunity. *Noise immunity* is a circuit's insensitivity or resistance to undesired voltages or noise. It is also called *noise margin* in digital circuits.

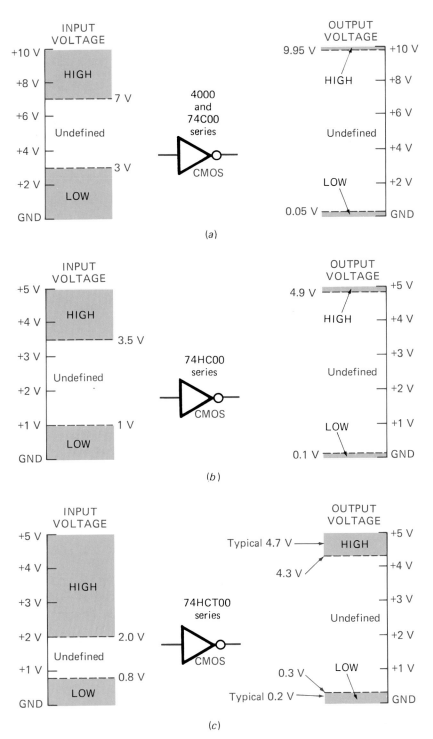

Fig. 5-2   Defining CMOS input and output voltage levels. (*a*) 4000 and 74C00 series voltage profile. (*b*) 74HC00 series voltage profile. (*c*) 74HCT00 series voltage profile.

The noise margins for typical TTL and CMOS families are compared in Fig. 5-3 on the next page. The noise margin is much better for the CMOS than for the TTL family. You may introduce almost 1.5 V of unwanted noise into the CMOS input before getting unpredictable results.

Noise in a digital system is *unwanted voltages* induced in the connecting wires and printed circuit board traces that might affect the input logic levels, thereby causing faulty output indications.

Consider the diagram in Fig. 5-4. The LOW, HIGH, and undefined regions are defined for

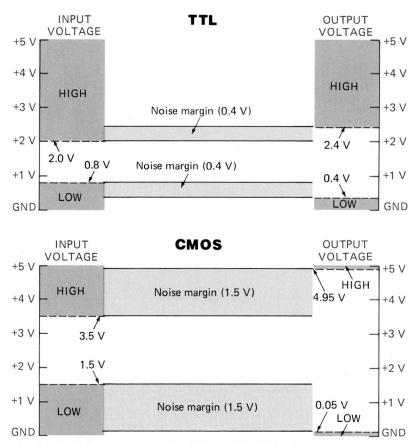

Fig. 5-3 **Defining and comparing TTL and CMOS noise margins.**

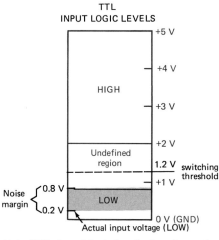

Fig. 5-4 **TTL input logic levels showing noise margin.**

TTL inputs. If the actual input voltage is 0.2 V, then the margin of safety between it and the undefined region is 0.6 V (0.8 − 0.2 = 0.6 V). This is the *noise margin*. In other words, it would take more than +0.6 V added to the LOW voltage (0.2 V in this example) to move the input into the undefined region.

In actual practice, the noise margin is even greater because the voltage must increase to the *switching threshold,* which is shown as 1.2 V in Fig. 5-4. With the actual LOW input at +0.2 V and the switching threshold at about +1.2 V, the actual noise margin is 1 V (1.2 − 0.2 = 1 V).

The switching threshold is *not* an absolute voltage. It does occur within the undefined region but varies widely because of manufacturer, temperature, and the quality of the components. However, the logic levels are guaranteed by the manufacturer.

Self-Test

Supply the missing word in each statement.

1. The design of interconnections between two circuits to make them compatible is called _____ .
2. An input of +3.1 V to a TTL IC would be considered _____ (HIGH, LOW, undefined).
3. An input of +0.7 V to a TTL IC would be considered _____ (HIGH, LOW, undefined).

Digital IC specifications

Drive capabilities

Fan-out

Fan-in

Propagation delay

Power dissipation

Worst-case conditions

4. An output of +2.0 V from a TTL IC would be considered _____ (HIGH, LOW, undefined).
5. An input of +6 V (10-V power supply) to a 4000 series CMOS IC would be considered _____ (HIGH, LOW, undefined).
6. A typical HIGH output (10-V power supply) from a CMOS IC would be about _____ V.
7. An input of +3 V (5-V power supply) to a 74HCT00 series CMOS IC would be considered _____ (HIGH, LOW, undefined).
8. The _____ (CMOS, TTL) family of ICs has better noise immunity.

5-2 OTHER DIGITAL IC SPECIFICATIONS

Digital logic voltage levels and noise margins were studied in the last section. In this section, other important specifications of digital ICs will be introduced. These include drive capabilities, fan-out and fan-in, propagation delay, and power dissipation.

DRIVE CAPABILITIES

A bipolar transistor has its maximum wattage and collector current ratings. These ratings determine its *drive capabilities*. One indication of output drive capability of a digital IC is called its fan-out. The *fan-out* of a digital IC is the number of "standard" inputs that can be driven by the gate's output. If the fan-out for standard TTL gates is 10, this means that the output of a single gate can drive up to 10 inputs *in the same subfamily*. A typical fan-out value for standard TTL ICs is 10. The fan-out for low-power Schottky TTL (LS-TTL) is 20 and for the 4000 series CMOS it is considered to be about 50.

Another way to look at the current characteristics of gates is to examine their output drive and input loading parameters. The diagram in Fig. 5-5(a), on the next page, is a simplified view of the output drive capabilities and input load characteristics of a standard TTL gate. A standard TTL gate is capable of handling 16 mA when the output is LOW (I_{OL}) and 400 µA when the output is HIGH (I_{OH}). This seems like a mismatch until you examine the input loading profile for a standard TTL gate. The input loading (worst-case conditions) is only 40 µA with the input HIGH (I_{IH}) and 1.6 mA when the input is LOW (I_{IL}). This means that the output of a standard TTL gate can drive 10 inputs (16 mA/1.6 mA = 10). Remember, these are *worst-case conditions* and in actual bench tests under static conditions these input load currents are much less than specified.

A summary of the *output drive* and *input loading* characteristics of several popular families of digital ICs is detailed in Fig. 5-5(b). Look over this chart of very useful information. You will need this data later when interfacing TTL and CMOS ICs.

The load represented by a single gate is called the *fan-in* of that family of ICs. The input loading column in Fig. 5-5(b) can be thought of as the fan-in of these IC families. Notice that the fan-in or input loading characteristics are different for each family of ICs.

Suppose you are given the interfacing problem in Fig. 5-6(a) on page 83. You are asked if the 74LS04 inverter has enough fan-out to drive the four standard TTL NAND gates on the right.

The voltage and current profiles for LS-TTL and standard TTL gates are sketched in Fig. 5-6(b). The voltage characteristics of all TTL families are compatible. The LS-TTL gate can drive 10 standard TTL gates when its output is HIGH (400 µA/40 µA = 10). However, the LS-TTL gate can drive only five standard TTL gates when it is LOW (8 mA/1.6 mA = 5). We could say that the fan-out of LS-TTL gates is only 5 when driving standard TTL gates.

PROPAGATION DELAY

Speed, or quickness of response to a change at the inputs is an important consideration in high-speed applications of digital ICs. Consider the waveforms in Fig. 5-7(a) on page 84. The top waveform shows the input to an inverter going from LOW to HIGH and then from HIGH to LOW. The bottom waveform shows the output response to the changes at the input. The slight delay between the time the input changes and the time the output changes is called the *propagation delay* of the inverter. Propagation delay is measured in seconds. The propagation delay for the LOW-to-HIGH transition of the input to the inverter is different from the HIGH-to-LOW delay. Propagation delays are shown in Fig. 5-7(a) for a standard TTL 7404 inverter IC.

The typical propagation delay for a standard TTL inverter (such as the 7404 IC) is about 12 ns for the LOW-to-HIGH change while only

Output drive and input loading characteristics

Propagation delays

High-speed CMOS

Power dissipation

Advanced low-power Schottky

TTL

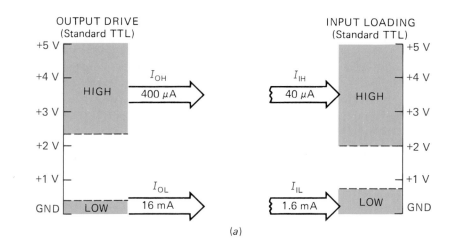

Fig. 5-5 (*a*) Standard TTL voltage and current profiles. (*b*) Output drive and input loading characteristics for selected TTL and CMOS logic families.

7 ns for the HIGH-to-LOW transition of the input.

Typical propagation delays are summarized on the graph in Fig. 5-7(*b*). The lower the propagation delay specifications for an IC, the higher the speed. Notice that the AS-TTL (advanced Schottky TTL) is the fastest, while the 4000 series CMOS family is the slowest. Generally, TTL ICs are faster than CMOS ICs. However, notice that the 74HC00 series (HC for high-speed CMOS) rivals standard TTL in speed. This is one of the reasons why the newer high-speed CMOS families are gaining popularity.

POWER DISSIPATION

Generally, as propagation delays decrease (increased speed), the power consumption and related heat generation increase. The very fast AS-TTL family dissipates about 19 mW per gate, while some slower CMOS gates consume as little as 0.01 mW per gate. Many designers have found the *advanced low-power Schottky TTL* family a very good compromise between speed and power dissipation. A typical advanced low-power Schottky TTL gate consumes 2 mW and has propagation delays of about 4 ns. Advanced low-power Schottky TTL gates consume one-fifth the power of standard TTL units but operate at approximately twice the speed.

Self-Test

Supply the missing word in each statement.

9. The number of "standard" input loads that can be driven by an IC is called its _____ (fan-in, fan-out).

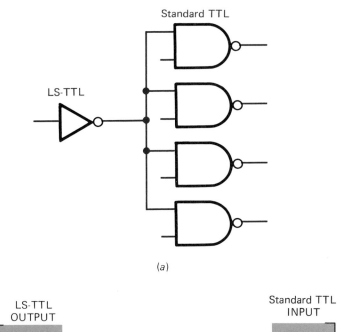

(a)

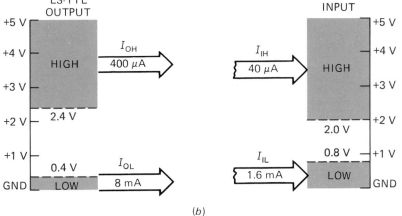

(b)

Fig. 5-6 **Interfacing LS-TTL to standard TTL problem.** (*a*) **Logic diagram of inter-facing problem.** (*b*) **Voltage and current profiles for visualizing the solution to the problem.**

10. The _____ (4000 series CMOS, standard TTL) gates have more output drive capabilities.

11. Refer to Fig. 5-5(*b*). The calculated fan-out when interfacing LS-TTL to LS-TTL is _____ .

12. The 4000 series CMOS gates have very low power dissipation, good noise immunity, and _____ (long, short) propagation delays.

13. Refer to Fig. 5-7(*b*). The fastest CMOS subfamily is _____ .

14. All TTL subfamilies have _____ (different, the same) voltage and different output drive and input loading characteristics.

5-3 MOS AND CMOS ICs

MOS ICs

The enhancement type of metal oxide semiconductor field-effect transistor (MOSFET) forms the primary component in MOS ICs. Because of their simplicity, MOS devices use less space on a silicon chip. Therefore, more functions per chip are typical in MOS devices than in bipolar ICs (such as TTL). Metal oxide semiconductor technology is widely used in large scale integration (LSI) and very large scale integration (VLSI) devices because of this packing density on the chip. Microprocessors, memory, and clock chips are typically fabricated us-

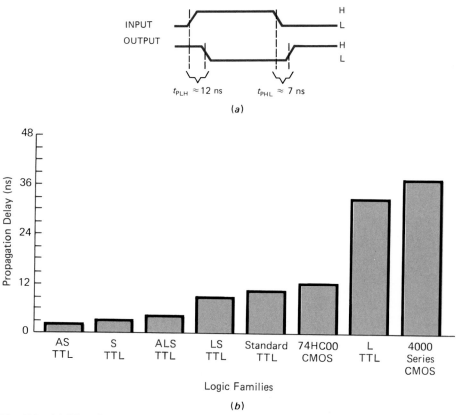

Fig. 5-7 (a) **Waveforms showing propagation delays for a standard TTL inverter.** (b) **Graph of propagation delays for selected TTL and CMOS IC families.**

ging MOS technology. Metal oxide semiconductor circuits are typically of either the PMOS (P-channel MOS) or the newer, faster NMOS (N-channel MOS) type. Metal oxide semiconductor chips are smaller, consume less power, and have a better noise margin and higher fan-out than bipolar ICs. The main disadvantage of MOS devices is their relative lack of speed.

CMOS ICs

Complementary symmetry metal oxide semiconductor (CMOS) devices use both P-channel and N-channel MOS devices connected end to end. Complementary symmetry metal oxide semiconductor ICs are noted for their *exceptionally low power consumption.* The CMOS family of ICs also has the advantages of low cost, simplicity of design, low heat dissipation, good fan-out, wide logic swings, and good noise performance. The CMOS family of digital ICs operates on a wide range of voltages (from +3 to +15 V dc).

The main disadvantage of CMOS ICs is that they are somewhat slower than bipolar digital ICs such as TTL devices. Also, extra care must be taken when handling CMOS ICs be-

cause they must be protected from static discharges. A static charge or transient voltage in a circuit can damage the very thin silicon dioxide layers inside the CMOS chip. The silicon dioxide layer acts like the dielectric in a capacitor and can be punctured by static discharge and transient voltages.

If you do work with CMOS ICs, manufacturers suggest preventing damage from static discharge and transient voltages by:

1. Storing CMOS ICs in special *conductive foam*
2. Using battery-powered soldering irons when working on CMOS chips or grounding the tips of ac-operated units
3. Changing connections or removing CMOS ICs only when the power is turned off
4. Ensuring that input signals do not exceed power supply voltages
5. Always turning off input signals before circuit power is turned off
6. Connecting *all unused input leads* to either the positive supply voltage or GND, whichever is appropriate (only unused CMOS *outputs* may be left unconnected)

The extremely low power consumption of CMOS ICs makes them ideal for battery-operated portable devices. Complementary symmetry metal oxide semiconductor ICs are widely used in electronic wristwatches, calculators, portable computers, and space vehicles.

A typical CMOS device is shown in Fig. 5-8. The top half is a P-channel MOSFET, while the bottom half is an N-channel MOSFET. Both are enhancement-mode MOSFETs. When the input voltage (V_{in}) is LOW, the top MOSFET is on and the bottom unit is off. The output voltage (V_{out}) is then HIGH. However, if (V_{in}) is HIGH, the bottom device is on and the top MOSFET is off. Therefore, (V_{out}) is LOW. The device in Fig. 5-8 acts as an inverter.

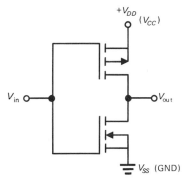

Fig. 5-8 CMOS structure using P-channel and N-channel MOSFETs in series.

Notice in Fig. 5-8 that the V_{DD} of the CMOS unit goes to the positive supply voltage. The V_{DD} lead is labeled V_{CC} (as in TTL) by some manufacturers. The "D" in V_{DD} stands for the *drain* supply in MOSFET. The V_{SS} lead of the CMOS unit is connected to the negative of the power supply. This connection is called GND (as in TTL) by some manufacturers. The "S" in V_{SS} stands for *source* supply in a MOSFET. CMOS ICs typically operate on 5-, 6-, 9-, or 12-V power supplies.

The CMOS technology is used in making several families of digital ICs. The most popular are the 4000, 74C00, and 74HC00 series ICs. The 4000 series is the oldest and probably the most widely used CMOS family. This family has all the customary logic functions plus a few devices that have no equivalent in TTL families. For instance, in CMOS it is possible to produce *transmission gates* or *bilateral switches*. These gates can conduct or allow a signal to pass in either direction like relay contacts.

The 74C00 series is an older CMOS logic family that is the pin-for-pin, function-for-function equivalent of the 7400 series of TTL ICs. As an example, a 7400 TTL IC is designated as a quadruple ("quad") two-input NAND gate as is the 74C00 CMOS IC.

The newer 74HC00 series CMOS logic family is designed to replace the 74C00 series and many 4000 series ICs. It has pin-for-pin, function-for-function equivalents for both 7400 and 4000 series ICs. It is a high-speed CMOS family with good drive capabilities. It operates on a 2- to 6-V power supply.

Self-Test

Supply the missing word in each statement.

15. Large-scale integration (LSI) and very large scale integration (VLSI) devices make extensive use of _____ (bipolar, MOS) technology.
16. The letters CMOS stand for _____ .
17. The most important advantage of using CMOS is its _____ .
18. The V_{SS} pin on a CMOS IC is connected to _____ (+5 V, GND) of the power supply.

5-4 INTERFACING TTL AND CMOS WITH SWITCHES

One of the most common means of entering information into a digital system is the use of switches or a keyboard. Examples might be the switches on a digital clock, the keys on a calculator, or the large keyboard on a microcomputer. This section will detail several methods of using a switch to enter data into either TTL or CMOS digital circuits.

Three simple switch interface circuits are depicted in Fig. 5-9 on the next page. Pressing the push-button switch in Fig. 5-9(*a*) will drop the input of the TTL inverter to ground level or LOW. Releasing the push-button switch in Fig. 5-9(*a*) opens the switch. The input to the TTL inverter now is allowed to "float." In TTL, inputs usually float at a HIGH logic level.

Floating inputs on TTL are not dependable. Figure 5-9(*b*) is a slight refinement of the switch input circuit in Fig. 5-9(*a*). The 10-kΩ resistor has been added to make sure the input to the TTL inverter goes HIGH when the switch is open. The 10-kΩ resistor is called a *pull-up resistor*. Its purpose is to pull the input voltage

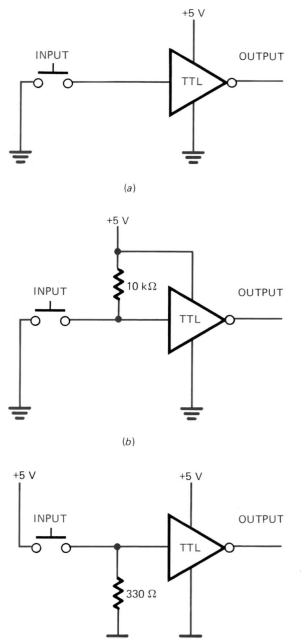

(a)

(b)

(c)

Fig. 5-9 Switch-to-TTL interfaces. (a) Simple active-LOW switch interface. (b) Active-LOW switch interface using pull-up resistor. (c) Active-HIGH switch interface using pull-down resistor.

(opened) the input is pulled LOW by the pull-down resistor. The value of the pull-down resistor is relatively low because the input current required by a standard TTL gate may be as high as 1.6 mA.

Two more switch-to-CMOS interface circuits are drawn in Fig. 5-10. An active LOW input switch is drawn in Fig. 5-10(a). The 100-kΩ pull-up resistor pulls the voltage to +5 V when the input switch is open. Figure 5-10(b) illustrates an active HIGH switch feeding a CMOS inverter. The 100-kΩ pull-down resistor makes sure the input to the CMOS inverter is near ground when the input switch is open. The resistance value of the pull-up and pull-down resistors is much greater than those in TTL interface circuits. This is because the input loading currents are much greater in TTL than in CMOS. The CMOS inverter illustrated in Fig. 5-10 could be from the 4000, 74C00, or the 74HC00 series of CMOS ICs.

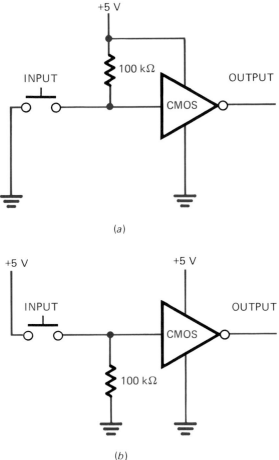

(a)

(b)

Fig. 5-10 Switch-to-CMOS interfaces. (a) Active-LOW switch interface with pull-up resistor. (b) Active-HIGH switch interface with pull-down resistor.

up to +5 V. Both circuits in Figs. 5-9(a) and (b) illustrate active LOW switches. They are called active LOW switches because the inputs go LOW only when the switch is activated.

An active HIGH input switch is sketched in Fig. 5-9(c). When the input switch is activated, the +5 V is connected directly to the input of the TTL inverter. When the switch is released

The switch interface circuits in Figs. 5-9 and 5-10 work well for some applications. However, none of the switches in Figs. 5-9 and 5-10 were *debounced*. The lack of a debouncing circuit can be demonstrated by operating the counter shown in Fig. 5-11(*a*). Each press of the input switch should cause the decade (0–9) counter to increase by 1. However, in practice each press of the switch increases the count by 1, 2, 3, or sometimes more. This means that several pulses are being fed into the clock (CLK) input of the counter each time the switch is pressed. This is caused by switch bounce.

A switch debouncing circuit has been added to the counting circuit in Fig. 5-11(*b*). The decade counter will now count each HIGH–LOW cycle of the input switch. The cross-coupled NAND gates in the debouncing circuit are sometimes called an RS flip-flop or latch. Flip-flops will be studied in greater detail in Chap. 7.

Several other switch debouncing circuits are illustrated in Fig. 5-12 on the next page. The simple debouncing circuit drawn in Fig. 5-12(*a*) will only work on the slower 4000 series CMOS ICs. The 40106 CMOS IC is a special inverter. The 40106 is a Schmitt trigger inverter, which means it has a "snap action" when changing to either HIGH or LOW. A Schmitt trigger can also change a slow-rising signal (such as a sine wave) into a square wave.

The switch debouncing circuit in Fig. 5-12(*b*) will drive 4000 or 74HC00 series CMOS or TTL ICs. Another general-purpose switch debouncing circuit is illustrated in Fig. 5-12(*c*). This debouncing circuit can drive either CMOS or TTL inputs. The 7403 is an *open-collector* TTL IC and needs pull-up resistors as shown in Fig. 5-12(*c*). The external pull-up resistors make it possible to have an output voltage of just about +5 V for a HIGH. Open-collector TTL gates with external pull-up resistors are useful when driving CMOS with TTL.

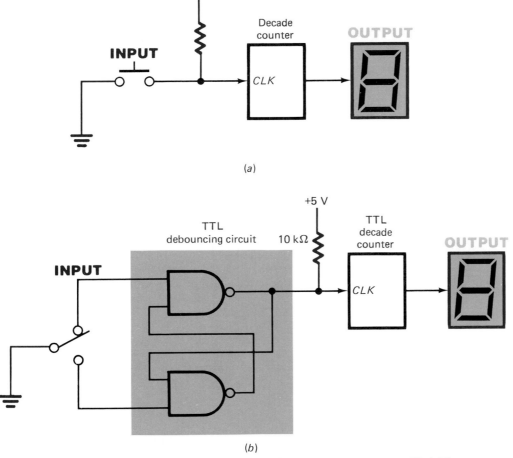

Fig. 5-11 (*a*) Block diagram of switch interfaced to a decimal counter system. (*b*) Adding a switch debouncing circuit to make the decimal counter work properly.

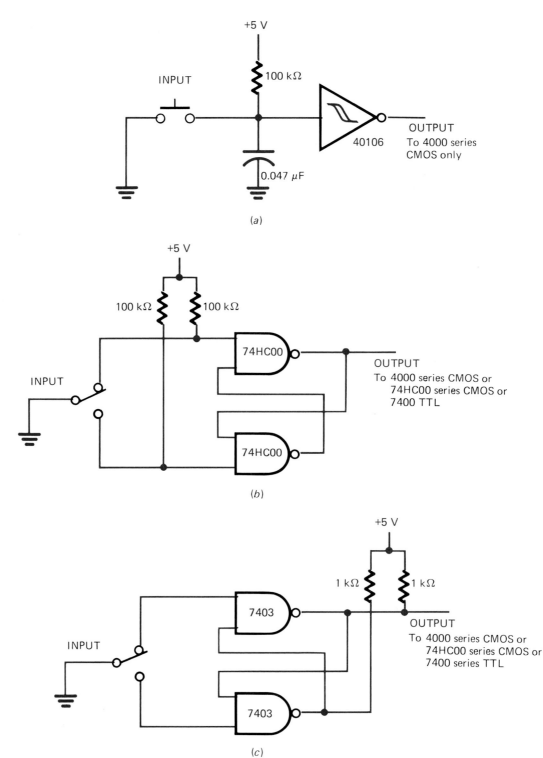

Fig. 5-12 Switch debouncing circuits. (*a*) A **4000 series switch debouncing circuit.**
(*b*) **General purpose switch debouncing circuit that will drive CMOS or TTL inputs.**
(*c*) **Another general-purpose switch debouncing circuit that will drive CMOS or TTL inputs.**

Self-Test

Answer the following questions.

19. Refer to Fig. 5-9(*c*). Pressing the switch causes the input of the inverter to go _____ (HIGH, LOW) while the output goes _____ (HIGH, LOW).

20. Refer to Fig. 5-10. The inverters and associated resistors form switch debouncing circuits. (T or F)

21. Refer to Fig. 5-11(*a*). This decade counter circuit lacks what circuit?

22. Refer to Fig. 5-12(*c*). The 7403 is a TTL inverter with a(n) _____ output.
  a. Open collector
  b. Totem pole
  c. Tri state

5-5 INTERFACING TTL AND CMOS WITH LEDs

Many of the lab experiments you will perform using digital ICs require an output indicator. The LED (light-emitting diode) is perfect for this job because it operates at low currents and voltages. The maximum current required by many LEDs is about 20 to 30 mA with about 2 V applied. An LED will light dimly on only 1.7 to 1.8 V and 2 mA.

Interfacing 4000 series CMOS devices with simple LED indicator lamps is easy. Figure 5-13(*a–f*) shows six examples of CMOS ICs driving LED indicators. Figures 5-13(*a*) and (*b*) show the CMOS supply voltage at +5 V. At this low voltage, no limiting resistors are needed in series with the LEDs. In Fig. 5-13(*a*), when the output of the CMOS inverter goes HIGH, the LED output indicator lights. The opposite is true in Fig. 5-13(*b*): when the CMOS output goes LOW, the LED indicator lights.

Figures 5-13(*c*) and (*d*) show the 4000 series CMOS ICs being operated on a higher supply voltage (+10 to +15 V). Because of the higher voltage, a 1-kΩ limiting resistor is placed in series with the LED output indicator lights. When the output of the CMOS inverter in Fig. 5-13(*c*) goes HIGH, the LED output indicator lights. In Fig. 5-13(*d*), however, the LED indicator is activated by a LOW at the CMOS output.

Figures 5-13(*e*) and (*f*) show CMOS buffers being used to drive LED indicators. The circuits may operate on voltages from +5 to +15 V. Figure 5-13(*e*) shows the use of an inverting CMOS buffer (like the 4049 IC), while Fig. 5-13(*f*) uses the noninverting buffer (like the 4050 IC). In both cases, a 1-kΩ limiting resistor must be used in series with the LED output indicator.

Standard TTL gates are sometimes used to drive LEDs directly. Two examples are illustrated in Figs. 5-13(*g*) and (*h*). When the output of the inverter in Fig. 5-13(*g*) goes HIGH, current will flow through the LED causing it to light. The indicator light in Fig. 5-13(*h*) only lights when the output of the 7404 inverter goes

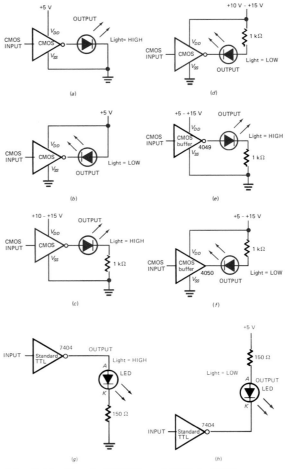

Fig. 5-13 Simple CMOS- and TTL-to-LED interfacing. (*a*) CMOS active-HIGH. (*b*) CMOS active-LOW. (*c*) CMOS active-HIGH, supply voltage = 10 to 15 V. (*d*) CMOS active-LOW, supply voltage = 10 to 15 V. (*e*) CMOS inverting buffer to LED interfacing. (*f*) CMOS noninverting buffer to LED interfacing. (*g*) TTL active-HIGH. (*h*) TTL active-LOW.

LOW. The circuits in Fig. 5-13 are not recommended for critical uses because they exceed the output current ratings of the ICs. However, they have been tested and work properly as simple output indicators.

Three improved LED output indicator designs are diagrammed in Fig. 5-14 on the next page. Each of the circuits uses transistor drivers and can be used with either CMOS or TTL. The LED in Fig. 5-14(*a*) lights when the output of the inverter goes HIGH. The LED in Fig. 5-14(*b*) lights when the output of the inverter goes LOW. Notice that the indicator in Fig. 5-14(*b*) uses a PNP instead of an NPN transistor.

The LED indicator circuits in Figs. 5-14(*a*) and (*b*) are combined in Fig. 5-14(*c*). The red light (LED1) will light when the inverter's out-

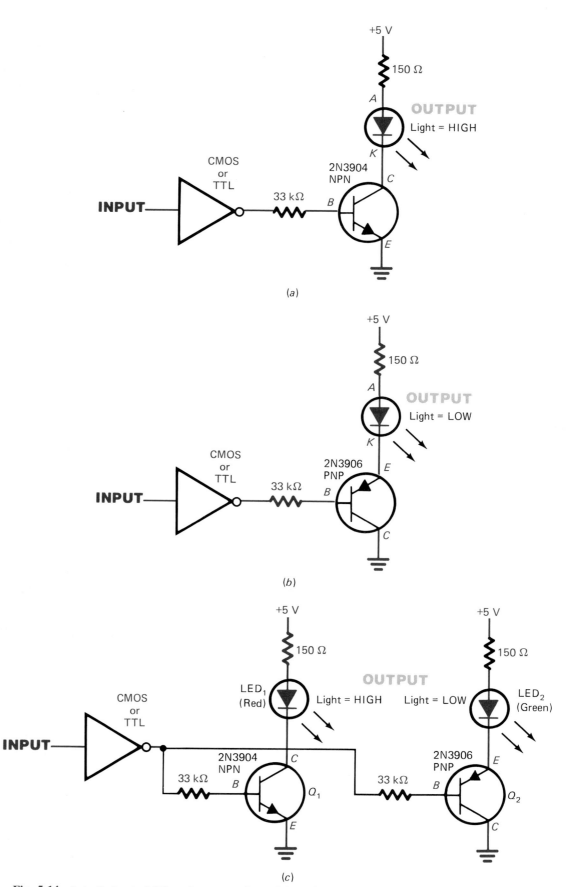

Fig. 5-14 Interfacing to LEDs using a transistor driver circuit. (*a*) Active-HIGH output using an NPN transistor driver. (*b*) Active-LOW output using a PNP transistor driver. (*c*) HIGH-LOW indicator circuit (simplified logic probe).

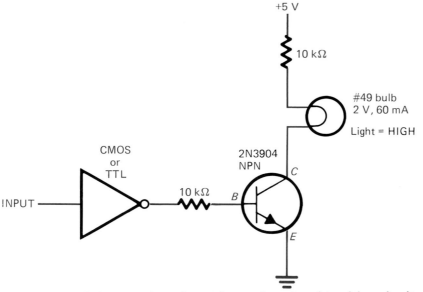

Fig. 5-15 Interfacing to an incandescent lamp using a transistor driver circuit.

put is HIGH. During this time LED2 will be off. When the output of the inverter goes LOW, transistor Q_1 turns off while Q_2 turns on. The green light (LED2) lights when the output of the inverter is LOW.

The circuit is Fig. 5-14(c) is a very basic logic probe. However, its accuracy is less than most logic probes.

The indicator light shown in Fig. 5-15 uses an incandescent lamp. When the output of the inverter goes HIGH, the transistor is turned on and the lamp lights. When the inverter's output is LOW, the lamp does not light.

Self-Test

Supply the missing word(s) in each statement.

23. Refer to Fig. 5-13(*a–f*). The _____ (4000, 74HCT00) series CMOS ICs are being used to drive the LEDs in these circuits.
24. Refer to Fig. 5-13(*h*). When the output of the inverter goes HIGH, the LED _____ (goes out, lights).
25. Refer to Fig. 5-14(*a*). When the output of the inverter goes LOW, the transistor is turned _____ (off, on) and the LED _____ (does not light, lights).
26. Refer to Fig. 5-14(*c*). When the output of the inverter goes HIGH, transistor _____ (Q_1, Q_2) is turned on and the _____ (green, red) LED lights.

5-6 INTERFACING TTL AND CMOS ICs

CMOS and TTL logic levels (voltages) are defined differently. These differences are illustrated in the voltage profiles for TTL and CMOS shown in Fig. 5-16(*a*) on the next page. Because of the differences in voltage levels, CMOS and TTL ICs usually cannot simply be connected together. Just as important, current requirements for CMOS and TTL ICs are different.

Look at the voltage and current profile in Fig. 5-16(*a*). Note that the output drive currents for the standard TTL are more than adequate to drive CMOS inputs. However, the voltage profiles do not match. The LOW outputs from the TTL are compatible because they fit within the wider LOW input band on the CMOS IC. There is a range of possible HIGH outputs from the TTL IC (2.4 to 3.5 V) that do not fit within the HIGH range of the CMOS IC. This incompatibility could cause problems. These problems can be solved by using a pull-up resistor between gates to pull the HIGH output of the standard TTL up closer to +5 V. A completed circuit for interfacing standard TTL to CMOS is shown in Fig. 5-16(*b*). Note the use of the 1-kΩ pull-up resistor. This circuit works for driving either 4000 series or 74HC00 series CMOS ICs.

Several other examples of TTL-to-CMOS and CMOS-to-TTL interfacing using a common 5-V power supply are detailed in Fig. 5-17. Figure 5-17(*a*) shows the popular LS-TTL driving

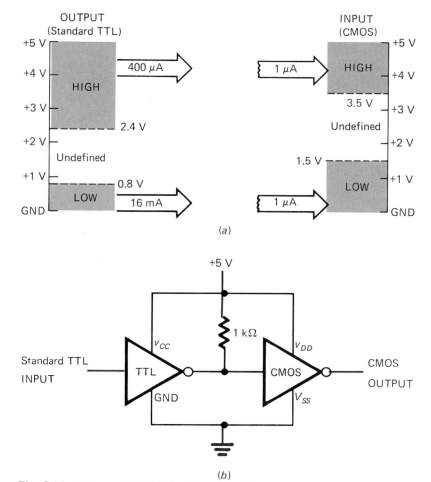

Fig. 5-16   TTL-to-CMOS interfacing. (*a*) TTL output and CMOS input profiles for visualizing checking compatibility. (*b*) TTL-to-CMOS interface using a pull-up resistor.

any CMOS gate. Notice the use of a 2.2-kΩ pull-up resistor. The pull-up resistor is being used to pull the TTL HIGH up near +5 V so that it will be compatible with the input voltage characteristics of CMOS ICs.

In Fig. 5-17(*b*), a CMOS inverter (any series) is driving an LS-TTL inverter *directly*. Complementary symmetry metal oxide semiconductor ICs can drive LS-TTL and ALS-TTL (advanced low-power Schottky) inputs; CMOS ICs cannot drive standard TTL inputs without special interfacing.

Manufacturers have made interfacing easier by designed special buffers and other interface chips for designers. One example is the use of the 4050 noninverting buffer in Fig. 5-17(*c*). The 4050 buffer allows the CMOS inverter to have enough drive current to operate up to two standard TTL inputs.

The problem of voltage incompatibility from TTL (or NMOS) to CMOS was solved in Fig. 5-16 using a pull-up resistor. Another method

of solving this problem is illustrated in Fig. 5-17(*d*). The 74HCT00 series of CMOS ICs is specifically designed as a convenient interface between TTL (or NMOS) and CMOS. Such an interface is implemented in Fig. 5-17(*d*) using the 74HCT34 noninverting IC.

The 74HCT00 series of CMOS ICs is widely used when interfacing between NMOS devices and CMOS. The NMOS output characteristics are almost the same as for LS-TTL.

Interfacing CMOS devices with TTL devices takes some added components when each operates on a *different voltage power supply*. Figure 5-18 shows three examples of TTL-to-CMOS and CMOS-to-TTL interfacing. Figure 5-18(*a*) shows the TTL inverter driving a general-purpose NPN transistor. The transistor and associated resistors translate the lower-voltage TTL outputs to the higher-voltage inputs needed to operate the CMOS inverter. The CMOS output has a voltage swing from about 0 to almost +10 V. Figure 5-18(*b*) shows an open-collector TTL buffer and

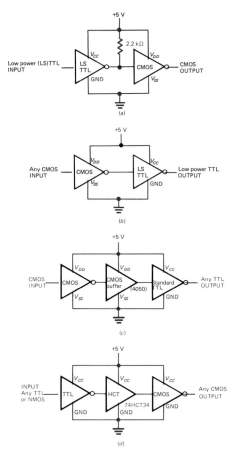

Fig. 5-17 Interfacing TTL and CMOS when both use a common +5-V power supply. (*a*) Low power Schottky TTL to CMOS interfacing using a pull-up resistor. (*b*) CMOS to low power Schottky TTL interfacing. (*c*) CMOS to standard TTL interfacing using a CMOS buffer IC. (*d*) TTL to CMOS interfacing using a 74HCT00 series IC.

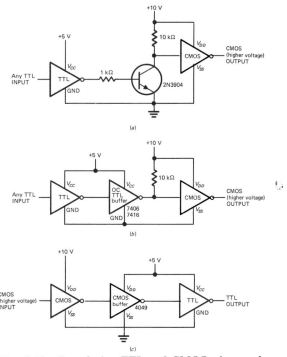

Fig. 5-18 Interfacing TTL and CMOS when each use a difference power supply voltage. (*a*) TTL-to-CMOS interfacing using a driver transistor. (*b*) TTL-to-CMOS interfacing using an open collector TTL buffer IC. (*c*) CMOS-to-TTL interfacing using a CMOS buffer IC.

a 10-kΩ pull-up resistor being used to translate the lower TTL to the higher CMOS voltages. The 7406 and 7416 TTL ICs are two inverting, open-collector buffers.

Interfacing between a higher-voltage CMOS inverter and a lower-voltage TTL inverter is shown in Fig. 5-18(*c*). The 4049 CMOS buffer is used between the higher-voltage CMOS inverter and the lower-voltage TTL IC. Note that the CMOS buffer is powered by the lower-voltage (+5 V) power supply in Fig. 5-18(*c*).

Looking at the voltage and current profiles [such as in Fig. 5-16(*a*)] is a good starting point when learning about or designing an interface. Manufacturers' manuals are also very helpful. Several techniques are used to interface between different logic families. These include the use of pull-up resistors and special interface ICs. Sometimes no extra parts are needed.

Self-Test

Supply the missing word in each statement.

27. Refer to Fig. 5-16(*a*). According to this profile of TTL output and CMOS input characteristics, the logic devices _____ (are, are not) voltage-compatible.

28. Refer to Fig. 5-17(*a*). The 2.2-kΩ resistor in this circuit is called a _____ resistor.

29. Refer to Fig. 5-17(*c*). The 4050 buffer is a special interface IC that solves the _____ (current drive, voltage) incompatibility between the logic families.

30. Refer to Fig. 5-18(*a*). The _____ (NMOS IC, transistor) translates the TTL logic levels to the higher-voltage CMOS logic levels.

5-7 INTERFACING WITH BUZZERS, RELAYS, MOTORS, AND SOLENOIDS

The objective of many electromechanical systems is to control a simple output device. This device might be as simple as a light, buzzer, relay, electric motor, or solenoid. Interfacing to LEDs and lamps has been explored. Simple interfacing between logic elements and buzzers, relays, motors, and solenoids will be investigated in this section.

The piezo buzzer is a modern signaling device drawing much less current than older buzzers and bells. The circuit in Fig. 5-19 below shows the interfacing necessary to drive a piezo buzzer with digital logic elements. A standard TTL inverter is shown driving a piezo buzzer *directly*. The standard TTL output can sink up to 16 mA. The piezo buzzer draws about 3 to 5 mA when sounding. Notice that the piezo buzzer has polarity markings. The diode across the buzzer is to suppress any transient voltages that might be induced in the system by the buzzer.

Most logic families do not have the current capacity to drive a buzzer directly. A transistor has been added to the output of the inverter

in Fig. 5-19(*b*) to drive the piezo buzzer. When the output of the inverter goes HIGH, the transistor is turned on and the buzzer sounds. A LOW at the output of the inverter turns the transistor off, switching the buzzer off. The diode protects against transient voltages. The interface circuit sketched in Fig. 5-19(*b*) will work for both TTL and CMOS.

A relay is an excellent method of isolating a logic device from a high-voltage circuit. Figure 5-20 shows how a TTL or CMOS inverter could be interfaced with a relay. When the output of the inverter goes HIGH, the transistor is turned on and the relay is activated. When activated, the normally open (NO) contacts of the relay close as the armature clicks downward. When the output of the inverter in Fig. 5-20 goes LOW, the transistor stops conducting and the relay is deactivated. The armature springs upward to its normally closed (NC) position. The clamp diode across the relay coil

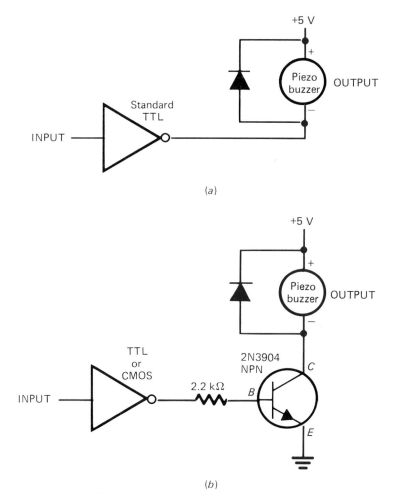

(*a*)

(*b*)

Fig. 5-19   Logic device to buzzer interfacing. (*a*) Standard TTL inverter driving a piezo buzzer directly. (*b*) TTL or CMOS interfaced with buzzer using a transistor driver.

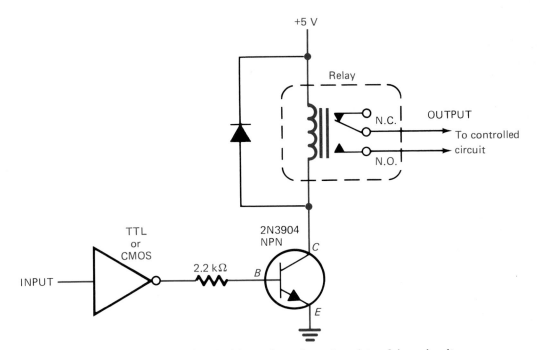

Fig. 5-20 TTL or CMOS interfaced with a relay using a transistor driver circuit.

prevents voltage spikes which might be induced in the system.

The circuit in Fig. 5-21(a) on page 96 uses a relay to isolate an electric motor from the logic devices. Notice that the logic circuit and dc motor have separate power supplies. When the output of the inverter goes HIGH, the transistor is turned on and the NO contacts of the relay snap closed. The dc motor operates. When the output of the inverter goes LOW, the transistor stops conducting and the relay contacts spring back to their NC position. This turns off the motor. The electric motor in Fig. 5-21(a) on the next page produces rotary motion. A solenoid is an electrical device that can produce linear motion. A solenoid is being driven by a logic gate in Fig. 5-21(b). Note the separate power supplies. This circuit works the same as the motor interface circuit in Fig. 5-21(a).

Voltage and current characteristics of most buzzers, relays, electric motors, and solenoids are radically different from those of logic circuits. Most of these electrical devices need special interfacing circuits to drive and isolate the devices from the logic circuits.

Self-Test

Supply the missing word(s) in each statement.

31. Refer to Fig. 5-19(a). If the piezo buzzer draws only 6 mA, it _____ (is, is not) possible for a 4000 series CMOS IC to

drive the buzzer directly [see Fig. 5-5(b) for 4000 series data].

32. Refer to Fig. 5-19(b). When the input to the inverter goes LOW, the transis-tor turns _____ (off, on) and the buzzer _____ (does not sound, sounds).

33. Refer to Fig. 5-20. The purpose of the diode across the coil of the relay is to suppress _____ (sound, transient voltages) induced in the circuit.

34. Refer to Fig. 5-21. The dc motor will run only when a _____ (HIGH, LOW) appears at the input of the inverter.

5-8 TROUBLESHOOTING SIMPLE LOGIC CIRCUITS

One test equipment manufacturer suggests that about three-quarters of all faults in digital circuits occur because of open input or output circuits. Many of these faults can be isolated in a logic circuit using a logic probe.

Consider the combinational logic circuit mounted on a printed circuit board in Fig. 5-22(a) on page 97. The equipment manual might include a schematic similar to the one shown in Fig. 5-22(b). Look at the circuit and schematic and determine the logic diagram. From that you can determine the Boolean expression and truth table. You will find that in this example, two NAND gates are feeding an OR gate. This is equivalent to the four-input NAND function. Its

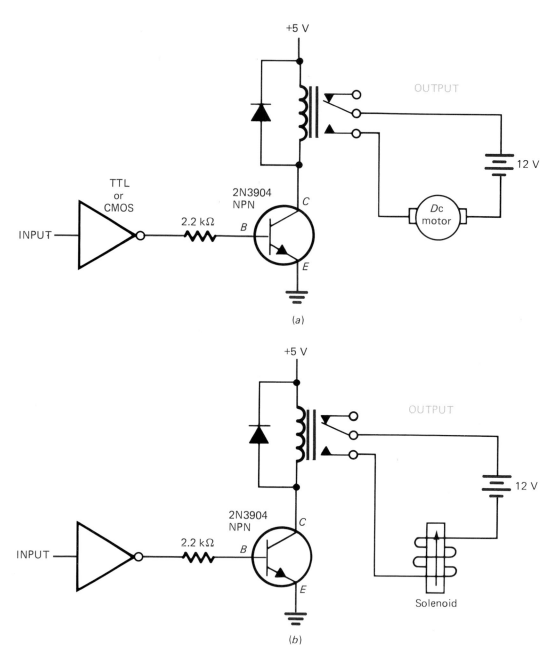

Fig. 5-21 Using a relay to isolate higher voltage/current circuits from logic gates.
(a) Interfacing TTL or CMOS with an electric motor. (b) Interfacing TTL or CMOS with a solenoid.

logic symbol diagram is the same as the one on the right side in Fig. 3-26.

The fault in the circuit in Fig. 5-22(a) is shown as an open circuit in the input to the OR gate. Now let's troubleshoot the circuit to see how we find this fault.

1. Set the logic probe to TTL and connect the power.
2. Test nodes 1 and 2 [see Fig. 5-22(a)]. *Result:* Both are HIGH.
3. Test nodes 3 and 4. *Result:* Both are LOW. *Conclusion:* Both ICs have power.

4. Test the four-input NAND circuit's unique state (inputs *A*, *B*, *C*, and *D* are all HIGH). Test at pins 1, 2, 4, and 5 of the 7400 IC. *Results:* All inputs are HIGH but the LED still glows and indicates a HIGH output. *Conclusion:* The unique state of the four-input NAND circuit is faulty.
5. Test the outputs of the NAND gates at pins 3 and 6 of the 7400 IC. *Results:* Both outputs are LOW. *Conclusions:* The NAND gates are working.
6. Test the inputs to the OR gate at pins 1 and 2 of the 7432 IC. *Results:* Both inputs are

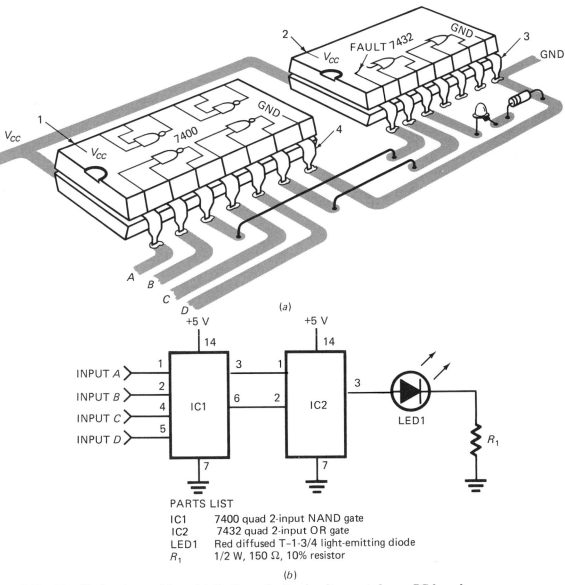

(a)

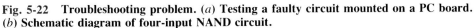

(b)

PARTS LIST

IC1   7400 quad 2-input NAND gate
IC2   7432 quad 2-input OR gate
LED1   Red diffused T–1-3/4 light-emitting diode
R_1 1/2 W, 150 Ω, 10% resistor

Fig. 5-22 Troubleshooting problem. (*a*) **Testing a faulty circuit mounted on a PC board.**
(*b*) **Schematic diagram of four-input NAND circuit.**

LOW. *Conclusions:* The OR gate inputs at pins 1 and 2 are correct but the output is still incorrect. Therefore the OR gate is faulty and the 7432 IC needs to be replaced.

Self-Test

Supply the missing word in each statement.

35. Most faults in digital circuits occur because of _____ (open, short) circuits in the inputs and outputs.

36. A simple piece of test equipment, such as a _____ , can be used for checking a digital logic circuit for open circuits in the inputs and outputs.

37. Refer to Fig. 5-22. With inputs *A*, *B*, *C*, and *D* all HIGH, the output (pin 3 of IC2) should be _____ (HIGH, LOW).

SUMMARY

1. Interfacing is the design of circuitry between devices that shifts voltage and current levels to make them compatible.

2. Interfacing between members of the same logic family is usually as simple as connecting one gate's output to the next input, etc.

3. In interfacing between logic families or between logic devices and the "outside world," the voltage and current characteristics are very important factors.

4. Noise margin is the amount of unwanted induced voltage that can be tolerated by a logic family. Complementary symmetry metal oxide semiconductor ICs have better noise margins than TTL families.

5. The fan-out and fan-in characteristics of a digital IC are determined by its output drive and input loading specifications.

6. Propagation delay (or speed) and power dissipation are important IC family characteristics. Transistor-transistor logic is faster, while CMOS ICs have lower power dissipation.

7. The LS-TTL (also ALS-TTL) and 74HC00 series CMOS IC subfamilies are very popular due to a combination of low power consumption, high speed, and good drive capabilities.

8. CMOS ICs are sensitive to static electricity and must be stored and handled properly.

Other precautions to be observed include turning off an input signal before circuit power and connecting all unused inputs.

9. Simple switches can drive logic circuits using pull-up and pull-down resistors. Switch debouncing is usually accomplished using latch circuits.

10. Driving LEDs and incandescent lamps with logic devices usually requires a driver transistor.

11. Most TTL-to-CMOS and CMOS-to-TTL interfacing requires some additional circuitry. This may take the form of a simple pull-up resistor, special interface IC, or transistor driver.

12. Interfacing digital logic devices with buzzers and relays usually requires a transistor driver circuit. Electric motors and solenoids can be controlled by logic elements using a relay to isolate them from the logic circuit.

13. Each logic family has its own definition of logical HIGH and LOW. Logic probes test for these levels.

CHAPTER REVIEW QUESTIONS

Answer the following questions.

5-1. How would you define interfacing?

5-2. Applying 3.1 V to a TTL input is interpreted by the IC as a(n) _____ (HIGH, LOW, undefined) logic level.

5-3. A TTL output of 2.0 V is considered a(n) _____ (HIGH, LOW, undefined) output.

5-4. Applying 2.4 V to a CMOS input (10-V power supply) is interpreted by the IC as a(n) _____ (HIGH, LOW, undefined) logic level.

5-5. Applying 3.0 V to a 74HC00 series CMOS input (5-V power supply) is interpreted by the IC as a(n) _____ (HIGH, LOW, undefined) logic level.

5-6. A "typical" HIGH output voltage for a TTL gate would be about _____ (0.1, 0.8, 3.5) V.

5-7. A "typical" LOW output voltage for a TTL gate would be about _____ (0.1, 0.8, 3.5) V.

5-8. A "typical" HIGH output voltage for a CMOS gate (10-V power supply) would be about _____ V.

5-9. A "typical" LOW output voltage for a CMOS gate (10-V power supply) would be about _____ V.

5-10. Applying 3.0 V to a 74HCT00 series CMOS input (5-V power supply) is interpreted by the IC as a(n) _____ (HIGH, LOW, undefined) logic level.

5-11. Applying 1.0 V to a 74HCT00 series CMOS input (5-V power supply) is interpreted by the IC as a _____ (HIGH, LOW, undefined) logic level.

5-12. How do you define *noise* in a digital system?

5-13. The _____ (CMOS, TTL) logic family has better noise immunity.

5-14. Refer to Fig. 5-3. The noise margin for the TTL family is _____ V.

5-15. Refer to Fig. 5-3. The noise margin for the CMOS family is about _____ V.

5-16. Refer to Fig. 5-4. The *switching threshold* for TTL is always exactly 1.2 V. (T or F)

5-17. The fan-out for standard TTL is said to be _____ when driving other standard TTL gates.

5-18. Refer to Fig. 5-5(*b*). A simple ALS-TTL output will drive _____ standard TTL inputs.

5-19. Refer to Fig. 5-5(*b*). A single 74HC00 series CMOS output has the capacity to drive at least _____ LS-TTL inputs.

5-20. Refer to Fig. 5-5(*b*). A single 4000 series CMOS output has the capacity to drive at least _____ LS-TTL inputs.

5-21. Refer to Fig. 5-23. If both family *A* and *B* are TTL, the inverter _____ (can, may not be able to) drive the AND gates.

5-22. Refer to Fig. 5-23. If family *A* is ALS-TTL and family *B* is standard TTL, the inverter _____ (can, may not be able to) drive the AND gates.

5-23. Refer to Fig. 5-23. If both families *A* and *B* are ALS-TTL, the inverter _____ (can, may not be able to) drive the AND gates.

5-24. The _____ (4000, 74HC00) series CMOS ICs have greater output drive capabilities.

5-25. What is the propagation delay of a logic gate?

5-26. Refer to Fig. 5-7(*b*). The _____ logic family has the lowest propagation delays and is considered the _____ (fastest, slowest).

5-27. Refer to Fig. 5-7(*b*). The _____ logic family has the highest propagation delays and is considered the _____ (fastest, slowest).

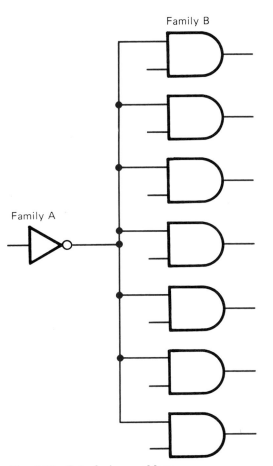

Fig. 5-23 **Interfacing problem.**

5-28. Refer to Fig. 5-7(b). The _____ is the fastest CMOS family.

5-29. Generally, _____ (CMOS, TTL) ICs are the fastest.

5-30. Generally, _____ (CMOS, TTL) ICs consume the least power.

5-31. List several advantages of CMOS logic elements.

5-32. List several precautions that should be observed when working with CMOS ICs.

5-33. The V_{DD} pin on a 4000 series CMOS IC is connected to _____ (ground, positive) of the dc power supply.

5-34. Refer to Fig. 5-9(b). With the switch open, the inverter's input is _____ (HIGH, LOW) while the output is _____ (HIGH, LOW).

5-35. Refer to Fig. 5-10(a). When the switch is open, the _____ resistor causes the input of the CMOS inverter to be pulled HIGH.

5-36. Refer to Fig. 5-24. Component R_1 is called a _____ resistor.

5-37. Refer to Fig. 5-24. Closing SW_1 causes the input to the inverter to go _____ (HIGH, LOW) and the LED _____ (goes out, lights).

5-38. Refer to Fig. 5-24. With SW_1 open, a _____ (HIGH, LOW) appears at the input of the inverter causing the output LED to _____ (go out, light).

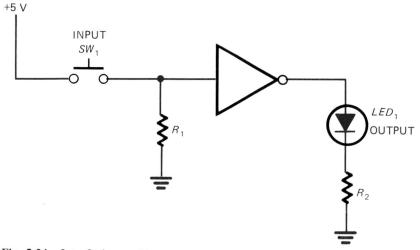

Fig. 5-24   Interfacing problem.

5-39. The common switch debouncing circuits in Figs. 5-12(b) and (c) are called RS flip-flops or _____ .

5-40. Refer to Fig. 5-14(c). With the input to the inverter HIGH, transistor _____ (Q_1, Q_2) will be turned on and the _____ (green, red) LED will light.

5-41. A TTL output can drive a CMOS input with the addition of a _____ resistor.

5-42. Any CMOS gate can drive at least one LS-TTL input. (T or F)

5-43. A 4000 series CMOS output can drive a standard TTL input with the addition of a(n) _____ .

5-44. Open-collector TTL gates require the use of _____ resistors at the outputs.

5-45. The 74HCT00 series of CMOS ICs are designed to serve as an interface between _____ and _____ logic families.

5-46. Refer to Fig. 5-20. Explain the circuit action when the inverter input is LOW.

5-47. Refer to Fig. 5-21(a). Explain the circuit action when the inverter input is HIGH.

Answers to Self-Tests

1. interfacing
2. HIGH
3. LOW
4. undefined
5. undefined
6. +10
7. HIGH
8. CMOS
9. fan-out
10. standard TTL
11. 20 (8 mA/400 µA = 20)
12. long
13. 74HC00 series CMOS
14. the same
15. MOS
16. complementary symmetry metal oxide semiconductor
17. low power consumption
18. GND
19. HIGH, LOW
20. False
21. switch debouncing circuit
22. open collector
23. 4000
24. goes out
25. off, does not light
26. Q_1, red
27. are not
28. pull-up
29. current drive
30. transistor
31. is not
32. on, sounds
33. transient voltages
34. LOW
35. open
36. logic probe or voltmeter
37. LOW (a) & (b)

CHAPTER 6

Encoding, Decoding, and Seven-Segment Displays

We humans use the decimal code to represent numbers. Digital electronic circuits in computers and calculators use mostly the binary code to represent numbers. Many other special codes are used in digital electronics to represent numbers, letters, punctuation marks, and control characters. This chapter covers several common codes used in digital electronic equipment. Electronic translators, which convert from one code to another, are widely used in digital electronics. In Chap. 2 we used an encoder to translate from decimal to binary numbers and a decoder to translate back from binary to decimal numbers. This chapter introduces you to several very common encoders and decoders used for translating from code to code. The construction and operation of three popular seven-segment displays are introduced. Some detail is given on liquid-crystal, vacuum fluorescent, and LED seven-segment displays. The topic of decoding and driving each type of display is detailed. A decoder/display troubleshooting problem will be solved.

6-1 THE 8421 BCD CODE

How would you represent the decimal number 926 in binary form? In other words, how would you convert 926 to the binary number 1110011110? The decimal-to-binary conversion would be done by using the method from Chap. 2 and illustrated in Fig. 6-1.

The binary number 1110011110 does not make much sense to most of us. A code that uses binary in a different way than in the pre-

ceding example is called the *8421 binary-coded decimal code*. This code is frequently referred to as just the *BCD code*.

The decimal number 926 is converted to the BCD (8421) code in Fig. 6-2(*a*). The result is that the decimal number 926 equals 1001 0010 0110 in the 8421 BCD code. Notice from Fig. 6-2(*a*) that each group of four binary digits represents a decimal digit. The right group (0110) represents the 1s place value in the decimal number. The middle group (0010) represents

Decimal number

926 ÷ 2 = 463	with a remainder of	0	1s
463 ÷ 2 = 231	with a remainder of	1	2s
231 ÷ 2 = 115	with a remainder of	1	4s
115 ÷ 2 = 57	with a remainder of	1	8s
57 ÷ 2 = 28	with a remainder of	1	16s
28 ÷ 2 = 14	with a remainder of	0	32s
14 ÷ 2 = 7	with a remainder of	0	64s
7 ÷ 2 = 3	with a remainder of	1	128s
3 ÷ 2 = 1	with a remainder of	1	256s
1 ÷ 2 = 0	with a remainder of	1	512s

binary number

Fig. 6-1 Converting decimal to binary numbers.

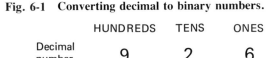

Fig. 6-2 (*a*) **Converting from decimal to 8421 BCD code.** (*b*) **Converting from BCD to decimal numbers.**

the 10s place value in the decimal number. The left group (1001) represents the 100s place value in the decimal number.

Suppose you are given the 8421 BCD number 0001 1000 0111 0001. What decimal number does this represent? Figure 6-2(*b*) shows how you translate from the BCD code to a decimal number. We find that the BCD number 0001 1000 0111 0001 is equal to the decimal number 1871. The 8421 code does not use the numbers 1010 1011 1100 1101 1110 1111. These are considered forbidden numbers.

The 8421 BCD code is very widely used in digital systems. As pointed out, it is common practice to substitute the term "BCD code" to mean the 8421 BCD code. A word of caution, however: some BCD codes do have different

weightings of the place values, such as the 4221 code and the excess-3 code.

Self-Test

Supply the missing number in each statement.

1. The decimal number 29 is the same as _____ in binary.
2. The decimal number 29 is the same as _____ in the 8421 BCD code.
3. The 8421 BCD number 1000 0111 0110 0101 equals _____ in decimal.

6-2 THE EXCESS-3 CODE

The term "BCD" is a general term, usually referring to an 8421 code. Another code that is really a BCD code is the *excess-3 code*. To convert a decimal number to the excess-3 form we *add 3 to each digit of the decimal number* and convert to binary form. Figure 6-3 shows how the decimal number 4 is converted to the excess-3 code number 0111. Some decimal numbers are converted to excess-3 code in Table 6-1 on the next page. You probably have noticed that the excess-3 code for decimal numbers is rather difficult to figure out. This is because the binary digits are not weighted as they are in regular binary numbers and in the 8421 BCD code. The excess-3 code is used in many arithmetic circuits because it is self-complementing.

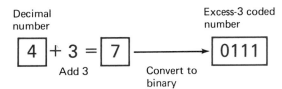

Fig. 6-3 Converting a decimal number to the excess-3 code.

The 8421 and excess-3 codes are but two of many BCD codes used in digital electronics. The 8421 code is by far the most widely used BCD code.

Self-Test

Supply the missing number in each statement.

4. The decimal number 18 equals _____ in excess-3 code.
5. The excess-3 code number 1001 0011 equals _____ in decimal.

From page 102:

BCD code

On this page:

Converting decimal to binary numbers

Converting decimal to 8421 BCD

Converting from BCD to decimal numbers

Excess-3 code

Converting a decimal number to the excess-3 code

Gray code

ASCII code

American Standard
Code for
Information
Interchage

Table 6-1 The Excess-3 Code

Decimal Number	Excess-3-Coded Number		
0			0011
1			0100
2			0101
3			0110
4			0111
5			1000
6			1001
7			1010
8			1011
9			1100
14		0100	0111
27		0101	1010
38		0110	1011
459	0111	1000	1100
606	1001	0011	1001
	Hundreds	Tens	Ones

Table 6-2 The Gray Code

Decimal Number	Binary Number	8421 BCD Coded Number		Gray-Coded Number
0	0000		0000	0000
1	0001		0001	0001
2	0010		0010	0011
3	0011		0011	0010
4	0100		0100	0110
5	0101		0101	0111
6	0110		0110	0101
7	0111		0111	0100
8	1000		1000	1100
9	1001		1001	1101
10	1010	0001	0000	1111
11	1011	0001	0001	1110
12	1100	0001	0010	1010
13	1101	0001	0011	1011
14	1110	0001	0100	1001
15	1111	0001	0101	1000
16	10000	0001	0110	11000
17	10001	0001	0111	1101

6-3 THE GRAY CODE

Table 6-2 compares the *Gray code* with some codes you already know. The important characteristic of the Gray code is that only *one digit changes* as you *count* from top to bottom, as shown in Table 6-2. The Gray code cannot be used in arithmetic circuits. The Gray code is used for input and output devices in digital systems. You can see from Table 6-2 that the Gray code is not classed as one of the many BCD codes. Also notice that it is quite difficult to translate from decimal numbers to the Gray code and back to decimals again. There is a method for making this conversion, but we usually use electronic decoders to do the job for us.

The Gray and excess-3 codes are not used extensively today. The purpose of mentioning them briefly is to make you aware that many codes exist in digital equipment. The codes you will probably encounter most often are binary, BCD (8421), and ASCII.

Self-Test

Answer the following questions.

6. The Gray code _____ (is, is not) a BCD-type code.
7. What characteristic is most important about the Gray code?

6-4 THE ASCII CODE

The ASCII code is widely used to send information to and from microcomputers. The ASCII code is a more complicated 7-bit code used in transferring coded information from keyboards and to computer displays and printers. The abbreviation ASCII (pronounced "ask-ee") stands for the *American Standard Code for Information Interchange*.

Table 6-3 is a summary of the ASCII code. The ASCII code is used to represent numbers, letters, punctuation marks, as well as control characters. For instance, the 7-bit ASCII code 111 1111 stands for DEL from the top chart. From the bottom chart we see that DEL means delete.

What is the coding for "A" in ASCII? Locate A on the top chart in Table 6-3. Assembling the 7-bit code gives 100 0001 = A. This is the code you would expect to be sent to a microcomputer's CPU if you pressed the A key on the keyboard.

Some care must be used in applying Table 6-3 to specific equipment. Be aware that the shaded control characters may have other meanings on specific computers or other equipment. However, common control characters such as BEL (bell), BS (backspace), LF

Table 6-3 The ASCII Code

Bit 7	Bit 6	Bit 5	Bit 4	Bit 3	Bit 2	Bit 1	0	0	0	0	1	1	1	1
							0	0	1	1	0	0	1	1
							0	1	0	1	0	1	0	1
			0	0	0	0	NUL	DLE	SP	0	@	P	\	P
			0	0	0	1	SOH	DC1	!	1	A	Q	a	q
			0	0	1	0	STX	DC2	''	2	B	R	b	r
			0	0	1	1	ETX	DC3	#	3	C	S	c	s
			0	1	0	0	EOT	DC4	$	4	D	T	d	t
			0	1	0	1	ENQ	NAK	%	5	E	U	e	u
			0	1	1	0	ACK	SYN	&	6	F	V	f	v
			0	1	1	1	BEL	ETB	'	7	G	W	g	w
			1	0	0	0	BS	CAN	(	8	H	X	h	x
			1	0	0	1	HT	EM	)	9	I	Y	i	y
			1	0	1	0	LF	SUB	*	:	J	Z	j	z
			1	0	1	1	VT	ESC	+	;	K	[	k	l
			1	1	0	0	FF	FS	,	<	L	\	l	|
			1	1	0	1	CR	GS	−	=	M	]	m	}
			1	1	1	0	SO	RS	·	>	N	∧	n	~
			1	1	1	1	S1	US	/	?	O	−	o	DEL

Control functions

NUL	Null	DLE	Data link escape
SOH	Start of heading	DC1	Device control 1
STX	Start of text	DC2	Device control 2
ETX	End of text	DC3	Device control 3
EOT	End of transmission	DC4	Device control 4
ENQ	Enquiry	NAK	Negative acknowledge
ACK	Acknowledge	SYN	Synchronous idle
BEL	Bell	ETB	End of transmission block
BS	Backspace	CAN	Cancel
HT	Horizontal tabulation (skip)	EM	End of medium
LF	Line feed	SUB	Substitute
VT	Vertical tabulation (skip)	ESC	Escape
FF	Form feed	FS	File separator
CR	Carriage return	GS	Group separator
SO	Shift out	RS	Record separator
SI	Shift in	US	Unit separator
DEL	Delete	SP	Space

(linefeed), CR (carriage return), DEL (delete), and SP (space) are used on most computers. The exact meaning of ASCII control codes should be looked up in your equipment manual.

The ASCII code is an *alphanumeric code*. It can represent both letters and numbers. Several other alphanumeric codes are *EBCDIC* (extended binary-coded decimal interchange code), *Baudot*, and *Hollerith*.

Self-Test

Answer the following questions.

8. ASCII is classified as an _____ code because it can represent both numbers and letters.
9. The letters ASCII stand for _____ .
10. The letter R is represented by the 7-bit ASCII code _____ .
11. The ASCII code 010 0100 represents what character?

6-5 ENCODERS

A digital system using an *encoder* is shown in Fig. 6-4. The encoder in this system must translate the decimal input from the keyboard to an 8421 BCD code. You used an encoder of this type in the experiments in Chap. 2. This encoder is called a *10-line-to-4-line priority encoder* by the manufacturer. Figure 6-5(*a*) is a block diagram of this encoder. If the decimal input 3 on the encoder is activated, then the logic circuit inside the unit outputs the BCD number 0011 as shown.

A more accurate description of a 10-line-to-4-line priority encoder is shown in Fig. 6-5(*b*). This is a connection diagram furnished by National Semiconductor and shows the 74147 10-line-to-4-line priority encoder. Note the bubbles at both the inputs (1 to 9) and the outputs (*A* to *D*). The bubbles mean that the 74147 priority encoder has both *active low inputs* and *active low outputs*. A truth table is given for the 74147 priority encoder in Fig. 6-5(*c*). Note that only low logic levels (L on the truth ta-

ble) activate the appropriate output. The active state for the outputs on this IC are also LOW. Notice that in the last line of the truth table in Fig. 6-5(*c*) the L (logical 0) at input 1 activates only the *A* output (the least significant bit of the four-bit group).

The 74147 TTL IC in Fig. 6-5(*c*) is packaged in a 16-pin DIP. Internally, the IC consists of circuitry equivalent to about 30 logic gates.

The 74147 encoder in Fig. 6-5 has a *priority* feature. This means that if two inputs are activated at the same time, only the larger number will be encoded. For instance, if both the 9 and the 4 inputs were activated (LOW), then the output would be LHHL, representing decimal 9. Note that the outputs need to be complemented (inverted) to form the true binary number of 1001.

Self-Test

Answer the following questions.

12. Refer to Fig. 6-5. The 74147 encoder IC has active _____ (HIGH, LOW) inputs and active _____ (HIGH, LOW) outputs.
13. Refer to Fig. 6-5. If only input 7 of the 74147 encoder is LOW, what is the logic state at each of the four outputs?
14. Refer to Fig. 6-5(*b*). What is the meaning of a bubble on the logic symbol at input 4 (pin 1 on the 74147 IC)?
15. Refer to Fig. 6-5. If both inputs 2 and 8 go LOW, what is the logic state at each of the four outputs?

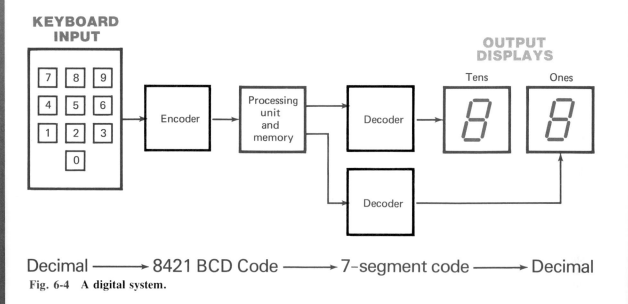

Decimal ⟶ 8421 BCD Code ⟶ 7-segment code ⟶ Decimal

Fig. 6-4 A digital system.

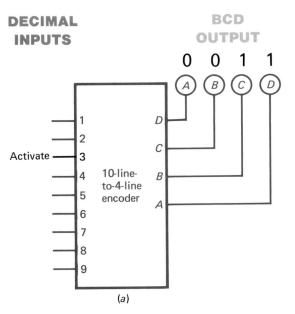

DECIMAL INPUTS

BCD OUTPUT

0 0 1 1

(a)

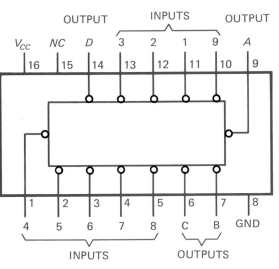

(b)

INPUTS									OUTPUTS			
1	2	3	4	5	6	7	8	9	D	C	B	A
H	H	H	H	H	H	H	H	H	H	H	H	H
X	X	X	X	X	X	X	X	L	L	H	H	L
X	X	X	X	X	X	X	L	H	L	H	H	H
X	X	X	X	X	X	L	H	H	H	L	L	L
X	X	X	X	X	L	H	H	H	H	L	L	H
X	X	X	X	L	H	H	H	H	H	L	H	L
X	X	X	L	H	H	H	H	H	H	L	H	H
X	X	L	H	H	H	H	H	H	H	H	L	L
X	L	H	H	H	H	H	H	H	H	H	L	H
L	H	H	H	H	H	H	H	H	H	H	H	L

H = HIGH logic level, L = LOW logic level, X = Don't care

(c)

Fig. 6-5 (*a*) **10-line-to-4-line encoder.** (*b*) **Pin diagram of 74147 encoder IC.** (*c*) **Truth table for 74147 encoder.** (*Courtesy of National Semiconductor Corporation.*)

6-6 SEVEN-SEGMENT LED DISPLAYS

The common task of decoding from machine language to decimal numbers is suggested in the system in Fig. 6-4. A very common output device used to display decimal numbers is the seven-segment display. The seven segments of the display are labeled *a* through *g* in Fig. 6-6(*a*). The displays representing decimal digits 0 through 9 are shown in Fig. 6-6(*b*). For instance, if segments *a*, *b*, and *c* are lit, a decimal 7 is displayed. If, however, all segments *a* through *g* are lit, a decimal 8 is displayed.

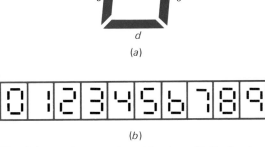

(a)

(b)

Fig. 6-6 (*a*) **Segment identification.** (*b*) **Decimal numbers on typical seven-segment display.**

Several common seven-segment display packages are shown in Fig. 6-7 on the next page. The seven-segment LED display in Fig. 6-7(*a*) fits a regular 14-pin DIP IC socket. Another single-digit seven-segment LED display is shown in Fig. 6-7(*b*). This display fits crosswise into a wider DIP IC socket. Finally, the unit in Fig. 6-7(*c*) is a multidigit LED display widely used in digital clocks.

The seven-segment display may be constructed with each of the segments being a thin filament which glows. This type of unit is called an *incandescent* display and is similar to a regular lamp. Another type of display is the *gas-discharge tube*, which operates at high voltages. It gives off an orange glow. The modern *vacuum fluorescent* (VF) display gives off a blue-green glow when lit and operates at low voltages. The newer

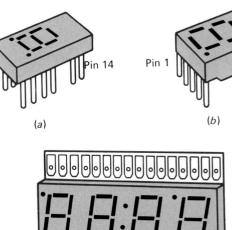

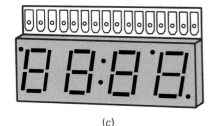

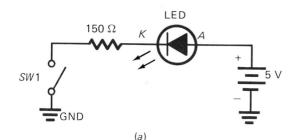

(a)   (b)

(a)

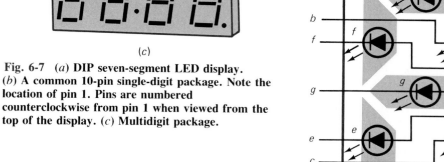

(c)

Fig. 6-7 (a) **DIP seven-segment LED display.** (b) **A common 10-pin single-digit package. Note the location of pin 1. Pins are numbered counterclockwise from pin 1 when viewed from the top of the display.** (c) **Multidigit package.**

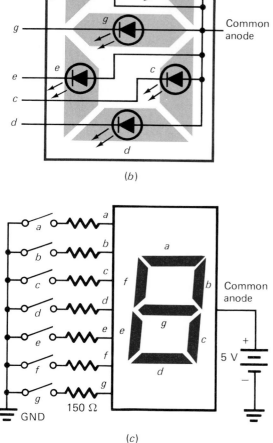

(b)

liquid-crystal display (LCD) creates numbers in a black or silvery color. The common LED display gives off a characteristic reddish glow when lit.

The LED is basically a PN-junction diode. When the diode is forward-biased, current flows through the PN junction and the light emitted is focused by a plastic lens, allowing the user to see the light. Many LEDs are fabricated from gallium arsenide (GaAs) and several related materials.

A single LED is being tested in Fig. 6-8(a). When the switch (SW1) is closed, current flows from the 5-V power supply through the LED, causing it to light. The series resistor limits current to about 20 mA. Without the limiting resistor the LED would burn out. Typically, LEDs can accept only about 1.7 to 2.1 V across their terminals when lit. Being a diode, the LED is sensitive to polarity. Hence, the cathode (K) must be toward the negative (GND) terminal while the anode (A) must be toward the positive terminal of the power supply.

A seven-segment LED display is shown in Fig. 6-8(b). Each segment (a through g) contains an LED, as shown by the seven symbols. The display shown has all the anodes tied together and coming out the right side as a single connection (common anode). The inputs on the left go to the various segments of the display. The device in Fig. 6-8(b) is referred to as

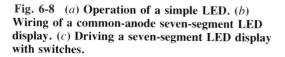

(c)

Fig. 6-8 (a) **Operation of a simple LED.** (b) **Wiring of a common-anode seven-segment LED display.** (c) **Driving a seven-segment LED display with switches.**

a common-anode seven-segment LED display. These units can also be purchased in common-cathode form.

To understand how segments on the display are activated and lit, consider the circuit in Fig.

6-8(c). If switch b is closed, current flows from GND through the limiting resistor to the b-segment LED and out the common-anode connection to the power supply. Only segment b will light.

Suppose you wanted the decimal 7 to light on the display in Fig. 6-8(c). Switches a, b, and c would be closed, lighting the LED segments a, b, and c. The decimal 7 would light on the display. Likewise, if the decimal 5 were to be lit, switches a, c, d, f, and g would be closed. These five switches would ground the correct segments, and a decimal 5 would appear on the display. Note that it takes a GND voltage (LOW logic level) to activate the LED segments on this display.

Mechanical switches are used in Fig. 6-8(c) to drive the seven-segment display. Usually power for the LED segments is provided by an IC. The IC is called a *display driver*. In practice, the display driver is usually packaged in the same IC as the decoder. Therefore, it is common to speak of *seven-segment decoder/drivers*.

Self-Test

Supply the missing word or words in each statement.

16. Refer to Fig. 6-6(a). If segments a, c, d, f, and g are lit, the decimal number _____ will appear on the seven-segment display.

17. The seven-segment unit that gives off a blue-green glow is a(n) _____ (vacuum fluorescent, incandescent, LCD, LED) display.

18. The letters "LED" stand for _____ , whereas "LCD" stand for _____ .

19. Refer to Fig. 6-8(c). If switches b and c are closed, segments _____ and _____ will light. This _____ (LCD, LED) seven-segment unit will display the decimal number _____ .

6-7 DECODERS

A *decoder*, like an encoder, is a code translator. Figure 6-4 shows two decoders being used in the system. The decoders are translating the 8421 BCD code to a seven-segment display code that lights the proper segments on the displays. The display will be a decimal number. Figure 6-9 shows the BCD number 0101 at the input of the *BCD-to-seven-segment decoder/ driver*. The decoder activates outputs a, c, d, f, and g to light the segments shown in Fig. 6-9. The decimal number 5 lights up on the display.

Decoders come in several varieties, such as the ones illustrated in Fig. 6-10 on the next page. Notice in Fig. 6-10 that the same block diagram is used for the 8421 BCD, the excess-3, and the Gray decoder.

Other decoders are available such as BCD converters, BCD-to-binary converters, 4-to-16-

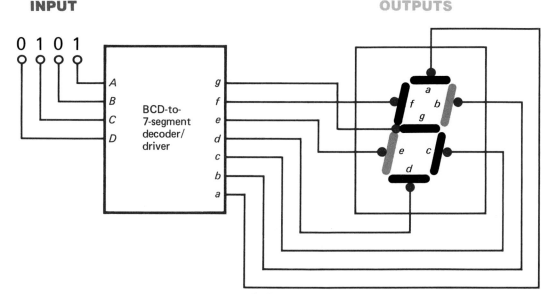

Fig. 6-9 Decoder driving a seven-segment display.

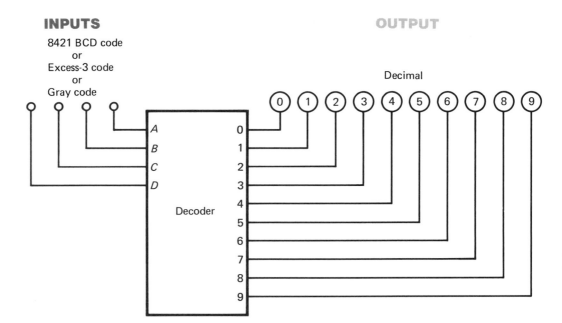

Fig. 6-10   A typical decoder block diagram. Note that inputs may be 8421 BCD, excess-3, or Gray code.

line decoders, and 2-to-4-line decoders. Other encoders available are a decimal-to-octal and 8-to-3-line priority encoder.

Decoders, like encoders, are combinational logic circuits with several inputs and outputs. Most decoders contain from 20 to 50 gates. Most decoders and encoders are packaged in single IC packages.

Self-Test

Answer the following questions.

20. Refer to Fig. 6-9. If the BCD input to the decoder/driver is 1000, which segments on the display will light? The seven-segment LED display will read what decimal number?

21. List at least three types of decoders.

6-8 BCD-TO-SEVEN-SEGMENT DECODER/DRIVERS

A logic symbol for a commercial TTL 7447A BCD-to-seven-segment decoder/driver is shown in Fig. 6-11(*a*). The BCD number to be decoded is applied to the inputs labeled *D*, *C*, *B*, and *A*. When activated with a LOW, the lamp-test (LT) input activates all outputs (*a* to *g*). When activated with a LOW, the blanking input (BI) causes all outputs HIGH, turning all attached displays OFF. When activated with a LOW, the

ripple-blanking input (RBI) blanks the display *only if it contains a 0*. When the RBI input becomes active, the BI/RBO pin temporarily becomes the ripple-blanking *output* (RBO) and drops to a LOW. Remember that "blanking" means to cause no LEDs on the display to light.

The seven outputs on the 7447A IC are all active LOW outputs. In other words, the outputs are normally HIGH and drop to a LOW when activated.

The exact operation of the 7447A decoder/driver IC is detailed in the truth table furnished by Texas Instruments and reproduced in Fig. 6-11(*b*). The decimal displays generated by the 7447A decoder are shown in Fig. 6-11(*c*). Note that invalid BCD inputs (decimals 10, 11, 12, 13, 14, and 15) do generate a unique output on the 7447A decoder.

The 7447A decoder/driver IC is typically connected to a common-anode seven-segment LED display. Such a circuit is shown in Fig. 6-12 on page 112. It is especially important that the seven 150-Ω limiting resistors be wired between the 7447A IC and the seven-segment display.

Assume that the BCD input to the 7447A decoder/driver in Fig. 6-12 is 0001 (LLLH). This is equal to line 2 of the truth table in Fig. 6-11(*b*). This input combination causes segments *b* and *c* on the seven-segment display to light (outputs *b* and *c* drop to LOW). Decimal 1 is displayed. The LT and two BIs are not shown in Fig. 6-12. When not connected,

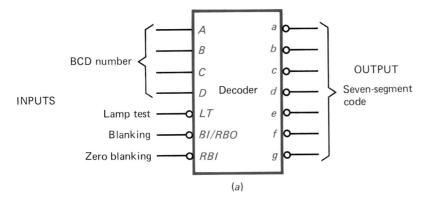

(a)

Decimal or function	INPUTS						BI/RBO	OUTPUTS							Note
	LT	RBI	D	C	B	A		a	b	c	d	e	f	g	
0	H	H	L	L	L	L	H	ON	ON	ON	ON	ON	ON	OFF	
1	H	X	L	L	L	H	H	OFF	ON	ON	OFF	OFF	OFF	OFF	
2	H	X	L	L	H	L	H	ON	ON	OFF	ON	ON	OFF	ON	
3	H	X	L	L	H	H	H	ON	ON	ON	ON	OFF	OFF	ON	
4	H	X	L	H	L	L	H	OFF	ON	ON	OFF	OFF	ON	ON	
5	H	X	L	H	L	H	H	ON	OFF	ON	ON	OFF	ON	ON	
6	H	X	L	H	H	L	H	OFF	OFF	ON	ON	ON	ON	ON	
7	H	X	L	H	H	H	H	ON	ON	ON	OFF	OFF	OFF	OFF	
8	H	X	H	L	L	L	H	ON	ON	ON	ON	ON	ON	ON	1
9	H	X	H	L	L	H	H	ON	ON	ON	OFF	OFF	ON	ON	
10	H	X	H	L	H	L	H	OFF	OFF	OFF	ON	ON	OFF	ON	
11	H	X	H	L	H	H	H	OFF	OFF	ON	ON	OFF	OFF	ON	
12	H	X	H	H	L	L	H	OFF	ON	OFF	OFF	OFF	ON	ON	
13	H	X	H	H	L	H	H	ON	OFF	OFF	ON	OFF	ON	ON	
14	H	X	H	H	H	L	H	OFF	OFF	OFF	ON	ON	ON	ON	
15	H	X	H	H	H	H	H	OFF	OFF	OFF	OFF	OFF	OFF	OFF	
BI	X	X	X	X	X	X	L	OFF	OFF	OFF	OFF	OFF	OFF	OFF	2
RBI	H	L	L	L	L	L	L	OFF	OFF	OFF	OFF	OFF	OFF	OFF	3
LT	L	X	X	X	X	X	H	ON	ON	ON	ON	ON	ON	ON	4

H = HIGH level, L = LOW level, X = Irrelevant

Notes:
1. The blanking input (BI) must be open or held at a HIGH logic level when output functions 0 through 15 are desired. The ripple-blanking input (RBI) must be open or HIGH if blanking of a decimal zero is not desired.
2. When a LOW logic level is applied directly to the blanking input (BI), all segment outputs ore OFF regardless of the level of any other input.
3. When ripple-blanking input (RBI) and inputs A, B, C, and D are at a LOW level with the lamp test (LT) input HIGH, all segment outputs go OFF and the ripple-blanking output (RBO) goes to a LOW level (response condition).
4. When the blanking input/ripple-blanking output (BI/RBO) is open or held HIGH and a LOW is applied to the lamp test (LT) input, all segment outputs are ON.

(b)

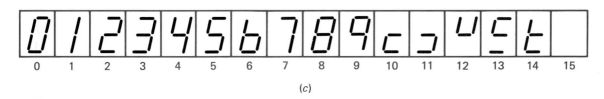

(c)

Fig. 6-11 (a) **Logic symbol for 7447A TTL decoder IC.** (b) **Truth table for 7447A decoder.** *(Courtesy of Texas Instruments, Inc.)* (c) **Format of readouts on seven-segment display using the 7447A decoder IC.**

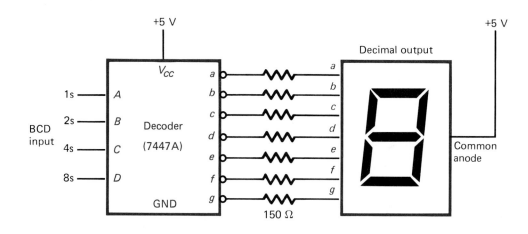

Fig. 6-12   Wiring a 7447A decoder and seven-segment LED display.

they are assumed to be "floating" HIGH and therefore disabled in this circuit. Good design practice suggests that these "floating" inputs should be connected to +5 V to make sure they stay HIGH.

The complexity of gating in the 7447A decoder is shown in Fig. 6-13. This logic diagram is reproduced from the 7447A decoder/driver data sheet furnished by Texas Instruments. Note that, even on the logic diagram in Fig. 6-13, active LOW inputs and outputs are designated by invert bubbles.

Many applications, such as a calculator or cash register, require that the *leading zeros be blanked*. The illustration in Fig. 6-14 (see page 114) shows the use of the 7447A decoder/drivers operating a group of displays as in a cash register. The six-digit display example details how the blanking of leading 0s would be accomplished using the 7447A IC driving LED displays.

The current inputs to the six decoders are shown across the bottom of the drawing in Fig. 6-14. The current BCD input is 0000 0000 0011 1000 0001 0000 (003810 in decimal). The two left 0s should be blanked, leaving the display reading 38.10. The blanking of the leading 0s is handled by wiring the RBI and RBO pins to each 7447A decoder IC together as shown in Fig. 6-14.

Working from left to right in Fig. 6-14, notice that the RBI input of IC6 is grounded. From the 7447A decoder's truth table in Fig. 6-11 it can be determined that when RBI is LOW and when all BCD inputs are LOW, then all segments of the display are blanked or off. Also the RBO is forced LOW. This LOW is passed to the RBI of IC5.

Continuing in Fig. 6-14, with the BCD input

to IC5 as 0000 and the RBI at LOW, the display is also blanked. The RBO of IC5 is forced LOW and is passed to the RBI input of IC4. Even with the RBI LOW, IC4 *does not* blank the display because the BCD is 0011. The RBO of IC4 remains HIGH, which is sent to IC3.

A question arises about the right LED display in Fig. 6-14. The BCD input to IC1 is 0000_{BCD} and a zero (0) appears on the display. The zero on the IC1 display is not blanked because the RBI input is not activated (RBI = HIGH). The first row of the truth table in Fig. 6-11(*b*) shows that the 7447A decoder/driver will display the zero when the RBI is HIGH.

Self-Test

Check your understanding by answering the following questions.

22. Refer to Fig. 6-11. The 7447A decoder/driver IC has active _____ (HIGH, LOW) BCD inputs and active _____ (HIGH, LOW) outputs.

23. Refer to Fig. 6-11. The lamp test, blanking, and zero-blanking inputs of the 7447A are active _____ (HIGH, LOW) inputs.

24. The RBI and RBO inputs of the 7447A are commonly used for blanking _____ on calculator and cash register multidigit displays.

25. What will the seven-segment display read for each input pulse shown in Fig. 6-15 (see page 114)?

26. List the segments on the seven-segment display that will light for each of the input pulses shown in Fig. 6-15.

**7447A
BCD-to-seven-
decoder IC**

**Liquid-crystal
displays**

Digital multimeter

Field-effect LCD

Nematic fluid

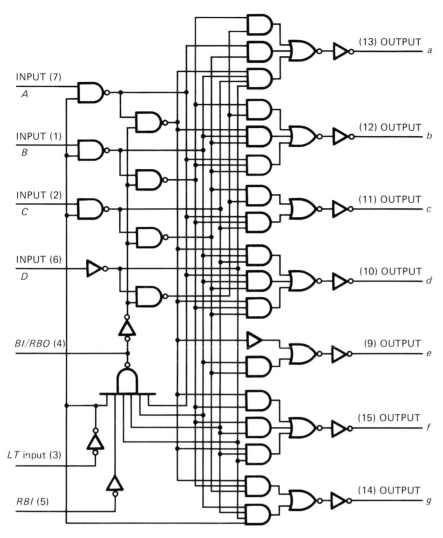

INPUT (7)
A

INPUT (1)
B

INPUT (2)
C

INPUT (6)
D

BI/RBO (4)

LT input (3)

RBI (5)

(13) OUTPUT a

(12) OUTPUT b

(11) OUTPUT c

(10) OUTPUT d

(9) OUTPUT e

(15) OUTPUT f

(14) OUTPUT g

Fig. 6-13   Detailed schematic of internal logic of the 7447A BCD-to-seven-segment decoder IC. *(Courtesy of Texas Instruments, Inc.)*

6-9 LIQUID-CRYSTAL DISPLAYS

The LED actually *generates* light, where the LCD simply *controls* available light. The LCD has gained wide acceptance recently because of its very low power consumption. The LCD is also well suited for use in sunlight or in other brightly lit areas. The DMM (digital multimeter) in Fig. 6-16 on the next page uses a modern LCD.

The LCD is also suited for more complex displays than just seven-segment decimal. The LCD display in Fig. 6-16 contains an analog scale across the bottom as well as the larger digital readout. In practice you will find that the DMM LCD has several other symbols which do not appear in Fig. 6-16 because they are not activated.

The construction of a common LCD unit is shown in Fig. 6-17(a) on page 115. This unit is called the *field-effect LCD*. When a segment is energized by a low-frequency square-wave signal, the LCD segment appears black while the rest of the surface remains shiny. Segment *e* is energized in Fig. 6-17(a). The nonenergized segments are nearly invisible.

The key to LCD operation is the liquid crystal, or *nematic fluid*. This nematic fluid is sandwiched between two glass plates. An ac voltage is applied across the nematic fluid, from the top metalized segments to the metalized back plane. When affected by the magnetic field of the ac voltage, the nematic fluid transmits light differently and the energized segment appears as black on a silvery background.

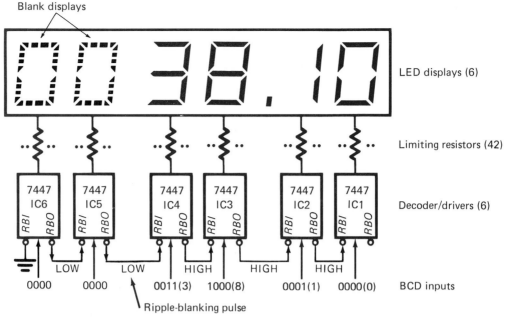

Blank displays

LED displays (6)

Limiting resistors (42)

Decoder/drivers (6)

LOW   LOW   HIGH   HIGH   HIGH

0000   0000   0011(3) 1000(8) 0001(1) 0000(0)   BCD inputs

Ripple-blanking pulse

Fig. 6-14   Using the ripple-blanking input (RBI) of the 7447A decoder/driver to blank leading zeros in a multidigit display.

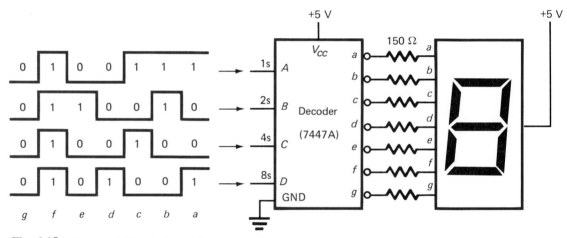

Fig. 6-15   Decoder-LED display pulse-train problem.

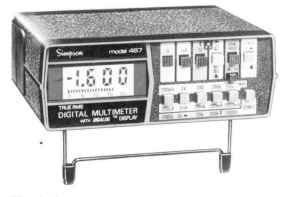

Fig. 6-16   Digital multimeter using a liquid-crystal display. *(Courtesy of Simpson Electric Company)*

The field-effect LCD uses a polarizing filter on the top and bottom of the display as shown in Fig. 6-17(*a*). The back plane and segments are internally wired to contacts on the edge of the LCD. Only two of the many contacts are shown in Fig. 6-17(*a*).

Decimal 7 is displayed on the LCD shown in Fig. 6-17(*b*). The BCD-to-seven-segment decoder on the left is receiving a BCD input of 0111. This input activates the *a*, *b*, and *c* outputs of the decoder (*a*, *b*, and *c* are HIGH in this example). The remaining outputs of the decoder are LOW (*d*, *e*, *f*, and *g* = LOW). The 100-Hz square-wave input is always applied to

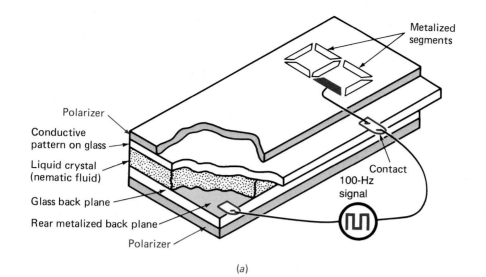

(a)

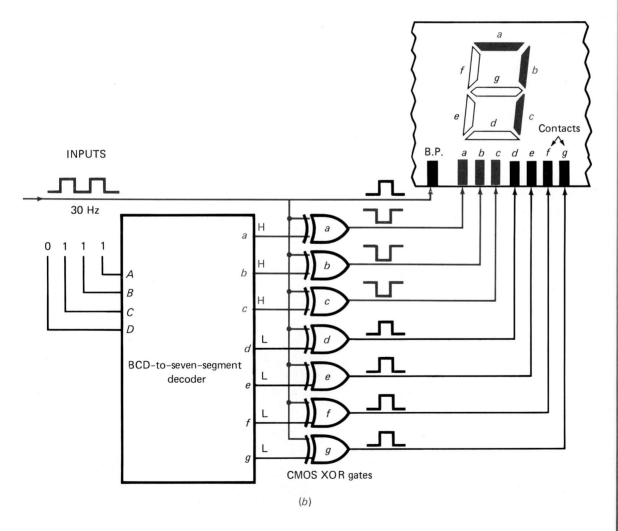

(b)

Fig. 6-17 (a) **Construction of a field-effect LCD.** (b) **Wiring of a CMOS decoder/driver system to an LCD.**

the back plane of the display. This signal is also applied to each of the CMOS XOR gates used to drive the LCD. Note that the XOR gates produce an inverted waveform when activated (*a*, *b*, and *c* XOR gates are activated). The 180° out-of-phase signals to the back plane and segments *a*, *b*, and *c* cause these areas of the LCD to turn black. The in-phase signals from XOR gates *d*, *e*, *f*, and *g* do not cause these segments to be activated. Therefore, these segments remain nearly invisible.

The XOR gates used as LCD drivers in Fig. 6-17(*b*) are CMOS units. TTL XOR gates are not used because they cause a small dc offset voltage to be developed across the LCD's nematic fluid. The *dc voltage* across the nematic fluid *will destroy the LCD* in a short time.

In actual practice, the decoder and XOR LCD display drivers pictured in Fig. 6-17(*b*) are usually packaged in a single CMOS IC. The 100-Hz square-wave signal is not critical and may range from 30 to 200 Hz. Liquid-crystal displays are sensitive to low temperatures. At below-zero temperatures the LCD display's turn-on and turn-off times become very slow. However, the long lifetime and extremely low power consumption make them ideal for battery or solar cell operation.

Figure 6-18 illustrates two examples of typical commercial LCD devices. Note that both have pins which can be soldered into a printed circuit board. In the lab, these LCD displays may be plugged into solderless mounting boards. However, this must be done with great care because there are many fragile pins. Most labs will have the LCDs mounted on a printed circuit board with a DIP (dual in-line package) jumper attached.

A simple two-digit seven-segment LCD is sketched in Fig. 6-18(*a*). Notice the use of two glass plates. Because of the thin glass used in LCDs, care must be taken not to drop or bend the display. Notice in Fig. 6-18(*a*) that two plastic headers with pins are fastened on each side of the glass back plane. The LCD illustrated in Fig. 6-18(*a*) has 18 pins. Only the common or back plane of the LCD is shown. Each segment and decimal point has a pin connection on this LCD package.

Another commercial LCD is illustrated in Fig. 6-18(*b*). This LCD has a more complex display, including symbols. This unit comes in a 40-pin package. All segments, decimal points, and symbols are assigned a pin number. Only the back plane or common pin is noted on the

drawing. Manufacturers' data sheets must be consulted for actual pin numbers.

Liquid-crystal displays that produce frosty white characters on a dark background are also available. This type is usually the *dynamic-scattering LCD*. The dynamic-scattering LCD uses a different nematic fluid and no polarizers. The dynamic-scattering LCD consumes more power than field-effect displays. Currently the field-effect LCD is the most popular. The black letters on a silvery background, as on the DMM in Fig. 6-16, tell you this is a field-effect LCD.

Self-Test

Supply the missing word or words in each statement.

27. Digits appear _____ (black, silver) on a _____ (black, silver) background on the field-effect LCD display.
28. The LCD display uses a liquid crystal, or _____ fluid, which transmits light differently when affected by a magnetic field from an ac voltage.
29. A(n) _____ (ac, dc) voltage applied to an LCD will destroy the unit.
30. The LCD unit consumes a _____ (large amount, moderate amount, very small amount) of power.

6-10 USING CMOS TO DRIVE AN LCD DISPLAY

A block diagram of an LCD decoder/driver system is sketched in Fig. 6-19(*a*) (see page 118). The input is 8421 BCD. The latch is a temporary memory to hold the BCD data. The BCD-to-seven-segment decoder operates somewhat like the 7447A decoder that was studied earlier. Note that the output from the decoder in Fig. 6-19(*a*) is in seven-segment code. The last block before the display is the LCD driver. This consists of XOR gates as in Fig. 6-17(*b*). The drivers and back plane (common) of the display must be driven with a 100-Hz ac square-wave signal. In actual practice the latch, decoder, and LCD driver are all available in a single CMOS package. The 74HC4543 and 4543 ICs described by the manufacturer as a *BCD-to-seven-segment latch/decoder/driver for LCDs* are such packages.

A wiring diagram for a single LCD driver circuit is shown in Fig. 6-19(*b*). The 74HC4543

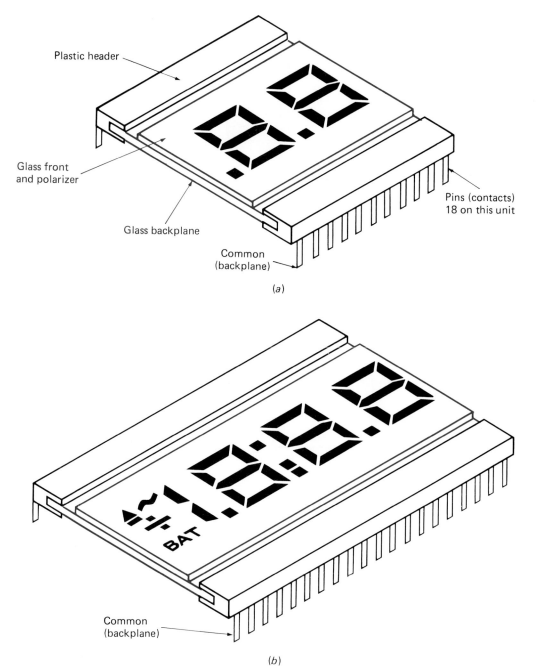

Plastic header

Glass front and polarizer

Glass backplane

Pins (contacts) 18 on this unit

Common (backplane)

(a)

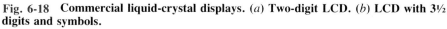

Common (backplane)

(b)

Fig. 6-18 **Commercial liquid-crystal displays.** (*a*) **Two-digit LCD.** (*b*) **LCD with 3½ digits and symbols.**

decoder/driver CMOS IC is being employed. The 8421 BCD input is 0011 (decimal 3). The 0011_{BCD} is decoded into seven-segment code. A separate 100-Hz clock feeds ac to both the LCD back plane (common) and the Ph (phase) input of the 74HC4543 IC. The driving signals in this example are sketched for each segment of the LCD. Note that *only out-of-phase signals will activate a segment*. In-phase signals (such as segments *e* and *f* in this example) do not activate LCD segments.

A pin diagram for the 74HC4543 BCD-to-seven-segment latch/decoder/driver CMOS IC is reproduced in Fig. 6-20(*a*) on page 119. Detailed information on the operation of the 74HC4543 IC is contained in the truth table in Fig. 6-20(*b*). On the output side of the truth table, an "H" means the segment is on while an "L" means the segment is off. The format of the decimal numbers generated by the decoder is shown in Fig. 6-20(*c*). Note especially the numbers 6 and 9. The 74HC4543 decoder

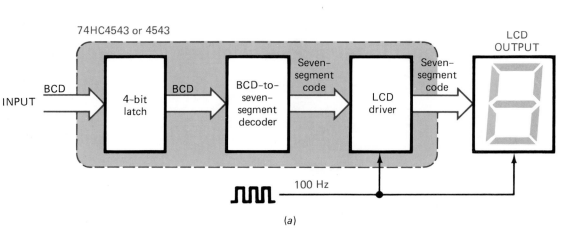

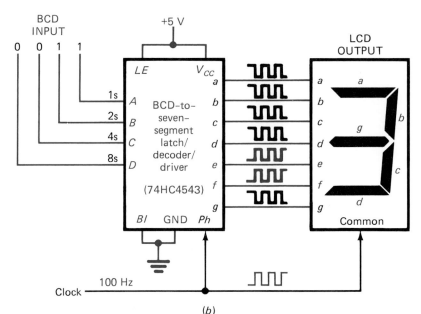

Fig. 6-19 **Driving seven-segment LCD. (a) Block diagram of system used to decode and drive LCD. (b) Using the 74HC4543 CMOS IC to decode and drive the LCD.**

forms the 6 and 9 differently from the 7447A TTL decoder studied earlier. Compare Fig. 6-20(c) with 6-11(c) to see the differences.

Self-Test

Answer the following questions.

31. Refer to Fig. 6-19(a). The job of the decoder block is to translate from _____ code to _____ code.
32. Refer to Fig. 6-19(b). All of the drive lines going from the driver to the LCD carry a(n) _____ (ac, dc) signal.
33. Refer to Fig. 6-21 on page 120. What is the decimal reading on the LCD for each input pulse (a to e)?
34. Refer to Fig. 6-21. For input pulse a only, which drive lines have an *out-of-phase* signal appearing on them?

6-11 VACUUM FLUORESCENT DISPLAYS

The *vacuum fluorescent (VF) display* is a modern relative of the triode vacuum tube. A schematic symbol for a triode vacuum tube is sketched in Fig. 6-22(a) (page 120). The three parts of the triode tube are called the *plate* (P), *grid* (G), and *cathode* (K). The cathode is also called the *filament* or the *heater*. The plate is also called the *anode*.

The cathode/heater is a fine tungsten wire coated with a material such as barium oxide. The cathode gives off electrons when heated. The grid is a stainless steel screen. The plate can be thought of as the "collector of electrons" in the triode tube.

Assume that the cathode (K) of the triode tube in Fig. 6-22(a) is heated and has "boiled off" some electrons into the vacuum surround-

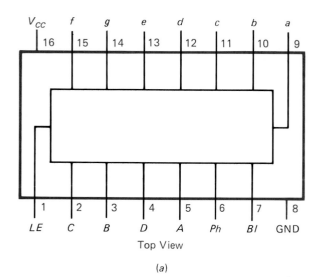

Top View

(a)

Truth Table

INPUTS							OUTPUTS							
LE	BI	Ph*	D	C	B	A	a	b	c	d	e	f	g	Display
X	H	L	X	X	X	X	L	L	L	L	L	L	L	Blank
H	L	L	L	L	L	L	H	H	H	H	H	H	L	0
H	L	L	L	L	L	H	L	H	H	L	L	L	L	1
H	L	L	L	L	H	L	H	H	L	H	H	L	H	2
H	L	L	L	L	H	H	H	H	H	H	L	L	H	3
H	L	L	L	H	L	L	L	H	H	L	L	H	H	4
H	L	L	L	H	L	H	H	L	H	H	L	H	H	5
H	L	L	L	H	H	L	H	L	H	H	H	H	H	6
H	L	L	L	H	H	H	H	H	H	L	L	L	L	7
H	L	L	H	L	L	L	H	H	H	H	H	H	H	8
H	L	L	H	L	L	H	H	H	H	H	L	H	H	9
H	L	L	H	L	H	L	L	L	L	L	L	L	L	Blank
H	L	L	H	L	H	H	L	L	L	L	L	L	L	Blank
H	L	L	H	H	L	L	L	L	L	L	L	L	L	Blank
H	L	L	H	H	L	H	L	L	L	L	L	L	L	Blank
H	L	L	H	H	H	L	L	L	L	L	L	L	L	Blank
H	L	L	H	H	H	H	L	L	L	L	L	L	L	Blank
L	L	L	X	X	X	X	**							**
†	†	H	†				Inverse of Output Combinations Above							Display as above

X = don't care

† = same as above combinations

* = for liquid crystal readouts, apply a square wave to Ph

** = depends upon the BCD code previously applied when LE = H

(b)

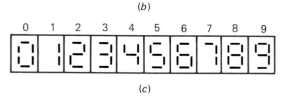

(c)

Fig. 6-20 The 74HC4543 BCD-to-seven-segment latch/decoder/driver CMOS IC. (a) **Pin diagram.** (b) **Truth table.** (c) **Format of digits formed by the 74HC4543 decoder IC.**

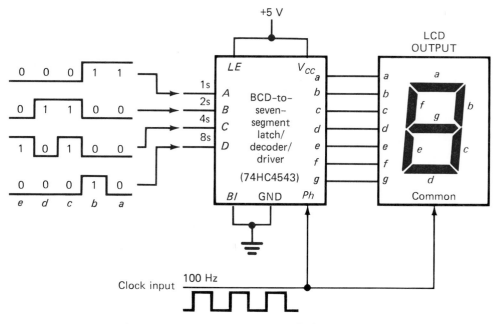

Fig. 6-21 Decoder-LCD display pulse-train problem (Self Test questions 33 and 34).

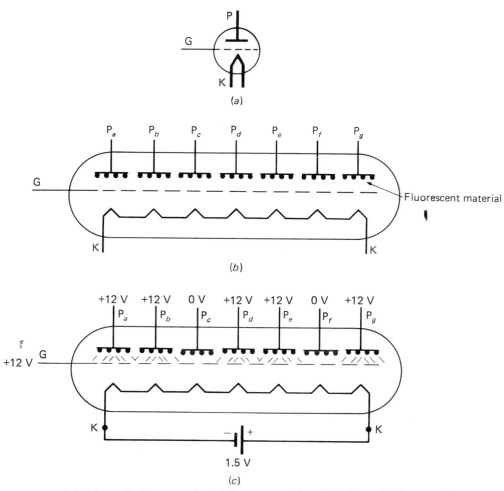

Fig. 6-22 (*a*) Schematic diagram of a triode vacuum tube. (*b*) Schematic diagram of single digit of a VF display. (*c*) Lighting plates on the VF display.

ing the cathode. Next, assume that the grid (G) becomes positive. The electrons are attracted to the grid. Next, assume that the plate (P) becomes positive. When the plate becomes positive, electrons will be attracted through the screenlike grid to the plate. Finally, the triode is conducting electricity from cathode to anode.

You can stop the triode tube from conducting in one of two ways. First, make the grid slightly negative (leave the plate positive). This will repel the electrons and they will not pass through the grid to the plate. Second, leave the grid positive and drop the plate voltage to 0. The plate will not attract electrons and the triode tube does not conduct electricity from cathode to anode.

The schematic symbol in Fig. 6-22(b) represents a single digit of a VF display. Notice the single cathode (K), single grid (G), and seven plates (P_a to P_g). Each of the seven plates has been coated with a *zinc oxide fluorescent* material. Electrons striking the fluorescent mate-

rial on the plate will cause it to glow a blue-green color. The seven plates in the schematic in Fig. 6-22(b) represent the seven segments of a normal numeric display. Note that the entire unit is held in a glass enclosure containing a vacuum.

A single-digit seven-segment display is being operated in Fig. 6-22(c). The cathode (heater) is being powered by direct current in this example, with +12 V applied to the grid (G). Two plates (P_c and P_f) are grounded. Each of the remaining five plates has +12 V applied. The high positive voltage on the five plates (P_a, P_b, P_d, P_e, and P_g) causes these plates to attract electrons. These five plates glow a blue-green color.

In actual practice, the plates of the VF display tube are shaped like segments of a number or other shape. Figure 6-23(a) is a view of the physical arrangement of cathode, grid, and plates. Notice that the plates are arranged in seven-segment format in this display unit. The

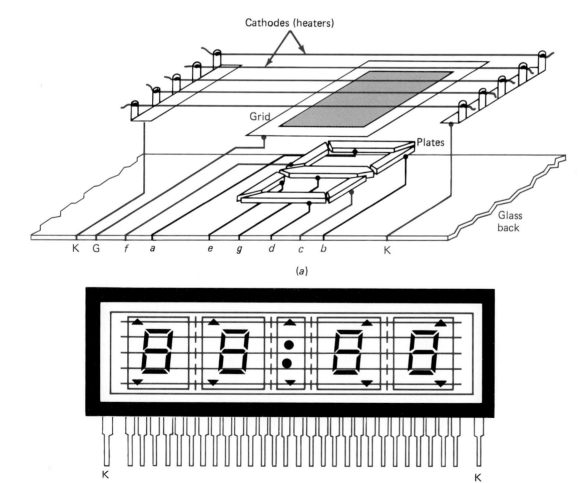

Fig. 6-23 (a) Typical construction of vacuum fluorescent display. (b) Commercial four-digit VF display.

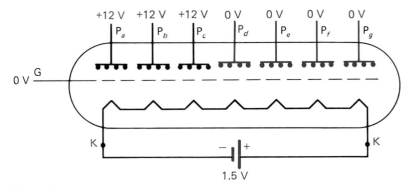

Fig. 6-24   VF display with no positive grid voltage.

screen above the segments is the grid. Above the grid are the cathodes (filaments or heaters). Each segment, grid, or cathode lead comes out the side of the sealed glass vacuum tube. The VF display shown in Fig. 6-23(a) would be viewed from above the unit looking downward. The fine wire cathodes and grid would be almost invisible. Lighted segments (plates) show through the mesh (grid).

A commercial vacuum fluorescent display is sketched in Fig. 6-23(b). This VF display contains 4 seven-segment numeric displays, a colon, and 10 triangle-shaped symbols. The internal parts of most VF displays are visible through the sealed glass package. Visible in the display are the cathodes (filaments or heaters) stretched horizontally across the display. These are very fine wires and are barely visible on a commercial display. Next, the grids are shown in five sections. Each of the five grids can be activated individually. Finally, the fluorescent-coated plates form the numeric segments, colons, and other symbols.

Vacuum fluorescent displays are based on an older technology but they have gained great favor in recent years. This is because they can operate at relatively low voltages and power and have an extremely long life and fast response. They can display in various colors (with filters), have good reliability, and low cost. Vacuum fluorescent displays are compatible with the popular 4000 series CMOS family of ICs. They are widely used in readouts found in automobiles, VCRs, TVs, household appliances, and digital clocks.

Self-Test

Answer the following questions.

35. A VF display glows with a _____ color when activated.

36. Refer to Fig. 6-24. Which plates of this VF display will glow?
37. Name the parts of the vacuum fluorescent display labeled *A*, *B*, and *C* in Fig. 6-25.
38. Refer to Fig. 6-25. What segments of the VF display glow, and what decimal number is lit?

6-12   DRIVING A VF DISPLAY

The voltage requirements for operating VF displays are somewhat higher than for LED or LCD units. This requirement makes them compatible with 4000 series CMOS ICs. Recall that 4000 series CMOS ICs can operate on voltages up to 18 V.

A wiring diagram of a simple BCD decoder/ driver circuit is detailed in Fig. 6-26. In this example, 1001_{BCD} is translated into the decimal 9 on the VF display. The circuit uses the 4511 BCD-to-seven-segment latch/decoder/driver CMOS IC. In this example, the *a*, *b*, *c*, *f*, and *g* output lines are HIGH (+12 V) with only the *d* and *e* lines LOW.

The +12-V supply is connected directly to the grid in Fig. 6-26. The cathode (filament or heater) circuit contains a resistor (R_1) to limit the current through the heaters to a safe level. The +12 V is also used to supply power for the 4511 decoder/driver CMOS IC. Note the labels on the power connections to the 4511 IC. The V_{DD} pin connects to +12 V while V_{SS} goes to ground (GND).

The pin diagram, truth table, and number formats for the 4511 CMOS IC are shown in Fig. 6-27 on page 124. The 4511 BCD-to-seven-segment latch/decoder/driver IC's pin diagram is shown in Fig. 6-27(a). This is the top view of this 16-pin DIP CMOS IC. Internally, the 4511 IC is organized like the 74HC4543 unit. The latch, decoder, and driver sections are illustrated in the shaded section of the block diagram in Fig. 6-19(a).

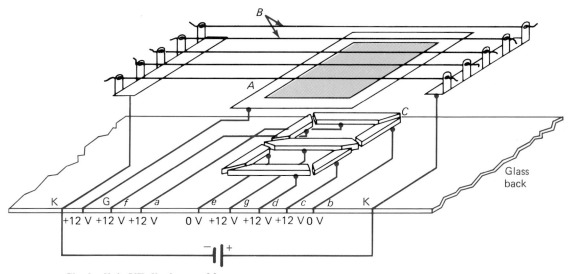

Fig. 6-25 **Single-digit VF display problem.**

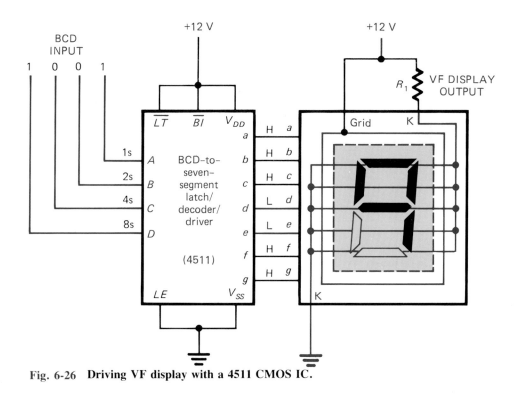

Fig. 6-26 **Driving VF display with a 4511 CMOS IC.**

The truth table in Fig. 6-27(*b*) shows seven inputs to the 4511 decoder/driver IC. The BCD data inputs are labeled *D, C, B,* and *A.* The $\overline{LT}$ input stands for lamp test. When activated by a LOW (row 1 on the truth table), all outputs go HIGH and light all segments of an attached display. The $\overline{BI}$ input stands for blanking input. When $\overline{BI}$ is activated with a LOW, all outputs go LOW and all segments of an attached display are blanked. The *LE* (latch enable) input can be used like a memory to hold data on display while the BCD in-

put data changes. If *LE* = 0, then data passes through the 4511 IC. However, if *LE* = 1 then the last data present at the data inputs (*D, C, B, A*) is latched and held on the display. The *LE,* $\overline{BI},$ *and* $\overline{LT}$ *inputs are all disabled in the circuit in Fig. 6-26.*

Next, refer to the output side of the truth table in Fig. 6-27. On the 4511 IC, a HIGH or 1 is an active output. In other words, an output of 1 turns on a segment on the attached display. Therefore, a 0 output means the display's segment is turned off.

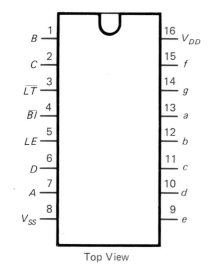

Top View

(a)

Truth Table

INPUTS							OUTPUTS							
LE	$\overline{BI}$	$\overline{LT}$	D	C	B	A	a	b	c	d	e	f	g	DISPLAY
X	X	0	X	X	X	X	1	1	1	1	1	1	1	B
X	0	1	X	X	X	X	0	0	0	0	0	0	0	
0	1	1	0	0	0	0	1	1	1	1	1	1	0	0
0	1	1	0	0	0	1	0	1	1	0	0	0	0	1
0	1	1	0	0	1	0	1	1	0	1	1	0	1	2
0	1	1	0	0	1	1	1	1	1	1	0	0	1	3
0	1	1	0	1	0	0	0	1	1	0	0	1	1	4
0	1	1	0	1	0	1	1	0	1	1	0	1	1	5
0	1	1	0	1	1	0	0	0	1	1	1	1	1	6
0	1	1	0	1	1	1	1	1	1	0	0	0	0	7
0	1	1	1	0	0	0	1	1	1	1	1	1	1	8
0	1	1	1	0	0	1	1	1	1	0	0	1	1	9
0	1	1	1	0	1	0	0	0	0	0	0	0	0	
0	1	1	1	0	1	1	0	0	0	0	0	0	0	
0	1	1	1	1	0	0	0	0	0	0	0	0	0	
0	1	1	1	1	0	1	0	0	0	0	0	0	0	
0	1	1	1	1	1	0	0	0	0	0	0	0	0	
0	1	1	1	1	1	1	0	0	0	0	0	0	0	
1	1	1	X	X	X	X				*				*

X = Don't care

*Depends upon the BCD code applied during the 0 to 1 transition of LE.

(b)

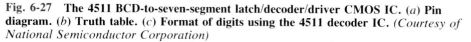

(c)

Fig. 6-27 **The 4511 BCD-to-seven-segment latch/decoder/driver CMOS IC.** (a) **Pin diagram.** (b) **Truth table.** (c) **Format of digits using the 4511 decoder IC.** (*Courtesy of National Semiconductor Corporation*)

The format of the digits generated by the 4511 BCD-to-seven-segment decoder IC are illustrated in Fig. 6-27(c). Especially note the formation of the decimal numbers 6 and 9.

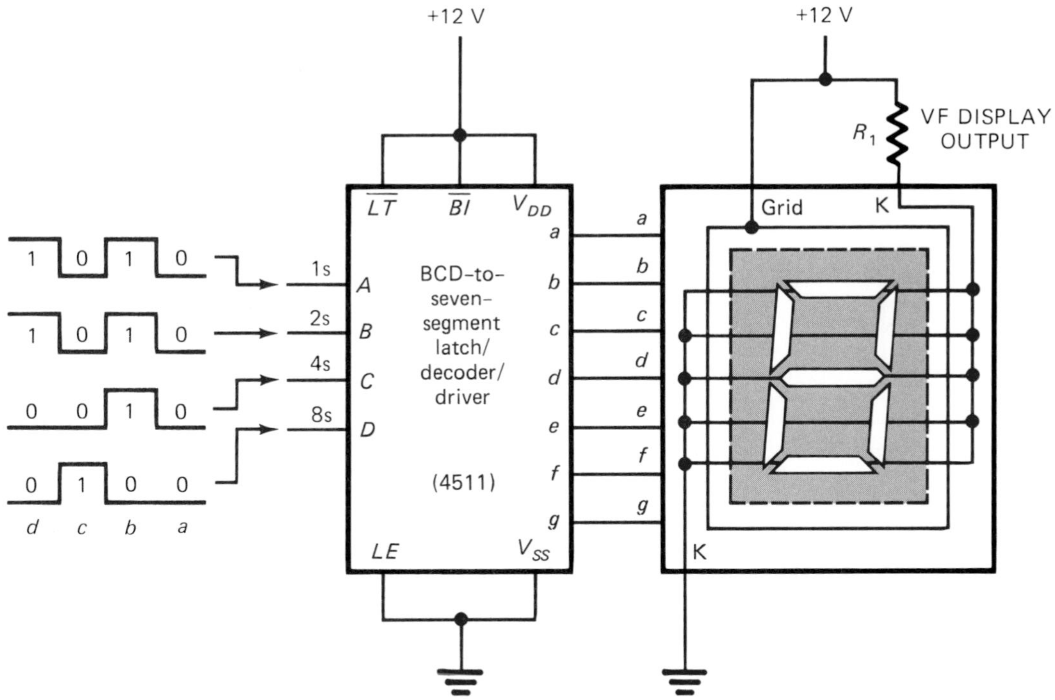

Fig. 6-28 Decoder–VF display pulse-train problem.

Self-Test

Answer the following questions.

39. Refer to Fig. 6-26. The +12-V power supply is being used because the _____ (CMOS, TTL) 4511 decoder/driver IC and the _____ (LCD, VF) display operate properly at this voltage.

40. Refer to Fig. 6-26. What is the purpose of resistor R_1 in this circuit?

41. Refer to Fig. 6-28. What is the decimal reading on the vacuum fluorescent display for each input pulse (*a* to *d*)?

42. Refer to Fig. 6-28. During pulse *a*, what voltages are applied to the seven plates (segments) of the VF display?

6-13 TROUBLESHOOTING A DECODING CIRCUIT

Consider the BCD-to-seven-segment decoder circuit in Fig. 6-29 on the next page. The problem is that segment *a* of the display does not light. The technician first checks the circuit visually. Then the IC is checked for signs of excessive heat. The V_{CC} and GND voltages are checked with a DMM or logic probe. In this

example, the results of these tests did not locate the problem. Next, the temporary jumper wire from GND to the LT input of the 7447A IC should cause all segments on the display to light, giving a decimal 8 indication. Segment *a* on the display still does not light. The logic probe is used to check the logic levels at the outputs (*a* to *g*) of the 7447A decoder. They are all L (LOW) in Fig. 6-29, as required. Next, the logic levels are checked on the display side of the resistors. They are all H (HIGH) except the faulty line, which is LOW. The LOW and HIGH pattern in Fig. 6-29 indicates a voltage drop across each of the bottom six resistors. The LOW indications on both ends of the top resistor in Fig. 6-29 indicate an open circuit in the segment *a* section of the seven-segment display. Segment *a* of the display must be faulty. The entire seven-segment LED display is replaced. The replacement must have the same pin diagram and be a common-anode LED display. After replacement, the circuit is checked for proper operation.

The circuit shown in Fig. 6-30 produces no display. The hurried technician checks the V_{CC} and GND pins with a logic probe. The readings as shown on Fig. 6-30 seem all right. The test jumper wire from the LT to the GND should light all segments of the LED display. No segments on the display light. The logic

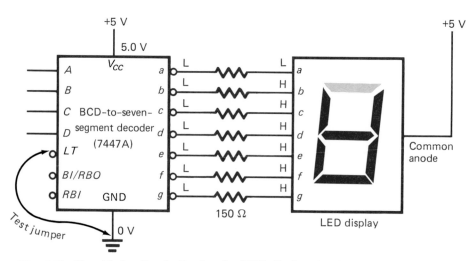

Fig. 6-29 Troubleshooting faulty decoder/LED display circuit.

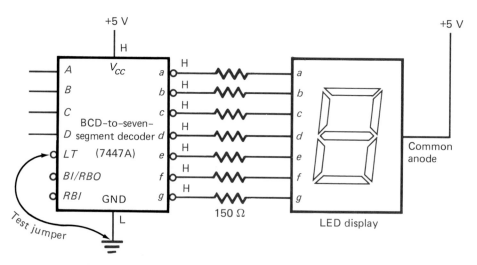

Fig. 6-30 Troubleshooting faulty decoder circuit with blank LED display.

probe shows faulty HIGH readings at all the outputs (*a* to *g*) of the 7447A IC. The technician checks the voltage at V_{CC} with a DMM. The reading is 4.65 V. This is quite low. The technician now touches the top of the 7447A IC. It is very hot. The chip (7447A) has an internal short circuit and must be replaced. The 7447A IC is replaced, and the circuit is checked for proper operation.

In this example, the technician forgot to use his or her own senses first. A simple touch of the top of the DIP IC circuit would have suggested a bad 7447A chip. Note that the HIGH reading on the V_{CC} pin did not give the technician an accurate picture. The voltage was actually 4.65 V instead of the normal 5.0 V. In

this case the voltmeter reading gave the technician a clue as to the difficulty in the circuit. The short circuit was dropping the power supply voltage to 4.65 V.

Self-Test

Answer the following questions.

43. What is the first step in troubleshooting a digital logic circuit?
44. An internal _____ (open, short) circuit in a TTL IC will many times cause the IC to become excessively hot.

SUMMARY

1. Many codes are used in digital equipment. You should now be familiar with decimal, binary, octal, hexadecimal, 8421 BCD, excess-3, Gray, and ASCII codes.
2. Converting from code to code is essential for your work in digital electronics. Table 6-4 will aid you in converting from several of the codes.
3. The most popular alphanumeric code is the 7-bit ASCII code. The ASCII code is widely used in microcomputer keyboard and display interfacing.
4. Electronic translators are called encoders and decoders. These complicated logic circuits are manufactured in single IC packages.
5. Seven-segment displays are very popular devices for reading out numbers. Light-emitting diode (LED), liquid-crystal display (LCD), and vacuum fluorescent (VF) types are popular displays.
6. The BCD-to-seven-segment decoder/driver is a common decoding device. It translates from BCD machine language to decimal numbers. The decimal numbers appear on seven-segment LED, LCD, or VF displays.

Table 6-4 The Gray Code

Decimal Number	Binary Number	BCD codes 8421		Excess-3		Gray code
0	0000	0000		0111		0000
1	0001	0001		0100		0001
2	0010	0010		0101		0011
3	0011	0011		0110		0010
4	0100	0100		0111		0110
5	0101	0101		1000		0111
6	0110	0110		1001		0101
7	0111	0111		1010		0100
8	1000	1000		1011		1100
9	1001	1001		1100		1101
10	1010	0001 0000		0100 0011		1111
11	1011	0001 0001		0100 0100		1110
12	1100	0001 0010		0100 0101		1010
13	1101	0001 0011		0100 0110		1011
14	1110	0001 0100		0100 0111		1001
15	1111	0001 0101		0100 1000		1000
16	10000	0001 0110		0100 1001		11000
17	10001	0001 0111		0100 1010		1101
18	10010	0001 1000		0100 1011		11011
19	10011	0001 1001		0100 1100		11010
20	10100	0010 0000		0101 0011		11110

CHAPTER REVIEW QUESTIONS

Answer the following questions.

6-1. Write the binary numbers for the decimal numbers in *a* to *f*:
   a. 17
   b. 31
   c. 42
   d. 75
   e. 150
   f. 300

6-2. Write the 8421 BCD numbers for the decimal numbers in *a* to *f*:
   a. 17
   b. 31
   c. 150
   d. 1632
   e. 47,899
   f. 103,926

6-3. Write the decimal numbers for the 8421 BCD numbers in *a* to *h*:
   a. 0010
   b. 1111
   c. 0011 0000
   d. 1110 0000 1111
   e. 0111 0001 0110 0000
   f. 0001 0001 0000 0000 0000
   g. 0101 1001 1000 1000 0101
   h. 0011 0010 0001 0100 0101 0110

6-4. Write the binary numbers for the 8421 BCD numbers in question 3.

6-5. Write the excess-3 code numbers for the decimal numbers in *a* to *f*:
   a. 7
   b. 27
   c. 59
   d. 318
   e. 4063
   f. 5533

6-6. Why is the excess-3 code used in some arithmetic circuits?

6-7. List two codes you learned about that are classified as BCD codes.

6-8. Write the Gray code numbers for the decimal numbers in *a* to *f*.

    a. 1   **d.** 4

    b. 2   **e.** 5

    c. 3   **f.** 6

6-9. As you count in the Gray code, what is the most important characteristic of this code?

6-10. The letters "ASCII" stand for _____ .

6-11. ASCII is a _____ -bit _____ (alphanumeric, BCD) code that is used to represent numbers, letters, punctuation marks, and control characters.

6-12. List two general names for code translators, or electronic code converters.

6-13. A(n) _____ (decoder, encoder) is the electronic device used to convert the decimal input of a calculator keypad to the BCD code used by the central processing unit.

6-14. A(n) _____ (decoder, encoder) is the electronic device used to convert the BCD of the central processing unit of a calculator to the decimal display output.

6-15. Which segments of the seven-segment display *will light* when the following decimal numbers appear? Use the letters *a, b, c, d, e, f,* and *g* as answers.

    a. 0   **f.** 5

    b. 1   **g.** 6

    c. 2   **h.** 7

    d. 3   **i.** 8

    e. 4   **j.** 9

6-16. The seven-segment displays that you will use give off a red glow and are of what type?

6-17. The _____ (LCD, LED) seven-segment display is used where battery operation demands low power consumption.

6-18. The _____ (LCD, LED) display is used where the unit must be read in bright light.

6-19. The _____ display generates light while the LCD controls available light.

6-20. The _____ (LED, VF) seven-segment display operates on a slightly higher voltage power supply.

6-21. Refer to Fig. 6-31. With the 1000 (*HLLL*) input, the display should read decimal _____ .

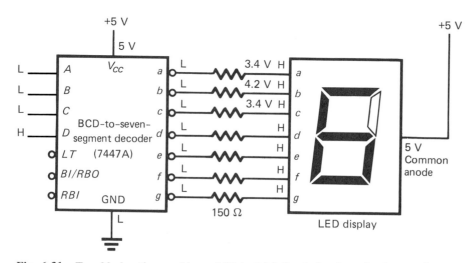

Fig. 6-31 **Troubleshooting problems 6-21 to 6-24. Logic levels and voltages given on faulty decoder/LED display circuit.**

6-22. Refer to Fig. 6-31. All the outputs from the 7447A decoder are _____ (HIGH, LOW). This is _____ (correct, not correct) for this circuit.

6-23. Both a voltmeter and a _____ _____ are used to troubleshoot the circuit in Fig. 6-31.

6-24. Refer to Fig. 6-31. Segment *b* of the LED display appears to be _____ (open, partially short circuited). The display should be replaced with a common- _____ LED display having the same pin diagram as the one in the circuit.

6-25. Refer to Fig. 6-32. With the BCD input shown, the six-digit display reads _____ .

6-26. List the condition (HIGH or LOW) of each of the ripple-blanking lines *A* to *F* in Fig. 6-32.

6-27. The front and back panels of a _____ (LCD, LED) seven-segment display are made of glass and can be broken by rough handling.

6-28. Refer to Fig. 6-33 on the next page. List the three functions of the 74HC4543 CMOS IC.

6-29. Refer to Fig. 6-33. With the driving signals shown, the LCD will display the decimal number _____ . The input must be the BCD number _____ .

6-30. Vacuum fluorescent displays can operate on 12 V, which makes them very compatible with (CMOS, TTL) ICs and automotive applications.

6-31. Refer to Fig. 6-34 on the next page. What will the VF seven-segment display read for each input pulse?

6-32. Refer to Fig. 6-34. List the approximate voltages at each of the seven plates and the grid of the VF display during pulse *b*.

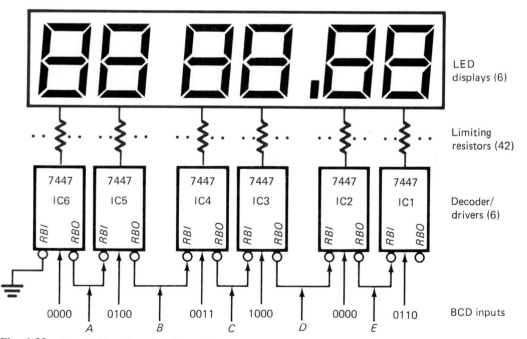

Fig. 6-32 **Ripple-blanking circuit problem.**

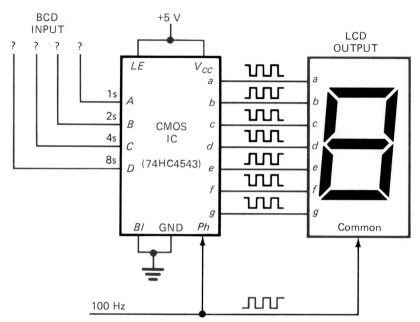

Fig. 6-33 Decoder/LCD display circuit problems 6-28 and 6-29.

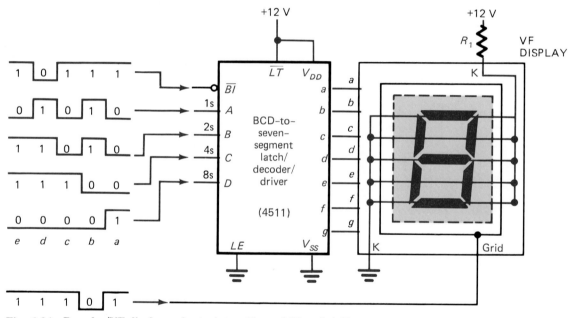

Fig. 6-34 Decoder/VF display pulse-train problems 6-31 and 6-32.

Answers to Self-Tests

1. 11101
2. 0010 1001
3. 8765
4. 0100 1011
5. 60
6. is not
7. Only one digit changes as you count in the Gray code.
8. alphanumeric
9. American Standard Code for Information Interchange
10. 101 0010
11. $
12. LOW, LOW
13. output D = HIGH
    output C = LOW
    output B = LOW
    output A = LOW

14. The invert bubble means that input 4 is an active LOW input; it is activated by a logical 0.
15. output D = LOW
    output C = HIGH
    output B = HIGH
    output A = HIGH
16. 5
17. vacuum fluorescent
18. light-emitting diode, liquid-crystal display
19. b, c, LED, 1
20. all segments, 8
21. 1. BCD-to-seven-segment
    2. 8421-BCD-to-decimal
    3. Excess-3-to-decimal
    4. Gray-code-to-decimal
    5. BCD-to-binary
    6. Binary-to-BCD
22. HIGH, LOW
23. LOW
24. leading zeros
25. pulse a = 9
    pulse b = 3
    pulse c = 5
    pulse d = 8
    pulse e = 2
    pulse f = blank display (not a BCD number)
    pulse g = 0
26. pulse a = a, b, c, f, g
    pulse b = a, b, c, d, g
    pulse c = a, c, d, f, g
    pulse d = a, b, c, d, e, f, g

    pulse e = a, b, d, e, g
    pulse f = blank display
    pulse g = a, b, c, d, e, f
27. black, silver
28. nematic
29. dc
30. very small amount
31. BCD, seven-segment
32. ac
33. pulse a = 1
    pulse b = 9
    pulse c = 6
    pulse d = 2
    pulse e = 4
34. b and c
35. blue-green (also blue or green)
36. none
37. part A = grid
    part B = cathode (heaters)
    part C = plates
38. a, c, d, f, g ; 5
39. CMOS, VF
40. limit current through cathodes to a safe level
41. pulse a = 0
    pulse b = 7
    pulse c = 8
    pulse d = 3
42. segment a–f = +12 V, segment g = GND
43. Use your senses to locate open or short circuits or ICs that are too hot.
44. short

CHAPTER 7

Flip-Flops

Engineers classify logic circuits into two groups. We have already worked with combination logic circuits using AND, OR, and NOT gates. The other group of circuits is classified as **sequential logic** *circuits. Sequential circuits involve timing and memory devices. The basic building block for combinational logic circuits is the logic gate. The basic building block for sequential logic circuits is the* **flip-flop** *(FF). This chapter covers several types of flip-flop circuits. In later chapters you will wire flip-flops together. Flip-flops are wired to form counters, shift registers, and various memory devices.*

7-1 THE R-S FLIP-FLOP

The logic symbol for the *R-S flip-flop* is drawn in Fig. 7-1. Notice that the R-S flip-flop has two inputs, labeled *S* and *R*. The two outputs are labeled Q and $\overline{Q}$. In flip-flops the outputs are always opposite, or complementary. In other words, if output $Q = 1$, then output $\overline{Q} = 0$, and so on. The letters "S" and "R" at the inputs of the R-S flip-flop are often referred to as the *set* and *reset* inputs.

The truth table in Table 7-1 details the operation of the R-S flip-flop. When the *S* and *R* inputs are both 0, both outputs go to a logical 1. This is called a *prohibited state* for the flip-flop and is not used. The second line of the truth table shows that when input *S* is 0 and *R* is 1, the *Q* output is set to logical 1. This is called the *set* condition. The third line shows that when input *R* is 0 and *S* is 1, output *Q* is reset (cleared) to 0. This is called the *reset* condition. Line 4 in the truth table shows both inputs (*R* and *S*) at 1. This is the idle or at rest condition and leaves Q and $\overline{Q}$ in their previous complementary states. This is called the *hold* condition.

From Table 7-1, it may be observed that it takes a logical 0 to activate the set (set *Q* to 1). It also takes a logical 0 to activate the reset, or clear (clear *Q* to 0). Because it takes a logical 0 to enable, or activate, the flip-flop, the

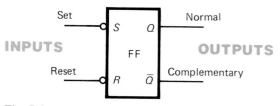

Fig. 7-1   Logic symbol for an R-S flip-flop.

Table 7-1   Truth table for R-S flip-flop

Mode of operation	INPUTS		OUTPUTS		
	S	R	Q	$\overline{Q}$	Effect on output Q
Prohibited	0	0	1	1	Prohibited - Do not use
Set	0	1	1	0	For setting Q to 1
Reset	1	0	0	1	For resetting Q to 0
Hold	1	1	Q	$\overline{Q}$	Depends on previous state

logic symbol in Fig. 7-1 has invert bubbles at the R and S inputs. These invert bubbles indicate that the set and reset inputs are activated by a logical 0.

R-S flip-flops can be purchased in an IC package, or they can be wired from logic gates, as shown in Fig. 7-2. The NAND gates in Fig. 7-2 form an R-S flip-flop. This NAND-gate R-S flip-flop operates according to the truth table in Table 7-1.

Many times *timing diagrams*, or *waveforms*, are given for sequential logic circuits. These diagrams show the voltage level and timing between inputs and outputs and are similar to what you would observe on an oscilloscope. The horizontal distance is *time*, and the vertical distance is *voltage*. Figure 7-3 shows the input waveforms (R, S) and the output waveforms (Q, $\overline{Q}$) for the R-S flip-flop. The bottom of the diagram lists the lines of the truth table from Table 7-1. The Q waveform shows the set and reset conditions of the output; the logic levels (0, 1) are on the right side of the waveforms. Waveform diagrams of the type

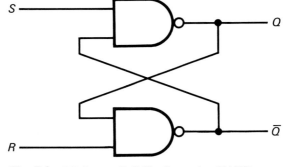

Fig. 7-2   Wiring an R-S flip-flop using NAND gates.

shown in Fig. 7-3 are very common when dealing with sequential logic circuits. Study this diagram to see what it tells you. The waveform diagram is really a type of truth table.

The R-S flip-flop is also called an *R-S latch* or a *set-reset flip-flop*. Do you know the logic symbol and truth table for the R-S flip-flop? Do you know the four modes of operation for the R-S flip-flop?

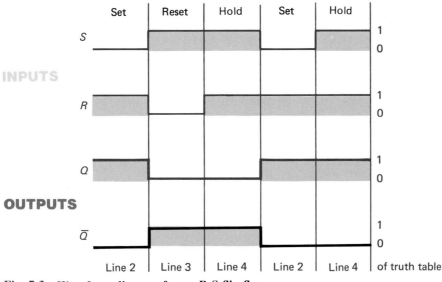

Fig. 7-3   Waveform diagram for an R-S flip-flop.

Self-Test

Answer the following questions.

1. The R-S flip-flop in Fig. 7-1 has active _____ (HIGH, LOW) inputs.
2. List the mode of operation of the R-S flip-flop for each input pulse shown in Fig. 7-4. Answer with the terms "set" and "reset," "hold," and "prohibited."
3. List the binary output at the normal output (Q) of the R-S flip-flop for each of the pulses shown in Fig. 7-4.

7-2 THE CLOCKED R-S FLIP-FLOP

The logic symbol for a *clocked R-S flip-flop* is shown in Fig. 7-5. Observe that it looks almost like an R-S flip-flop except that it has one extra input labeled *CLK* (for clock). Figure 7-6 diagrams the operation of the clocked R-S flip-flop. The *CLK* input is at the top of the diagram. Notice that the clock pulse (1) has no effect on output Q with inputs S and R in the 0 position. The flip-flop is in the *idle*, or *hold*, mode during clock pulse 1. At the preset S position, the S (set) input is moved to 1, but output Q is not yet set to 1. The rising edge of clock pulse 2 permits Q to go to 1. Pulses 3 and 4 have no effect on output Q. During pulse 3 the flip-flop is in its set mode, while during pulse 4 it is in its hold mode. Next, input R is preset to 1. On the rising edge of clock pulse 5 the Q output is reset (or cleared) to 0. The flip-flop is in the reset mode during both clock pulses 5 and 6. The flip-flop is in its hold mode during clock pulse 7; therefore, the normal output (Q) remains at 0.

Notice that the outputs of the clocked R-S flip-flop *change only on a clock pulse.* We say that this flip-flop operates *synchronously;* it operates *in step with* the clock. Synchronous operation is very important in calculators and computers, where each step must happen in a very exact order.

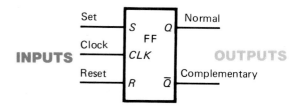

Fig. 7-5   Logic symbol for a clocked R-S flip-flop.

Another characteristic of the clocked R-S flip-flop is that once it is set or reset it stays that way even if you change some inputs. This is a *memory characteristic,* which is extremely valuable in many digital circuits. This characteristic is evident during the hold mode of operation. In the waveform diagram in Fig. 7-6, this flip-flop is in the hold mode during clock pulses 1, 4, and 7.

Figure 7-7(*a*) shows a truth table for the clocked R-S flip-flop. Notice that only the top three lines of the truth table are usable; the bottom line is prohibited and not used. Observe that the R and S inputs to the clocked R-S flip-flop are active HIGH inputs. That is, it takes a HIGH on input S while $R = 0$ to cause output Q to be set to 1.

Figure 7-7(*b*) shows a wiring diagram of a clocked R-S flip-flop. Notice that two NAND gates have been added to the inputs of the R-S flip-flop to add the clocked feature.

It is strongly suggested that you actually wire the R-S and clocked R-S flip-flop. Operating flip-flops in the laboratory will help you to better understand their operation.

Self-Test

Answer the following questions.

4. The set and reset inputs (S, R) of the clocked R-S flip-flop in Fig. 7-5 are active _____ (HIGH, LOW) inputs.
5. List the mode of operation of the clocked R-S flip-flop for each input pulse shown in

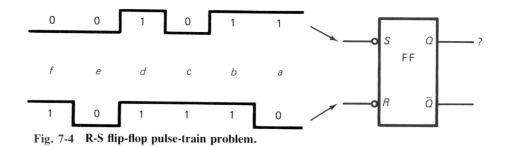

Fig. 7-4   R-S flip-flop pulse-train problem.

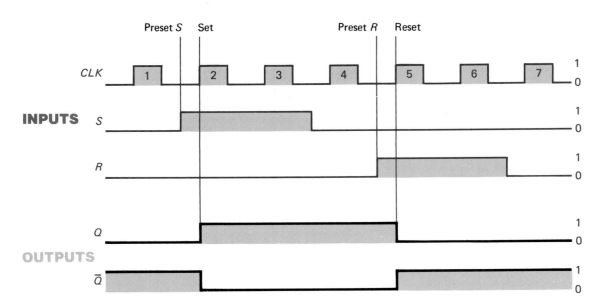

Fig. 7-6 **Waveform diagram for a clocked R-S flip-flop.**

TRUTH TABLE

Mode of operation	INPUTS			OUTPUTS		
	CLK	S	R	Q	$\overline{Q}$	Effect on output Q
Hold	⎍	0	0	No change	No change	No change
Reset	⎍	0	1	0	1	Reset or cleared to 0
Set	⎍	1	0	1	0	Set to 1
Prohibited	⎍	1	1	1	1	Prohibited— do not use

(a)

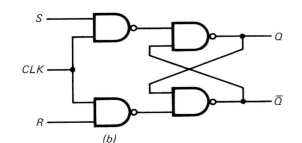

(b)

Fig. 7-7 (a) **Truth table for a clocked R-S flip-flop.** (b) **Wiring a clocked R-S flip-flop using NAND gates.**

Fig. 7-8 on the next page. Answer with the terms "set," "reset," "hold," and "prohibited."

6. List the binary output at the normal output (Q) of the clocked R-S flip-flop for each of the pulses shown in Fig. 7-8.

7-3 THE D FLIP-FLOP

The logic symbol for the D *flip-flop* is shown in Fig. 7-9(a). It has only one *data input* (D) and a clock input (CLK). The outputs are labeled Q and $\overline{Q}$. The D flip-flop is often called a

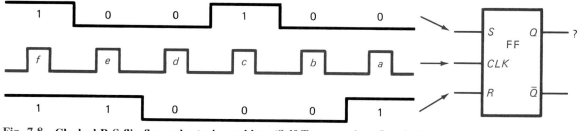

Fig. 7-8 Clocked R-S flip-flop pulse-train problem (Self-Test questions 5 and 6).

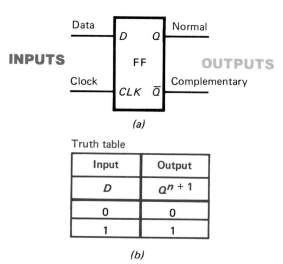

INPUTS   OUTPUTS

(a)

Truth table

Input	Output
D	Q^{n+1}
0	0
1	1

(b)

Fig. 7-9 D flip-flop. (a) Logic symbol. (b) Truth table.

delay flip-flop. The word "delay" describes what happens to the data, or information, at input D. The data (a 0 or 1) at input D is *delayed one clock pulse* from getting to output Q. A simplified truth table for the D flip-flop is shown in Fig. 7-9(b). Notice that output Q follows input D *after one clock pulse* (see Q^{n+1} column).

A D flip-flop may be formed from a clocked R-S flip-flop by adding an inverter, as shown in Fig. 7-10. More commonly you will use a D flip-flop contained in an IC. Figure 7-11(a)

shows a typical commercial D flip-flop. Two extra inputs [*PS* (preset) and *CLR* (clear)] have been added to the D flip-flop in Fig. 7-11(a). The *PS* input sets output Q to 1 when enabled by a logical 0. The *CLR* input clears output Q to 0 when enabled by a logical 0. The *PS* and *CLR* inputs will override the D and *CLK* inputs. The D and *CLK* inputs operate as they did in the D flip-flops in Fig. 7-9.

A more detailed truth table for the commercial 7474 TTL D flip-flop is shown in Fig. 7-11(b). Remember that the asynchronous (not synchronous) inputs (*PS* and *CLR*) override the synchronous inputs. The asynchronous inputs are in control of the D flip-flop in the first three lines of the truth table in Fig. 7-11(b). The synchronous inputs (D and *CLK*) are irrelevant as shown by the "X"'s on the truth table. The prohibited condition, line 3 on the truth table, should be avoided. With both asynchronous inputs disabled (*PS* = 1 and *CLR* = 1), the D flip-flop can be set and reset using the D and *CLK* inputs. The last two lines of the truth table use a clock pulse to transfer data from input D to output Q of the flip-flop. Being in step with the clock, this is called *synchronous operation*. Note that this flip-flop uses the LOW-to-HIGH transition of the clock pulse to transfer data from input D to output Q.

D flip-flops are wired together to form *shift registers* and *storage registers*. These registers

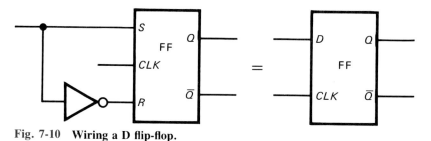

Fig. 7-10 Wiring a D flip-flop.

136   CHAPTER 7   FLIP-FLOPS

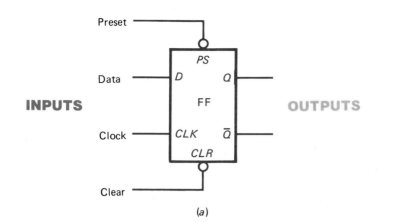

(a)

TRUTH TABLE

| Mode of operation | INPUTS | | | | OUTPUTS | |
| | Asynchronous | | Synchronous | | | |
	PS	CLR	CLK	D	Q	$\bar{Q}$
Asynchronous set	0	1	X	X	1	0
Asynchronous reset	1	0	X	X	0	1
Prohibited	0	0	X	X	1	1
Set	1	1	↑	1	1	0
Reset	1	1	↑	0	0	1

0 = LOW
1 = HIGH
X = Irrelevant
↑ = LOW–to–HIGH transition of clock pulse

(b)

Fig. 7-11 (a) Logic symbol for commercial D flip-flop. (b) Truth table for 7474 D flip-flop.

are widely used in digital systems. Remember that the D flip-flop *delays* data from reaching output Q one clock pulse and is called a delay flip-flop. D flip-flops are available in both TTL and CMOS ICs. Typical CMOS D flip-flops might be the 74HC74, 74HC273, 4013, or 40174 ICs.

Self-Test

Answer the following questions.

7. List the mode of operation of the 7474 D flip-flop for each input pulse shown in Fig. 7-12 on the next page. Answer with the terms "asynchronous set," "asynchronous reset," "prohibited," "set," and "reset."

8. List the binary output at the normal output (Q) of the D flip-flop for each of the pulses shown in Fig. 7-12.

7-4 THE J-K FLIP-FLOP

The *J-K flip-flop* is probably the most widely used and universal flip-flop, having the features of all the other types of flip-flops. The logic symbol for the J-K flip-flop is illustrated in Fig. 7-13(a). The inputs labeled *J* and *K* are the data inputs. The input labeled *CLK* is the clock input. Outputs Q and $\bar{Q}$ are the usual normal and complementary outputs on a flip-flop. A truth table for the J-K flip-flop is shown in Fig. 7-13(b). When the *J* and *K* inputs are both 0, the flip-flop is in the *hold* mode. In the hold mode the data inputs have no effect on the outputs. The outputs "hold" the last data present.

Lines 2 and 3 of the truth table show the reset and set conditions for the Q output. Line 4 illustrates the useful *toggle* position of the J-K flip-flop. When both data inputs *J* and *K* are at 1, repeated clock pulses cause the output to turn off-on-off-on-off-on, and so on. This off-

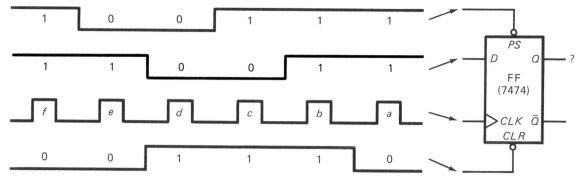

Fig. 7-12 D flip-flop pulse-train problems (Self-Test questions 7 and 8).

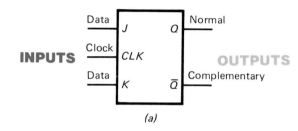

(a)

TRUTH TABLE

Mode of operation	INPUTS			OUTPUTS		
	CLK	J	K	Q	Q̄	Effect on output Q
Hold	⊓	0	0	No change		No change — disable
Reset	⊓	0	1	0	1	Reset or cleared to 0
Set	⊓	1	0	1	0	Set to 1
Toggle	⊓	1	1	Toggle		Changes to opposite state

(b)

Fig. 7-13 J-K flip-flop. (a) Logic symbol. (b) Truth table.

on action is like a toggle switch and is called *toggling*.

The logic symbol for the commercial 7476 TTL J-K flip-flop is shown in Fig. 7-14(a). Added to the symbol are two asynchronous inputs (preset and clear). The synchronous inputs are the J and K data and clock inputs. The customary normal (Q) and complementary (Q) outputs are also shown. A detailed truth table for the commercial 7476 J-K flip-flop is drawn in Fig. 7-14(b). Recall that asynchronous inputs (such as PS and CLR) override synchronous inputs. The asynchronous inputs are activated in the first three lines of the truth table. The synchronous inputs are irrelevant (overriden) in the first three lines in Fig. 7-14(b); therefore,

an "X" is placed under the J, K, and CLK inputs for these rows. The prohibited state occurs when both asynchronous inputs are activated at the same time. The prohibited state should be avoided.

When both asynchronous inputs (PS and CLR) are disabled with a 1, the synchronous inputs can be activated. The bottom four lines of the truth table in Fig. 7-14(b) detail the *hold*, *reset*, *set*, and *toggle* modes of operation for the 7476 J-K flip-flop. Note that the 7476 J-K flip-flop uses the entire pulse to transfer data from the J and K data inputs to the Q and Q̄ outputs.

J-K flip-flops are widely used in many digital circuits. You will use the J-K flip-flop es-

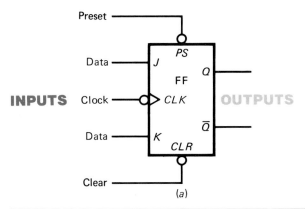

Mode of operation	INPUTS					OUTPUTS	
	Asynchronous		Synchronous				
	PS	CLR	CLK	J	K	Q	$\overline{Q}$
Asynchronous set	0	1	X	X	X	1	0
Asynchronous reset	1	0	X	X	X	0	1
Prohibited	0	0	X	X	X	1	1
Hold	1	1	⎍	0	0	No change	
Reset	1	1	⎍	0	1	0	1
Set	1	1	⎍	1	0	1	0
Toggle	1	1	⎍	1	1	Opposite state	

0 = LOW
1 = HIGH
X = Irrelevant
⎍ = Positive clock pulse

(b)

Fig. 7-14 (a) Logic symbol for commercial J-K flip-flop. (b) Truth table for 7476 J-K flip-flop.

pecially in *counters.* Counters are found in almost every digital system. J-K flip-flops are available in both TTL and CMOS. Typical CMOS J-K flip-flops are the 74HC76 and 4027 ICs.

Self-Test

Answer the following questions.

9. List the mode of operation of the 7476 J-K flip-flop for each input pulse shown in Fig. 7-15 on the next page. Answer with the terms "asynchronous set," "asynchronous reset," "prohibited," "hold," "reset," "set," and "toggle."
10. List the binary output at the normal output (*Q*) of the J-K flip-flop after each of the pulses shown in Fig. 7-15.

7-5 IC LATCHES

Consider the block diagram of the digital system in Fig. 7-16(*a*). Press and hold the decimal number 7 on the keyboard. A 7 will be observed on the seven-segment display. Release the 7 on the keyboard and the 7 disappears from the display. It is obvious that a *memory device* is needed to hold the BCD code for 7 at the inputs to the decoder. A device that serves as a temporary buffer memory is called a *latch.* A four-bit latch has been added to the system in Fig. 7-16(*b*). Now when the decimal number 7 on the keyboard is *pressed and released* the seven-segment display continues to show a 7.

The term "latch" refers to a digital storage device. The D flip-flop is a good example of a device used to latch data. However, other types of flip-flops are also used for the latching function.

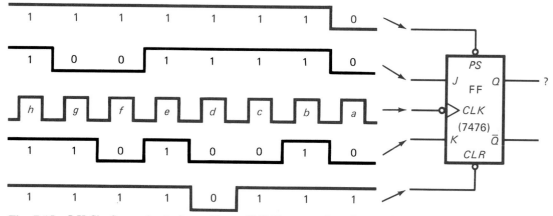

Fig. 7-15 J-K flip-flop pulse-train problems (Self-Test questions 9 and 10).

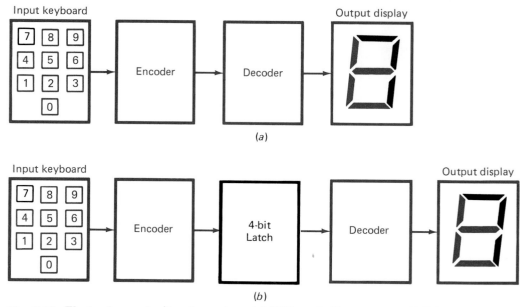

Fig. 7-16 Electronic encoder/decoder system. (*a*) Without buffer memory. (*b*) With buffer memory added.

Manufacturers have developed many latches in IC form. The logic diagram for the *7475 TTL four-bit transparent latch* is shown in Fig. 7-17(*a*). This unit has four D flip-flops enclosed in a single IC package. The D_0 data input and the normal Q_0 and complementary $\overline{Q}_0$ outputs form the first D flip-flop. The enable input (E_{0-1}) is similar to the clock input on the D flip-flop and activates both the D_0 and the D_1 flip-flop inside the 7475 latch IC. Data at D_1 is transferred to the normal Q_1 output, whereas its complement arrives at $\overline{Q}_1$.

A simplified truth table for the 7475 latch IC is shown in Fig. 7-17(*b*). If the enable input is at a logical 1, data is transferred, without a separate clock pulse, from the D input to the Q and $\overline{Q}$ outputs. As an example, if $E_{0-1} = 1$

and $D_1 = 1$, then without a clock pulse output Q_1 would be set to 1 while $\overline{Q}_1$ would be reset to 0. In the *data-enabled* mode of operation the Q outputs follow their respective D inputs on the 7475 latch.

Consider the last line of the truth table in Fig. 7-17(*b*). When the enable input drops to 0, the 7475 IC enters the *data-latched* mode. The data that was at Q remains the same even if the D inputs change. The data is said to be latched. The 7475 IC is called a *transparent* latch because when the enable input is HIGH, the normal outputs follow the data at the D inputs. Note that the D_0 and D_1 flip-flops in the 7475 IC are controlled by the E_{0-1} enable input whereas the E_{2-3} input controls the D_2 and D_3 flip-flops.

7475 4-bit transparent latch

Counters

Shift registers

Delay unit

Frequency dividers

Triggering flip-flops

Synchronous flip-flops

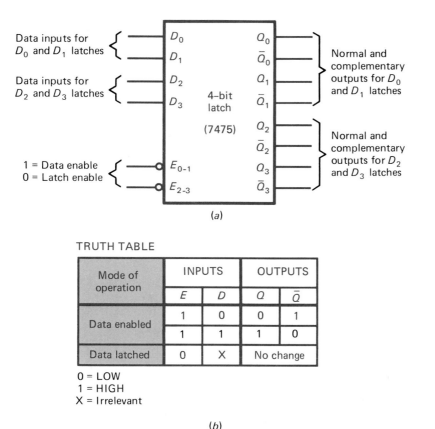

TRUTH TABLE

Mode of operation	INPUTS		OUTPUTS	
	E	D	Q	$\bar{Q}$
Data enabled	1	0	0	1
	1	1	1	0
Data latched	0	X	No change	

0 = LOW
1 = HIGH
X = Irrelevant

Fig. 7-17 (a) **Logic symbol for commercial 7475 4-bit transparent latch.** (b) **Truth table for 7475 D latch.**

One use of a flip-flop is to hold, or latch, data. When used for this purpose, the flip-flop is called a latch. Flip-flops have many other uses, including counters, shift registers, delay units, and frequency dividers.

Latches are available in all logic families. Several typical CMOS latches are the 4042, 4099, 74HC75, and 74HC373 ICs. Latches are sometimes built into other ICs such as the 4511 and 4543 BCD-to-seven segment latch/decoder/driver chips illustrated in Chap. 6.

One of the primary advantages of digital over analog circuitry is the availability of easy-to-use memory devices. The latch is the most fundamental memory device used in digital electronics. Almost all digital equipment contains simple memory devices called latches.

Self-Test

Supply the missing word in each statement.

11. When the 7475 latch IC is in its data-enabled mode of operation, the _____ outputs follow their respective D inputs.

12. A _____ (HIGH, LOW) at the enable inputs places the 7475 latch IC in the data-latched mode of operation.

13. In the data-latched mode, a change at any of the D inputs to the 7475 latch IC has _____ (an immediate effect on their respective outputs; no effect on the outputs).

14. When a flip-flop is used to temporarily hold data, it is sometimes called _____ .

7-6 TRIGGERING FLIP-FLOPS

We have classified flip-flops as synchronous or asynchronous in their operation. Synchronous flip-flops are all those that have a clock input. We found that the clocked R-S, the D, and the J-K flip-flop operate in step with the clock.

When using manufacturers' data manuals you will notice that many synchronous flip-flops are also classified as either *edge-triggered* or *master/slave*. Figure 7-18 on the next page shows two edge-triggered flip-flops in the toggle position. On clock pulse 1 the positive edge (positive-going edge) of the pulse is identified.

Positive-edge-
triggered flip-flop

Negative-edge-
triggered flip-flop

Flip-flop triggering

J-K master/slave
flip-flop

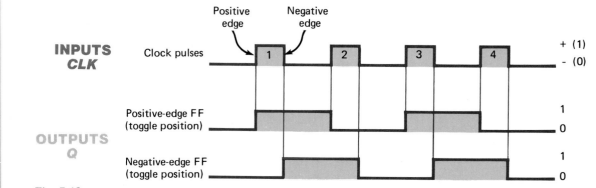

Fig. 7-18   Waveforms for positive- and negative-edge-triggered flip-flops.

The second waveform shows how the positive-edge-triggered flip-flop toggles each time a positive-going pulse comes along (see pulses 1 to 4). On pulse 1 in Fig. 7-18 the negative edge (negative-going edge) of the pulse is also labeled. The bottom waveform shows how the negative-edge-triggered flip-flop toggles. Notice that it changes state, or toggles, each time a negative-going pulse comes along (see pulses 1 to 4). Especially notice the difference in timing between the positive- and negative-edge-triggered flip-flops. This triggering time difference is quite important for some applications.

It is common to show the type of triggering on the flip-flop. The logic symbol for a D flip-flop with positive-edge triggering is shown in Fig. 7-19(a). Note the use of the small > inside the flip-flop near the clock input. This > symbol says data is transferred to the output on the edge of the pulse. A logic symbol for a D flip-flop using negative-edge triggering is shown in Fig. 7-19(b). The added invert bubble at the clock input shows that triggering occurs on the negative-going edge of the clock pulse. Finally, a typical D latch symbol is shown in Fig. 7-19(c). Note the lack of a > symbol next to the enable (similar to a clock) input. This means that this unit is not considered an edge-triggered unit. Like the R-S flip-flop, the D latch is considered asynchronous. Recall that the D latch normal output (Q) follows its input (D) when the enable (E) input is HIGH. The data is latched when the enable input drops to LOW. Several manufacturers label the enable input with a "G" on the D latch.

Another class of flip-flop triggering is the master/slave type. The J-K master/slave flip-flop uses the entire pulse (positive edge and negative edge) to trigger the flip-flop. Figure 7-20 shows the triggering of a master/slave flip-

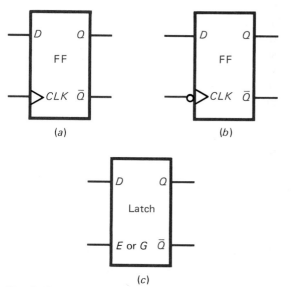

Fig. 7-19   (a) Logic symbol for positive-edge-triggered D flip-flop. (b) Logic symbol for negative-edge-triggered D flip-flop. (c) Logic symbol for D latch.

flop. Pulse 1 shows four positions (a to d) on the waveform. The following sequence of operation takes place in the master/slave flip-flop at each point on the clock pulse:

- Point a: leading edge—isolate input from output
- Point b: leading edge—enter information from J and K inputs
- Point c: trailing edge—disable J and K inputs
- Point d: trailing edge—transfer information from input to output

A very interesting characteristic of the master/slave flip-flop is shown on pulse 2, Fig. 7-20. Notice that at the beginning of pulse 2 the outputs are disabled. For a very brief moment the J and K inputs are moved to the tog-

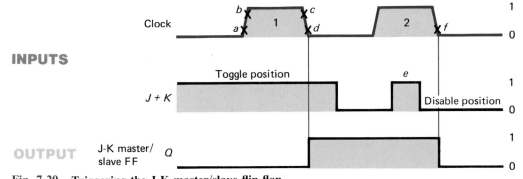

Fig. 7-20 Triggering the J-K master/slave flip-flop.

gle positions (see point e) and then disabled. The J-K master/slave flip-flop "remembers" that the J and K inputs were in the toggle positions, and it toggles at point f on the waveform diagram. This memory characteristic happens only while the clock pulse is high (at logical 1).

Self-Test

Supply the missing word in each statement.

15. A positive-edge-triggered flip-flop changes state on the _____ transition of the clock pulse.
16. A negative-edge-triggered flip-flop changes state on the _____ transition of the clock pulse.
17. The " $>$ " near the clock input inside a flip-flop logic symbol means _____ .
18. A(n) _____ J-K flip-flop uses both the positive and the negative edge of the clock pulse for triggering.

7-7 SCHMITT TRIGGER

Digital circuits prefer waveforms with fast rise and fall times. The waveform on the right side of the inverter symbol in Fig. 7-21 is an example of a good digital signal. The square wave's L-to-H and H-to-L edges are vertical. This means that the rise and fall times are very fast (almost instantaneous).

The waveform to the left of the inverter symbol in Fig. 7-21 has very slow rise and fall times. The poor waveform on the left in Fig. 7-21 might lead to unreliable operation if fed directly into counters, gates, or other digital circuitry. In this example, a *Schmitt trigger* inverter is being used to "square up" the input signal and make it more useful. The Schmitt trigger in Fig. 7-21 is reshaping the waveform. This is called *signal conditioning*. Schmitt triggers are widely used in signal conditioning.

A voltage profile of a typical TTL inverter (7404 IC) is reproduced in Fig. 7-22(a) on the next page. Of special interest is the *switching threshold* of the 7404 IC. The switching threshold may vary from chip to chip, but it is always in the undefined region. Figure 7-22(a) shows that a typical 7404 IC has a switching threshold of +1.2 V. In other words, when the voltage rises to +1.2 V, the output changes from HIGH to LOW. However, if the voltage drops below +1.2 V, the output switches from LOW to HIGH. Most regular gates have a single switching threshold voltage whether the input voltage is rising (L to H) or falling (H to L).

A voltage profile for a 7414 Schmitt trigger inverter TTL IC is sketched in Fig. 7-22(b). Note that the *switching threshold is different* for positive-going (V +) and negative-going (V −) voltages. The voltage profile for the 7414 IC shows that the switching threshold is 1.7 V for positive-going (V +) input voltage. However, the switching threshold is 0.9 V for a negative-going (V −) input voltage. The differ-

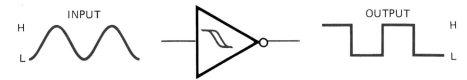

Fig. 7-21 Schmitt trigger used for wave shaping.

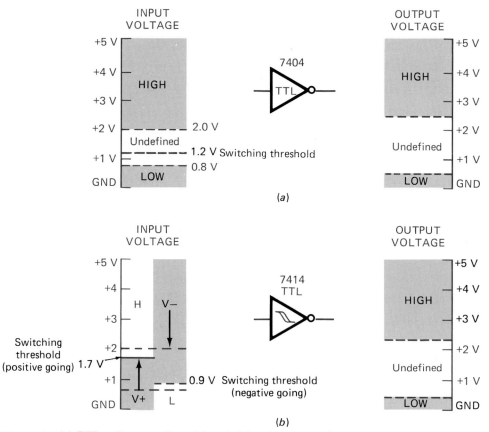

Fig. 7-22 (*a*) **TTL voltage profiles with switching threshold.** (*b*) **Voltage profiles for 7414 TTL Schmitt trigger IC showing switching thresholds.**

ence between these switching thresholds (1.7 V and 0.9 V) is called *hysteresis*. Hysteresis provides for excellent noise immunity and helps the Schmitt trigger square up waveforms with slow rise and fall times.

Schmitt triggers are also available in CMOS. These include the 40106, 4093, and 74HC14 ICs.

One of the characteristics of a bistable multivibrator (or flip-flop) is that its outputs are either HIGH or LOW. When changing states (H to L or L to H), they do so rapidly without the outputs being in the undefined region. This "snap action" of the output is also characteristic of Schmitt triggers.

Self-Test

Answer the following questions.

19. The _____ is a good device for squaring up a waveform with slow rise and fall times.
20. Draw the schematic symbol for a Schmitt trigger inverter.
21. A Schmitt trigger is said to have _____ because its switching thresholds are different for positive-going and negative-going inputs.
22. Schmitt triggers are commonly used for _____ (memory, signal conditioning).

SUMMARY

1. Logic circuits are classified as combinational or sequential. Combinational logic circuits use AND, OR, and NOT gates. Sequential logic circuits use flip-flops and involve a memory characteristic.

2. Flip-flops are wired together to form counters, registers, and memory devices.
3. Flip-flop outputs are always opposite, or complementary.
4. Table 7-2 summarizes some basic flip-flops.

Table 7-2 Summary of Basic Flip-Flops

Circuit	Logic Symbol	Truth table	Remarks:
R-S flip-flop	S Q / FF / R Q̄	S R Q 0 0 prohibited 0 1 1 set 1 0 0 reset 1 1 hold	R-S latch Set-reset flip-flop (asynchronous)
Clocked R-S flip-flop	S Q / FF / CLK / R Q̄	CLK S R Q ⎍ 0 0 hold ⎍ 0 1 0 reset ⎍ 1 0 1 set ⎍ 1 1 prohibited	(synchronous)
D flip-flop	D Q / FF / CLK Q̄	CLK D Q ⎍ 0 0 ⎍ 1 1	Delay flip-flop Data flip-flop (synchronous)
J-K flip-flop	J Q / FF / CLK / K Q̄	CLK J K Q ⎍ 0 0 hold ⎍ 0 1 0 ⎍ 1 0 1 ⎍ 1 1 toggle	Most universal FF (synchronous)

5. Waveform (timing) diagrams are used to describe the operation of sequential devices.

6. Flip-flops can be edge-triggered or master/slave types.

7. Special flip-flops called latches are widely used in most digital circuits as temporary buffer memories.

8. Schmitt triggers are special gates that are used for signal conditioning.

CHAPTER REVIEW QUESTIONS

7-1. Logic _____ are the basic building blocks of combinational logic circuits; the basic building blocks of sequential circuits are devices called _____ .

7-2. List one type of asynchronous and three types of synchronous flip-flops (mark the synchronous types).

7-3. List two other names sometimes given to an R-S flip-flop.

7-4. Draw a logic symbol for the following flip-flops:
   a. J-K
   b. D
   c. Clocked R-S
   d. R-S
7-5. Draw a truth table for the following flip-flops:
   a. J-K
   b. D
   c. Clocked R-S
   d. R-S
7-6. Draw the logic symbol for the following flip-flops that have *PS* and *CLR* asynchronous inputs:
   a. D
   b. J-K
7-7. Draw the truth table for the asynchronous inputs only (*PS* and *CLR*) on the following flip-flops:
   a. D
   b. J-K
7-8. If both synchronous and the asynchronous inputs on a J-K flip-flop are activated, which input will control the output?
7-9. When we say the flip-flop is in the set condition, we mean output _____ is at a logical _____ .
7-10. When we say the flip-flop is in the reset, or clear, condition, we mean output _____ is at a logical _____ .
7-11. On a timing, or waveform, diagram the horizontal distance stands for _____ and the vertical distance stands for _____ .
7-12. Refer to Fig. 7-3. Notice that line 4 is listed two times across the bottom. Why does output $Q = 0$ once and $\overline{Q} = 1$ on the right side when inputs R and S are both 1 in each case?
7-13. Refer to Fig. 7-6. This waveform diagram is for a _____ flip-flop. This flip-flop is _____ -edge triggered.
7-14. List two types of edge-triggered flip-flops.
7-15. The "D" in "flip-flop" stands for _____ , or data.
7-16. D flip-flops are widely used as temporary memories called _____ .
7-17. If a flip-flop is in its toggle position, what will the output act like upon repeated clock pulses?
7-18. Explain how a J-K master/slave flip-flop is triggered.
7-19. Identify these acronyms used on flip-flops:
   a. *CLK*
   b. *CLR*
   c. *D/*
   d. *FF*
   e. *PS*
   f. *R*
   g. *S*
7-20. Explain how a J-K master/slave flip-flop can still toggle even when *J* and *K* inputs are in the disable condition.
7-21. Give a descriptive name for the following TTL ICs:
   a. 7474
   b. 7475
   c. 7476
7-22. The 7474 IC is a _____ -edge-triggered unit.
7-23. List the modes of operation of the 7474 IC.
7-24. List the mode of operation of the 7476 J-K flip-flop for each input pulse shown in Fig. 7-23.

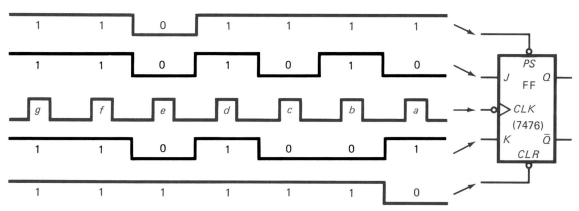

Fig. 7-23 **Pulse-train problem.**

7-25. List the binary outputs at the normal output (Q) of the J-K flip-flop for each of the pulses shown in Fig. 7-23.

7-26. List the mode of operation of the 7475 four-bit latch for each time period (t_1 through t_7) shown in Fig. 7-24.

7-27. List the binary output (four-bit) at the output indicators of the 7475 four-bit latch for each time period (t_1 through t_7) shown in Fig. 7-24.

7-28. Refer to Fig. 7-25. The output waveform on the right of the logic symbol will be a(n) _____ (sine, square) wave.

7-29. The inverter in Fig. 7-25 is being used as a signal _____ (conditioner, multiplexer) in this circuit.

7-30. The logic symbol in Fig. 7-25 is for a _____ (two words) inverter IC.

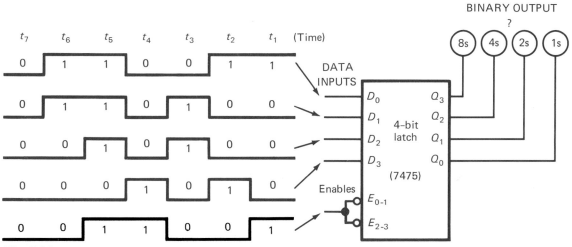

Fig. 7-24 **Pulse-train problem.**

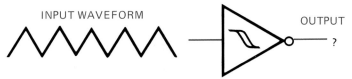

Fig. 7-25 **Sample problem.**

Answers to Self-Tests

1. LOW
2. pulse a = reset
   pulse b = hold
   pulse c = set
   pulse d = hold
   pulse e = prohibited
   pulse f = set
3. pulse a = 0
   pulse b = 0
   pulse c = 1
   pulse d = 1
   pulse e = 1
   pulse f = 1
4. HIGH
5. pulse a = reset
   pulse b = hold
   pulse c = set
   pulse d = hold
   pulse e = reset
   pulse f = prohibited
6. pulse a = 0
   pulse b = 0
   pulse c = 1
   pulse d = 1
   pulse e = 0
   pulse f = 1
7. pulse a = asynchronous reset (or clear)
   pulse b = set
   pulse c = reset
   pulse d = asynchronous set (or preset)
   pulse e = prohibited
   pulse f = asynchronous reset (or clear)
8. pulse a = 0
   pulse b = 1
   pulse c = 0
   pulse d = 1
   pulse e = 1
   pulse f = 0
9. pulse a = asynchronous set (or preset)
   pulse b = toggle
   pulse c = set
   pulse d = asynchronous reset (or clear)
   pulse e = toggle
   pulse f = hold
   pulse g = reset
   pulse h = toggle
10. pulse a = 1
    pulse b = 0
    pulse c = 1
    pulse d = 0
    pulse e = 1
    pulse f = 1
    pulse g = 0
    pulse h = 1
11. Q (normal)
12. LOW
13. no effect on the outputs
14. latch
15. LOW-to-HIGH
16. HIGH-to-LOW
17. edge-triggering
18. master/slave
19. Schmitt trigger
20. See figure below

21. hysteresis
22. signal conditioning

CHAPTER 8

Counters

Almost any complex digital system contains several counters. A counter's job is the obvious one of counting events or periods of time or putting events into sequence. Counters also do some not so obvious jobs: dividing frequency, addressing, and serving as memory units. This chapter discusses several types of counters and their uses. Flip-flops are wired together to form circuits that count. Because of the wide use of counters, manufacturers also make self-contained counters in IC form. Many counters are available in all TTL and CMOS families.

8-1 RIPPLE COUNTERS

Counting in binary and decimal is illustrated in Fig. 8-1. With four binary places (D, C, B, and A) we can count from 0000 to 1111 (0 to 15 in decimal). Notice that column A is the 1s binary place, or least significant digit (LSD). The term "least significant bit" (LSB) is usually used. Column D is the 8s binary place, or most significant digit (MSD). The term "most significant bit" (MSB) is usually used. Notice that the 1s column changes state the most often. If we design a counter to count from binary 0000 to 1111, we need a device that has 16 different output states: a *modulo* (mod)-*16 counter*. The *modulus* of a counter is the number of different states the counter must go through to complete its counting cycle.

A mod-16 counter using four J-K flip-flops is diagramed in Fig. 8-2(*a*) on the next page. Each J-K flip-flop is in its toggle position (*J* and *K* both at 1). Assume the outputs are cleared to 0000. As clock pulse 1 arrives at the clock (*CLK*) input of flip-flop 1 (FF 1), it toggles (on the negative edge) and the display shows 0001.

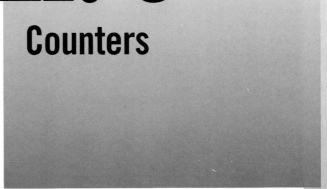

BINARY COUNTING				DECIMAL COUNTING
D	C	B	A	
8s	4s	2s	1s	
0	0	0	0	0
0	0	0	1	1
0	0	1	0	2
0	0	1	1	3
0	1	0	0	4
0	1	0	1	5
0	1	1	0	6
0	1	1	1	7
1	0	0	0	8
1	0	0	1	9
1	0	1	0	10
1	0	1	1	11
1	1	0	0	12
1	1	0	1	13
1	1	1	0	14
1	1	1	1	15

Fig. 8-1   Counting sequence for an electronic counter.

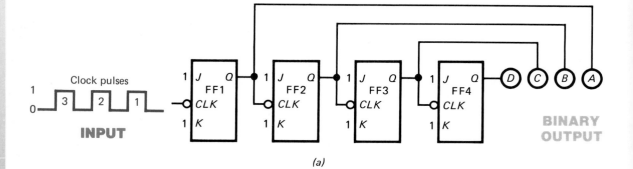

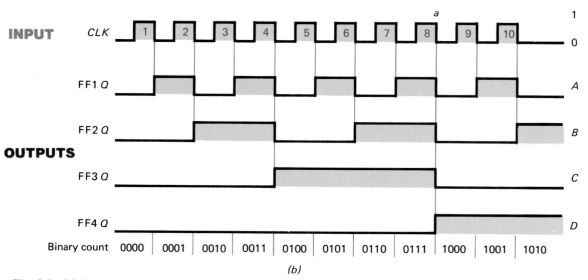

Fig. 8-2 Mod-16 counter. (*a*) **Logic diagram.** (*b*) **Waveform diagram.**

Clock pulse 2 causes FF 1 to toggle again, returning output *Q* to 0, which causes FF 2 to toggle to 1. The count on the display now reads 0010. The counting continues, with each flip-flop output triggering the next flip-flop on its negative-going pulse. Look back at Fig. 8-1 and see that column *A* (1s column) must change state on every count. This means that FF 1 in Fig. 8-2(*a*) must toggle each pulse. FF 2 must toggle only half as often as FF 1, as seen from column *B* in Fig. 8-1. Each more significant bit in Fig. 8-1 toggles less often.

The counting of the mod-16 counter is shown up to a count of decimal 10 (binary 1010) by waveforms in Fig 8-2(*b*). The *CLK* input is shown on the top line. The state of each flip-flop (FF 1, FF 2, FF 3, FF 4) is shown on the waveforms below. The binary count is shown across the bottom of the diagram. Especially note the vertical lines on Fig. 8-2(*b*); these lines show that the clock triggers only FF 1. FF 1 triggers FF 2, FF 2 triggers FF 3, and so on.

Because one flip-flop affects the next one, it takes some time to toggle all the flip-flops. For instance, at point *a* on pulse 8, Fig. 8-2(*b*), notice that the clock triggers FF 1, causing it to go to 0. This in turn causes FF 2 to toggle from 1 to 0. This in turn causes FF 3 to toggle from 1 to 0. As output *Q* of FF 3 reaches 0 it triggers FF 4, which toggles from 0 to 1. We see that the changing of states is a chain reaction that *ripples* through the counter. For this reason this counter is called a *ripple counter.*

The counter we studied in Fig. 8-2 could be described as a ripple counter, a mod-16 counter, a 4-bit counter, or an asynchronous counter. All these names describe something about the counter. The ripple and asynchronous labels mean that all the flip-flops do not trigger at one time. The mod-16 description comes from the number of states the counter goes through. The 4-bit label tells how many binary places there are at the output of the counter.

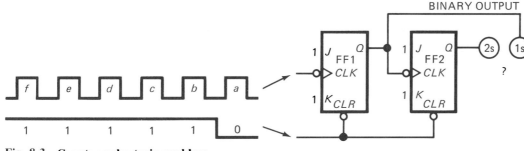

Fig. 8-3   Counter pulse-train problem.

Self-Test

Answer the following questions.

1. The unit in Fig. 8-3 is a _____ -bit ripple counter.
2. The unit in Fig. 8-3 is a mod- _____ up counter.
3. Each J-K flip-flop in Fig. 8-3 is in the _____ (hold, reset, set, toggle) mode because inputs *J* and *K* are both HIGH.
4. List the binary output after each of the six input pulses shown in Fig. 8-3.

8-2 MOD-10 RIPPLE COUNTERS

The counting sequence for a mod-10 counter is from 0000 to 1001 (0 to 9 in decimal). This is down to the heavy line in Fig. 8-1. This mod-10 counter, then, has four place values: 8s, 4s, 2s, and 1s. This takes four flip-flops connected as a ripple counter in Fig. 8-4. We must *add* a NAND gate to the ripple counter to clear all the flip-flops back to zero *immediately after* the 1001 (9) count. The *trick* is to look at Fig. 8-1

and determine what the *next count will be after 1001*. You will find it is 1010 (decimal 10). You must feed the two 1s in the 1010 into a NAND gate as shown in Fig. 8-4. The NAND gate then clears the flip-flop back to 0000. The counter then starts its count from 0000 up to 1001 again. We say we are using the NAND gate to reset the counter to 0000. By using a NAND gate in this manner we can make several other modulo counters. Fig. 8-4 illustrates a mod-10 ripple counter. This type of counter might also be called a *decade* (meaning 10) *counter.*

Ripple counters can be constructed from individual flip-flops. Manufacturers also produce ICs with all four flip-flops inside a single package. Some IC counters even contain the reset NAND gate, such as the one you used in Fig. 8-4.

Self-Test

Answer the following questions.

5. Refer to Fig. 8-4. This is the logic diagram for a mod-10 _____ (ripple, synchronous) counter. Because it has 10 states

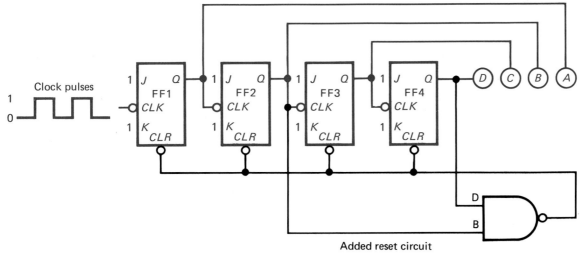

Fig. 8-4   Logic diagram of a mod-10 ripple counter.

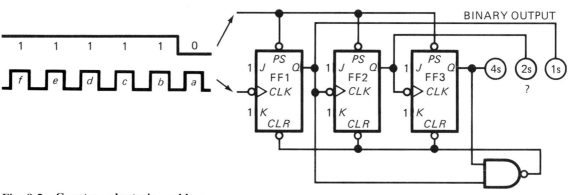

Fig. 8-5 Counter pulse-train problem.

(counts from 0 through 9), it is also called a _____ counter.

6. The circuit in Fig. 8-5 is a _____ (ripple, synchronous) mod-_____ counter.

7. List the binary output after each of the six input pulses shown in Fig. 8-5.

8-3 SYNCHRONOUS COUNTERS

The ripple counters we have studied are asynchronous counters. Each flip-flop does not trigger exactly in step with the clock pulse. For some high-frequency operations it is necessary to have all stages of the counter trigger together. There is such a counter: _a synchronous counter._

A rather complicated-looking synchronous counter is shown in Fig. 8-6(_a_). This logic diagram is for a 3-bit (mod-8) counter. First notice the _CLK_ connections. The clock is connected directly to the _CLK_ input of each flip-flop. We say that the _CLK_ inputs are connected in _parallel_. Figure 8-6(_b_) gives the counting sequence this counter goes through. Column _A_ is the binary 1s column, and FF 1 does the counting for this column. Column _B_ is the binary 2s column, and FF 2 counts this column. Column _C_ is the binary 4s column, and FF 3 counts this column.

Let us go through the counting sequence of this mod-8 counter by referring to Fig. 8-6(_a_) and (_b_):

Pulse 1—row 2
 Circuit action: Each flip-flop is pulsed by clock.
  Only FF 1 can toggle because it is the only one with 1s applied to both _J_ and _K_ inputs.
  FF 1 goes from 0 to 1.
 Output result: 001 (decimal 1).

Pulse 2—row 3
 Circuit action: Each flip-flop is pulsed.
  Two flip-flops toggle because they have 1s applied to both _J_ and _K_ inputs.
  FF 1 and FF 2 both toggle.
  FF 1 goes from 1 to 0.
  FF 2 goes from 0 to 1.
 Output result: 010 (decimal 2).

Pulse 3—row 4
 Circuit action: Each flip-flop is pulsed.
  Only one flip-flop toggles.
  FF 1 toggles from 0 to 1.
 Output result: 011 (decimal 3).

Pulse 4—row 5
 Circuit Action: Each flip-flop is pulsed.
  All flip-flops toggle to opposite state.
  FF 1 goes from 1 to 0.
  FF 2 goes from 1 to 0.
  FF 3 goes from 0 to 1.
 Output result: 100 (decimal 4).

Pulse 5—row 6
 Circuit action: Each flip-flop is pulsed.
  Only one flip-flop toggles.
  FF 1 goes from 0 to 1.
 Output result: 101 (decimal 5).

Pulse 6—row 7
 Circuit action: Each flip-flop is pulsed.
  Two flip-flops toggle.
  FF 1 goes from 1 to 0.
  FF 2 goes from 0 to 1.
 Output result: 110 (decimal 6).

Pulse 7—row 8
 Circuit action: Each flip-flop is pulsed.
  Only one flip-flop toggles.
  FF 1 goes from 0 to 1.
 Output result: 111 (decimal 7).

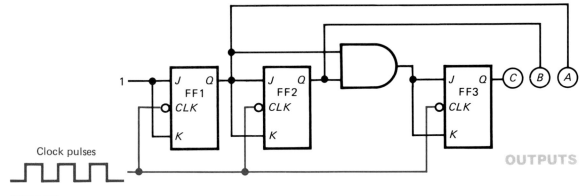

1

FF1 · J · Q · CLK · K

FF2 · J · Q · CLK · K

FF3 · J · Q · CLK · K

C · B · A

Clock pulses

INPUT

OUTPUTS

(a)

ROW	NUMBER OF CLOCK PULSES	BINARY COUNTING SEQUENCE			DECIMAL COUNT
		C	B	A	
1	0	0	0	0	0
2	1	0	0	1	1
3	2	0	1	0	2
4	3	0	1	1	3
5	4	1	0	0	4
6	5	1	0	1	5
7	6	1	1	0	6
8	7	1	1	1	7
9	8	0	0	0	0

(b)

Fig. 8-6 **A 3-bit synchronous counter.** (*a*) **Logic diagram.** (*b*) **Counting sequence.**

Pulse 8—row 9

   Circuit action: Each flip-flop is pulsed. All three flip-flops toggle. All flip-flops change from 1 to 0.

   Output result: 000 (decimal 0).

We now have completed the explanation of how the 3-bit synchronous counter works. Notice that the J-K flip-flops are used in their toggle mode (*J* and *K* at 1) or disable mode (*J* and *K* at 0) only.

Because of their complexity, synchronous counters are most often purchased in IC form. Synchronous counters are available in both TTL and CMOS.

Self-Test

Supply the missing word in each statement.

8. A counter that triggers all the flip-flops at the same instant is called a _____ (ripple, synchronous) counter.

9. Clock inputs are connected in _____ (parallel, series) on a synchronous counter.

10. Refer to Fig. 8-6(*a*). FF 1 is always in the _____ (hold, reset, set, toggle) mode in this circuit.

11. Refer to Fig. 8-6. On clock pulse 4, _____ (only FF 1 toggles; both FF 1 and FF 2 toggle; only FF 3 toggles; all the flip-flops toggle), leaving a binary count of 100 at the outputs of the counter.

8-4 DOWN COUNTERS

To now we have used counters that count upward (0, 1, 2, 3, 4, . . .). Sometimes, however, we must count downward (9, 8, 7, 6, . . .) in digital systems. A counter that counts from higher to lower numbers is called a *down counter*.

A logic diagram of a mod-8 asynchronous down counter is shown in Fig. 8-7(*a*); the counting sequence for this counter is listed in Fig. 8-7(*b*). Note how much the down counter

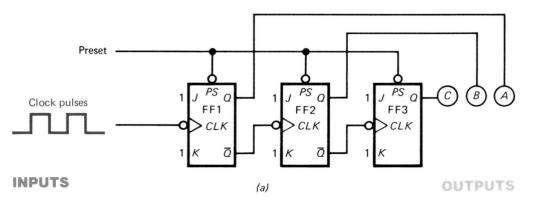

(a)

NUMBER OF CLOCK PULSES	BINARY COUNTING SEQUENCE			DECIMAL COUNT
	C	B	A	
0	1	1	1	7
1	1	1	0	6
2	1	0	1	5
3	1	0	0	4
4	0	1	1	3
5	0	1	0	2
6	0	0	1	1
7	0	0	0	0
8	1	1	1	7
9	1	1	0	6

(b)

Fig. 8-7 **A 3-bit ripple down counter.** (*a*) **Logic diagram.** (*b*) **Counting sequence.**

in Fig. 8-7(*a*) looks like the up counter in Fig. 8-2(*a*). The only difference is in the "carry" from FF 1 to FF 2 and the carry from FF 2 to FF 3. The up counter carries from Q to the *CLK* input of the next flip-flop. The down counter carries from $\overline{Q}$ (not Q) to the *CLK* input of the next flip-flop. Notice that the down counter has a preset (*PS*) control to preset the counter to 111 (decimal 7) to start the downward count. FF 1 is the binary 1s place (column *A*) counter. FF 2 is the 2s place (column *B*) counter. FF 3 is the 4s place (column *C*) counter.

Self-Test

Answer the following questions.

12. Refer to Fig. 8-7(*a*). All flip-flops are in the _____ (hold, reset, set, toggle) mode in this counter.
13. Refer to Fig. 8-7(*a*). It takes a (HIGH-to-LOW, LOW-to-HIGH) transition of the clock pulse to trigger these J-K flip-flops.

14. Refer to Fig. 8-7. On clock pulse 1, _____ (only FF 1 toggles; both FF 1 and FF 2 toggle; only FF 3 toggles; all the flip-flops toggle) leaving a binary count of 110 at the outputs of the counter.
15. List the binary output for each of the six input pulses shown in Fig. 8-8.

8-5 SELF-STOPPING COUNTERS

The down counter shown in Fig. 8-7(*a*) *recirculates*. That is, when it gets to 000 it starts at 111, then 110, and so forth. However, sometimes you want a counter to *stop* when a sequence is finished. Figure 8-9 illustrates how you could stop the down counter in Fig. 8-7 at the 000 count. The counting sequence is shown in Fig. 8-7(*b*). In Fig. 8-9 we add an *OR* gate to place a logical 0 on the *J* and *K* inputs of FF 1 when the count at outputs *C*, *B*, and *A* reaches 000. The preset must be enabled (*PS* to 0) again to start the sequence at 111 (decimal 7).

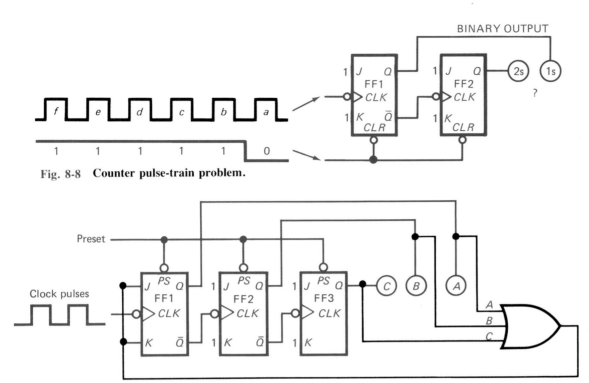

Fig. 8-8 Counter pulse-train problem.

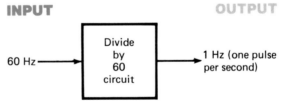

Fig. 8-9 A 3-bit down counter with self-stopping feature.

Up or down counters can be stopped after any sequence of counts by using a logic gate or combination of gates. The output of the gate is fed back to the *J* and *K* inputs of the first flip-flop in a ripple counter. The logical 0s fed back to the *J* and *K* inputs of FF 1 in Fig. 8-9 place it in the hold mode. This stops FF 1 from toggling, thereby stopping the count at 000.

Self-Test

Supply the missing word or words in each statement.

16. Refer to Fig. 8-9. This is the logic diagram for a self-stopping 3-bit _____ (down, up) counter.
17. Refer to Fig. 8-9. With an output count of 111, the OR gate outputs a _____ (HIGH, LOW). This places FF 1 in the _____ (hold, toggle) mode.
18. Refer to Fig. 8-9. With an output count of 000, the OR gate outputs a _____ (HIGH, LOW). This places FF 1 in the _____ (hold, toggle) mode.

8-6 COUNTERS AS FREQUENCY DIVIDERS

An interesting and common use of counters is for *frequency division*. An example of a

simple system using a frequency divider is shown in Fig. 8-10. This system is the basis for an electric clock. The 60-Hz input frequency is from the power line (formed into a square wave). The circuit must divide the frequency by 60, and the output will be one pulse per second (1 Hz). This is a seconds timer.

INPUT   **OUTPUT**

```
                    ┌──────────┐
                    │  Divide  │
 60 Hz ───────────▶ │    by    │ ─────────▶ 1 Hz (one pulse
                    │    60    │              per second)
                    │  circuit │
                    └──────────┘
```

Fig. 8-10 A 1-second timer system.

A block diagram of a decade counter is drawn in Fig. 8-11(*a*) on the next page. In Fig. 8-11(*b*) the waveforms at the *CLK* input and the binary 8s place (output Q_D) are shown. Notice that it takes 30 input pulses to produce 3 output pulses. Using division, we find that $30 \div 3 = 10$. Output Q_D of the decade counter in Fig. 8-11(*a*) is a *divide-by-10* counter. In other words, the output frequency at Q_D is only one-tenth the frequency at the input of the counter.

If we use the decade counter (divide-by-10

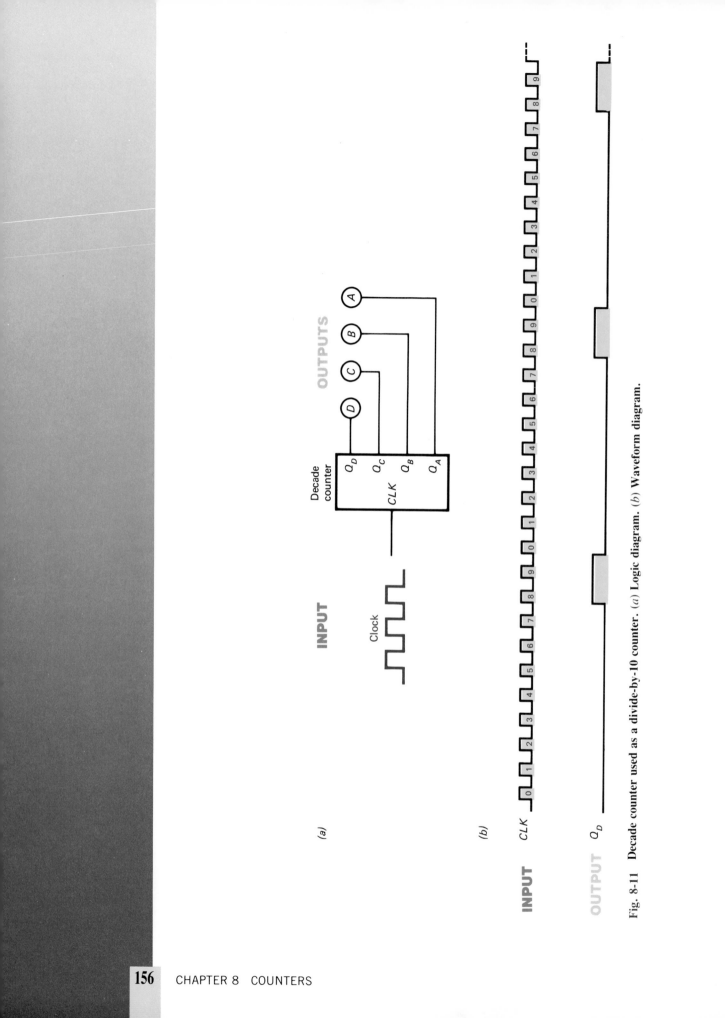

Fig. 8-11 Decade counter used as a divide-by-10 counter. (*a*) Logic diagram. (*b*) Waveform diagram.

counter) from Fig. 8-10 and a mod-6 counter (divide-by-6 counter) in series, we get the divide-by-60 circuit we need in Fig. 8-10. A diagram of such a system is illustrated in Fig. 8-12. The 60-Hz square wave enters the divide-by-6 counter and comes out at 10 Hz. The 10 Hz then enters the divide-by-10 counter and comes out at 1 Hz.

You are already aware that counters are used as frequency dividers in digital time-pieces, such as electronic digital clocks, automobile digital clocks, and digital wristwatches. Frequency division is also used in frequency counters, oscilloscopes, and television servicing dot-and-bar generators.

Self-Test

Supply the missing word in each statement.

19. Refer to Fig. 8-12. If the input frequency on the left is 60,000 Hz, the output frequency from the decade counter is _____ Hz.
20. Refer to Fig. 8-11(*a*). Output *A* divides the input clock frequency by _____ (number).

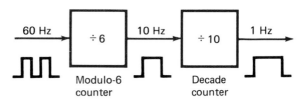

Fig. 8-12 Practical divide-by-60 circuit used as a 1-second timer.

8-7 TTL IC COUNTERS

Manufacturers' IC data manuals contain long lists of counters. This section covers only two representative types of TTL IC counters.

Signetics, a manufacturer of ICs, provided the diagrams and tables in Fig. 8-13 (see page 158). The diagrams are for a 7493 *four-bit binary counter.* Look carefully at the logic diagram in Fig. 8-13(*a*); you will see that the 7493 is a ripple counter. You will notice also that the top J-K flip-flop does not have its output (Q_A) connected to the *CLK* input of the second flip-flop. To operate this counter as a mod-16, you must *externally connect* Q_A to input *B* on the 7493. The counting sequence is shown in Fig. 8-13(*c*). The pin diagram for the 7493 counter is drawn in Fig. 8-13(*b*). To clear, or reset, the counter to 0000, inputs $R_0(1)$ and

$R_0(2)$ must be connected to a logical 1, as shown in Fig. 8-13(*d*). If these reset inputs are left floating (not connected to anything), the counter will not operate. Inputs $R_0(1)$ and $R_0(2)$ float HIGH, and therefore the counter remains in the reset condition and does not count. It is understood that the *J* and *K* inputs to each flip-flop are held at a logical 1. This places each flip-flop in its toggle mode.

The second IC counter we shall discuss is the 74192 *synchronous decade up/down counter.* Figure 8-14 (see page 159) gives information on the 74192 decade counter. Read the manufacturer's description of the IC counter in Fig. 8-14(*a*). Notice that the counter has many features. Because the counter is a synchronous counter, its circuitry is quite complex, as seen in Fig. 8-14(*b*). Figure 8-14(*d*) diagrams in waveforms some typical sequences used on the 74192 counter. You will find upon using this IC that most of the inputs float high. This causes a problem with the *CLR* input; if it is left unconnected, it floats HIGH and clears the output to 0000. The *CLR* input is an active HIGH input on the 74192 counter IC. The pin diagram for the 74192 IC is shown in Fig. 8-14(*c*).

You probably have already figured out that some of the features are not used on these IC counters for some applications. Figure 8-15(*a*) (see page 160) shows the 7493 IC counter being used as a mod-8 counter. Look back at Fig. 8-13 and notice that several inputs and an output are not being used. Figure 8-15(*b*) shows the 74192 counter being used as a decade down counter. Six inputs and two outputs are not being used in this circuit. Simplified logic diagrams similar to those in Fig. 8-15 are more common than the complicated diagrams in Figs. 8-13(*a*) and 8-14(*b*).

Self-Test

Answer the following questions.

21. Refer to Fig. 8-13. If both inputs to the NAND gate (pins 2 and 3 on the 7493 IC) are HIGH, the output from the 7493 counter will be _____ (4 bits).
22. Refer to Fig. 8-13. The 7493 IC is a _____ -bit _____ (down, up) counter.
23. Refer to Fig. 8-14. The 74192 IC is a _____ (decade, mod-16) up/down _____ (ripple, synchronous) counter.

Digital clock

Frequency counter

Oscilloscopes

Television servicing dot-and-bar generators

7493 four-bit binary counter

74192 synchronous decade up/down counter

Divide-by-60 circuit

(a) BLOCK DIAGRAM

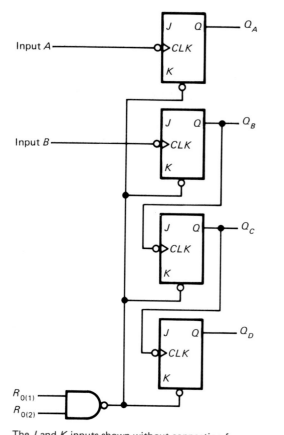

Input A —

Input B —

$R_{0(1)}$
$R_{0(2)}$

The J and K inputs shown without connection for reference only and are functionally at a high level.

(b) PIN CONFIGURATION

54/74
A, F, W package

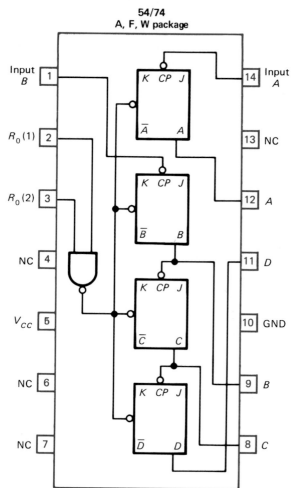

(c) COUNT SEQUENCE

COUNT	OUTPUT			
	Q_D	Q_C	Q_B	Q_A
0	L	L	L	L
1	L	L	L	H
2	L	L	H	L
3	L	L	H	H
4	L	H	L	L
5	L	H	L	H
6	L	H	H	L
7	L	H	H	H
8	H	L	L	L
9	H	L	L	H
10	H	L	H	L
11	H	L	H	H
12	H	H	L	L
13	H	H	L	H
14	H	H	H	L
15	H	H	H	H

Output Q_A is connected to input B.

(d) RESET/COUNT FUNCTION TABLE

RESET INPUTS		OUTPUT			
$R_0(1)$	$R_0(2)$	Q_D	Q_C	Q_B	Q_A
H	H	L	L	L	L
L	X	Count			
X	L	Count			

Fig. 8-13 **A 4-bit binary counter IC (7493).** (a) **Block diagram.** (b) **Pin configuration.**
(c) **Count sequence.** (d) **Reset/count function table.** (*Courtesy of Signetics Corporation*)

(a) DESCRIPTION

This monolithic circuit is a synchronous reversible (up/down) counter having a complexity of 55 equivalent gates. Synchronous operation is provided by having all flip-flops clocked simultaneously so that the outputs change coincidently with each other when so instructed by the steering logic. This mode of operation eliminates the output counting spikes which are normally associated with asynchronous (ripple-clock) counters.

The outputs of the four master-slave flip-flops are triggered by a low-to-high-level transition of either count (clock) input. The direction of counting is determined by which count input is pulsed while the other count input is high.

All four counters are fully programmable; that is, each output may be preset to either level by entering the desired data at the data inputs while the load input is low. The output will change to agree with the data inputs independently of the count pulses. This feature allows the counters to be used as modulo-N dividers by simply modifying the count length with the preset inputs.

A clear input has been provided which forces all outputs to the low level when a high level is applied. The clear function is independent of the count and load inputs. The clear, count, and load inputs are buffered to lower the drive requirements. This reduces the number of clock drivers, etc., required for long words.

These counters were designed to be cascaded without the need for external circuitry. Both borrow and carry outputs are available to cascade both the up- and down-counting functions. The borrow output produces a pulse equal in width to the count-down input when the counter underflows. Similarly, the carry output produces a pulse equal in width to the count-down input when an overflow condition exists. The counters can then be easily cascaded by feeding the borrow and carry outputs to the count-down and count-up inputs respectively of the succeeding counter.

(b) BLOCK DIAGRAM

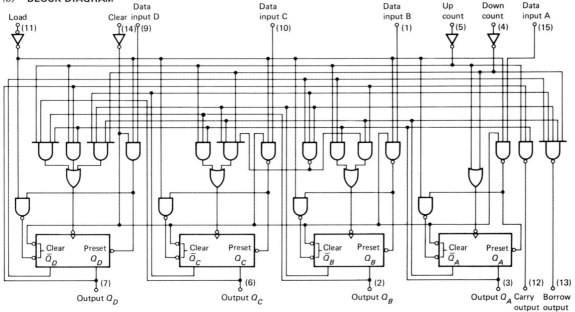

(c) PIN CONFIGURATION

(d) TYPICAL CLEAR, LOAD, AND COUNT SEQUENCE

Illustrated below is the following sequence:
1. Clear output to zero.   3. Count up to eight, nine, carry, zero, one, and two.
2. Load (preset) to BCD seven.   4. Count down to one, zero, borrow, nine, eight, and seven.

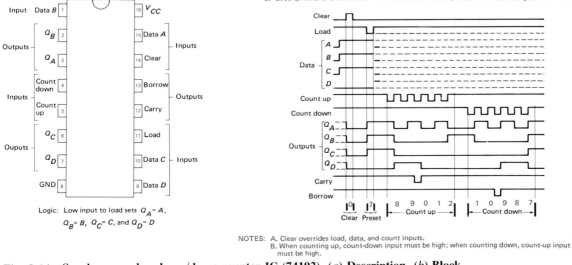

Logic: Low input to load sets $Q_A = A$,
$Q_B = B$, $Q_C = C$, and $Q_D = D$.

NOTES: A. Clear overrides load, data, and count inputs.
B. When counting up, count-down input must be high; when counting down, count-up input must be high.

Fig. 8-14 **Synchronous decade up/down counter IC (74192).** (*a*) **Description.** (*b*) **Block diagram.** (*c*) **Pin configuration.** (*d*) **Waveforms.** (*Courtesy of Signetics Corporation*)

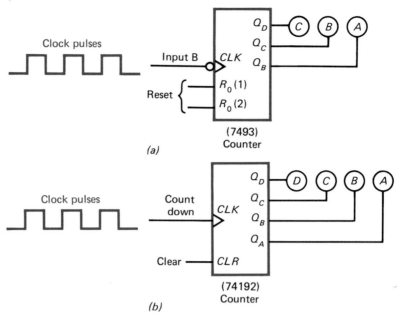

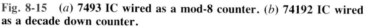

Fig. 8-15 (*a*) **7493 IC wired as a mod-8 counter.** (*b*) **74192 IC wired as a decade down counter.**

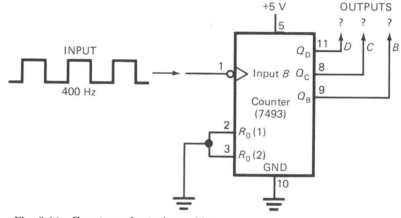

Fig. 8-16 **Counter pulse-train problem.**

24. Refer to Fig. 8-14. The clock input to the 74192 for counting upward is pin _____ (number) on the IC.

25. Refer to Fig. 8-14. The 74192 IC has an active _____ (HIGH, LOW) clear input.

26. List the output frequency at points *B*, *C*, and *D* in Fig. 8-16 (see page 161).

27. The 7493 IC is a ripple divide-by-2, divide-by-4, and divide-by- _____ unit in Fig. 8-16.

8-8 CMOS IC COUNTERS

Manufacturers of CMOS chips offer a variety of counters in IC form. This section covers only two types of CMOS ICs.

Signetics, a manufacturer of CMOS ICs, provided the diagrams and tables in Fig. 8-17 (see page 162). The diagrams are for a 74HC393 *dual 4-bit binary ripple counter*. A *functional diagram* (something like a logic diagram) of the 74HC393 counter IC is shown in Fig. 8-17(*a*). Note that the IC contains two 4-bit binary ripple counters. The table in Fig. 8-17(*b*) gives the names and functions of each input and output pin on the 74HC393 IC. Note that the clock inputs are labeled with the letters *CP* instead of *CLK*, as used earlier. Pin labels vary from manufacturer to manufacturer. For this reason, you must learn to use manufacturer's data manuals for exact information.

Each 4-bit counter in the 74HC393 IC package consists of four T flip-flops. A *T flip-flop* is any flip-flop that is in the toggle mode. This is shown in the detailed logic diagram drawn in Fig. 8-17(c). Note that the *MR* input is an asynchronous master reset pin. The *MR* pins are active HIGH inputs. In other words, a HIGH at the *MR* input will override the clock and reset the individual counter to 0000.

A pin diagram for the 74HC393 IC is reproduced in Fig. 8-17(d). This dual in-line package IC is being viewed from the top. The counting sequence for the 74HC393 counter is binary 0000 through 1111 (0 to 15 in decimal).

The functional diagram in Fig. 8-17(a) and logic diagram in Fig. 8-17(c) both suggest that the counters are triggered on the HIGH-to-LOW transition of the clock pulse. The outputs (Q_0, Q_1, Q_2, Q_3) of the ripple counter are asynchronous (not exactly in step with the clock). As with all ripple counters, there is a slight delay in outputs because the first flip-flop triggers the second, the second the third, and so forth. Note that the > symbol at the clock (*CP*) inputs has been omitted by this manufacturer. Again, many variations occur in both labels and logic diagrams from manufacturer to manufacturer.

The second CMOS IC counter we shall discuss is the 74HC193 *presettable synchronous 4-bit binary up/down counter IC*. The 74HC193 counter has more features than the 74HC393 IC. Manufacturer's information on the 74HC193 counter IC is detailed in Fig. 8-18 (see page 163).

A functional diagram of the 74HC193 IC is drawn in Fig. 8-18(a) with pin descriptions following in Fig. 8-18(b). The 74HC193 has two clock inputs (CP_U and CP_D). One clock input is used for counting up (CP_U) and the other when counting down (CP_D). Figure 8-18(b) notes that the clock inputs are edge-triggered on the LOW-to-HIGH transition of the clock pulse.

A truth table for the 74HC193 counter is drawn in Fig. 8-18(d). The operating modes for the counter on the left give an overview of the many functions of the 74HC193 counter. Its modes of operation are *reset, parallel load, count up,* and *count down*. The truth table in Fig. 8-18(d) also makes it clear which pins are inputs and which are outputs.

Typical clear (reset), preset (parallel load), count up, and count down sequences are shown in Fig. 8-18(e). Waveforms are useful when investigating an IC's typical operations or timing.

Figures 8-19 and 8-20 (see page 164) show two possible applications for the CMOS counter ICs studied in this section. Figure 8-19 shows a logic diagram for a 74HC393 IC wired as a simple 4-bit binary counter. The *MR* (master reset) pin must be tied to either 0 or 1. The *MR* input is an active HIGH input so a 1 clears the binary outputs to 0000. With a logical 0 at the reset pin (*MR*), the IC is allowed to count upward from binary 0000 to 1111.

The 74HC193 CMOS IC is a more sophisticated counter. Figure 8-20 diagrams a mod-6 counter which starts at binary 001 and counts up to 110 (1 to 6 in decimal). This might be useful in a game circuit where the rolling of dice is simulated. The NAND gate in the mod-6 counter activates the asynchronous parallel load ($\overline{PL}$) input with a LOW just after the highest required count of binary 0110. The counter is then loaded with 0001 which is permanently connected to the data inputs (D_0 to D_3). Clock pulses enter the count-up clock input (C_U). The count-down clock input (C_D) must be tied to +5 V and the master reset (*MR*) pin must be grounded to disable these inputs and allow the counter to operate. The mod-6 counter circuit in Fig. 8-20 shows the flexibility of the CMOS 74HC193 presettable 4-bit up/down counter IC.

Self-Test

Answer the following questions.

28. Refer to Fig. 8-17 (see page 162). The 74HC393 IC contains two _____ (4-bit binary, decade) counters.
29. Refer to Fig. 8-17. The reset pin (*MR*) on the 74HC393 counter is an active _____ (HIGH, LOW) input.
30. Refer to Fig. 8-17. The 74HC393 counter's clock inputs are triggered by the _____ (H-to-L, L-to-H) transition of the clock pulse.
31. The circuit drawn in Fig. 8-19 (page 164) is a mod-_____ (number) _____ (ripple, synchronous) counter.
32. Refer to Fig. 8-18. The 74HC193 is a presettable _____ (ripple, synchronous) 4-bit up/down counter IC.
33. Refer to Fig. 8-18. The reset pin (*MR*) is _____ (asynchronous, synchronous) and overrides all other inputs on the 74HC193 IC.
34. Refer to Fig. 8-18. The outputs of the 74HC193 are labeled _____ (D_0–D_3, Q_0–Q_3).

74HC193 CMOS IC

Mod-6 counter circuit

74HC193 presettable synchronous 4-bit binary up/down counter IC

Reset

Parallel load

Count up

Count down

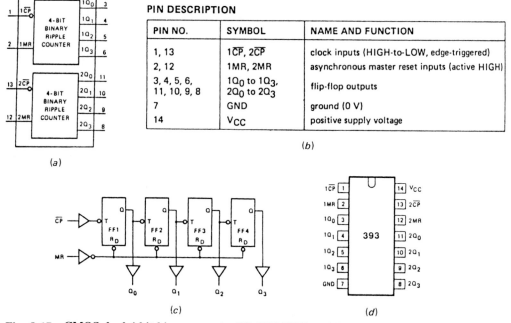

Fig. 8-17 **CMOS dual 4-bit binary counter IC (74HC393).** (*a*) **Function diagram.** (*b*) **Pin descriptions.** (*c*) **Detailed logic diagram.** (*d*) **Pin diagram.** (*Courtesy of Signetics Corporation*)

35. Refer to Fig. 8-20 (see page 164). List the binary counting sequence for this counter circuit.

36. Refer to Fig. 8-20. What is the purpose of the three-input NAND gate in this counter circuit?

37. Refer to Fig. 8-17(*a*). How do you explain the lack of the > symbol near the clock inputs even if the 74HC393 counters are edge-triggered?

8-9 USING A CMOS COUNTER IN AN ELECTRONIC GAME

This section will feature a CMOS counter being used in an electronic game. The game is the classic computer game of "guess the number." In the computer version, a random number is generated and the player tries to guess the unknown number. The computer responds with one of three responses: correct, too high, or too low. The player can then guess again until he or she zeros in on the unknown number. The player who uses the fewest guesses wins the game.

The schematic for a simple electronic version of this game is drawn in Fig. 8-21 (see page 165). To operate the game first press the push-button switch (SW_1). This allows the approx-

imately 1-kHz signal into the clock input of the binary counter. When the push button is released, a random binary number (from 0000 to 1111) is held at the B inputs to the 74HC85 4-bit magnitude comparator. The player's guess is entered at the A inputs to the comparator IC. If the random number (B inputs) and the guess (A inputs) are equal, then the $A = B_{\text{OUT}}$ output will be activated (HIGH) and the green LED will light. This means the guess was correct. After a correct guess a new random number should be generated by pressing SW_1.

If a player's guess (A inputs) is lower than the random number (B inputs), the comparator will activate the $A < B_{\text{OUT}}$ output. The yellow LED will light. This means that the guess was too low and the player should try again by entering a somewhat higher number.

Finally, if a player's guess (A inputs) is higher than the random number (B inputs), the comparator will activate the $A > B_{\text{OUT}}$ output. The red LED will light. This means that the guess was too high and the player should try again.

Figure 8-22 (see page 166) gives greater detail on the operation of the 74HC85 *4-bit magnitude comparator* IC. The pin diagram is shown in Fig. 8-22(*a*). This is the top view of the DIP 74HC85 CMOS IC. The truth table for the 74HC85 comparator is reproduced in Fig. 8-22(*b*).

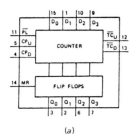

(a)

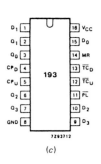

(c)

PIN DESCRIPTION

PIN NO.	SYMBOL	NAME AND FUNCTION
3, 2, 6, 7	Q_0 to Q_3	flip-flop outputs
4	CP_D	count down clock input*
5	CP_U	count up clock input*
8	GND	ground (0 V)
11	$\overline{PL}$	asynchronous parallel load input (active LOW)
12	$\overline{TC}_U$	terminal count up (carry) output (active LOW)
13	$\overline{TC}_D$	terminal count down (borrow) output (active LOW)
14	MR	asynchronous master reset input (active HIGH)
15, 1, 10, 9	D_0 to D_3	data inputs
16	V_{CC}	positive supply voltage

* LOW-to-HIGH, edge triggered

(b)

OPERATING MODE	INPUTS								OUTPUTS					
	MR	$\overline{PL}$	CP_U	CP_D	D_0	D_1	D_2	D_3	Q_0	Q_1	Q_2	Q_3	$\overline{TC}_U$	$\overline{TC}_D$
reset (clear)	H	X	X	L	X	X	X	X	L	L	L	L	H	L
	H	X	X	H	X	X	X	X	L	L	L	L	H	H
parallel load	L	L	X	L	L	L	L	L	L	L	L	L	H	L
	L	L	X	H	L	L	L	L	L	L	L	L	H	H
	L	L	L	X	H	H	H	H	H	H	H	H	L	H
	L	L	H	X	H	H	H	H	H	H	H	H	H	H
count up	L	H	↑	H	X	X	X	X	count up				H*	H
count down	L	H	H	↑	X	X	X	X	count down				H	H**

* $\overline{TC}_U$ = CP_U at terminal count up (HHHH)
** $\overline{TC}_D$ = CP_D at terminal count down (LLLL)

H = HIGH voltage level
L = LOW voltage level
X = don't care
↑ = LOW-to-HIGH clock transition

(d)

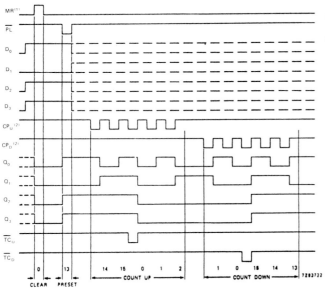

(1) Clear overrides load, data and count inputs.

(2) When counting up the count down clock input (CP_D) must be HIGH, when counting down the count up clock input (CP_U) must be HIGH.

Sequence

Clear (reset outputs to zero);

load (preset) to binary thirteen;

count up to fourteen, fifteen, terminal count up, zero, one and two;

count down to one, zero, terminal count down, fifteen, fourteen and thirteen.

(e)

Fig. 8-18 CMOS presettable 4-bit synchronous up/down counter IC (74HC193).
(a) Function diagram. (b) Pin descriptions. (c) Pin diagram. (d) Truth table. (e) Typical
clear, preset, and count sequence. (Courtesy of Signetics Corporation)

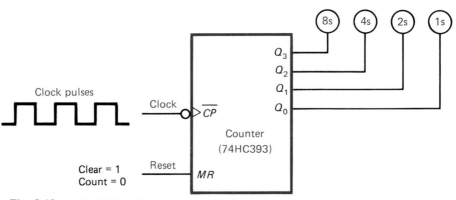

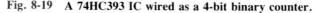

Fig. 8-19   A 74HC393 IC wired as a 4-bit binary counter.

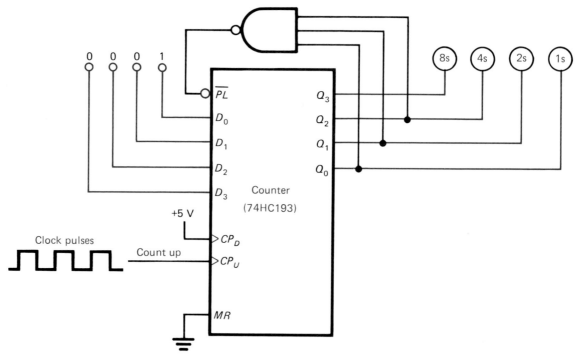

Fig. 8-20   A 74HC193 IC wired as a mod-6 counter.

The 74HC85 comparator has three "extra" inputs used for *cascading* comparators. Typical cascading of 74HC85 magnitude comparators is shown in Fig. 8-23 (see page 167). This circuit compares the magnitude of the two 8-bit binary words $A_7 A_6 A_5 A_4 A_3 A_2 A_1 A_0$ and $B_7 B_6 B_5 B_4 B_3 B_2 B_1 B_0$. The output from IC_2 is one of three responses ($A > B$, $A = B$, or $A < B$).

Self-Test

Answer the following questions.

38. Refer to Fig. 8-21. If the binary counter holds the number 1001 and your guess is 1011, the _____ (color) LED will light indicating your guess is _____ (correct, too high, too low).

39. Refer to Fig. 8-21. How do you generate a random number before guessing?

40. Refer to Fig. 8-21. The 555 timer is wired as a(n) _____ (astable, monostable) multivibrator.

41. Refer to Fig. 8-24 (see page 168). List the *color* of the output LED that is lit for each time period (t_1 to t_6).

8-10 SEQUENTIAL LOGIC TROUBLESHOOTING EQUIPMENT

Until now, troubleshooting has been performed on combinational logic circuits (gating circuits). Sequential logic circuits are composed of flip-flops and as such are somewhat more difficult

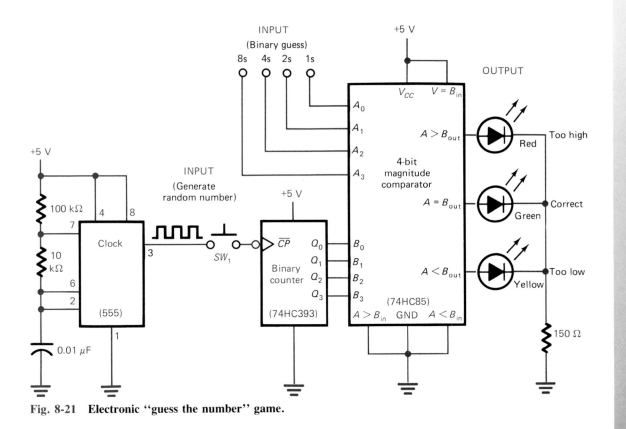

Fig. 8-21 Electronic "guess the number" game.

to troubleshoot. Your favorite pieces of test equipment to this time have been the logic probe and the voltmeter. Besides these, several other pieces of test equipment are handy when troubleshooting sequential logic circuits. It is common to have available a logic pulser, logic monitors, and an oscilloscope. You may also have available a signature analyzer, logic analyzer, and IC tester.

A commercial *logic probe* is pictured in Fig. 8-25 (*a*) (see page 169). Notice that this logic probe has a diode-transistor/transistor-transistor logic (DTL/TTL) or CMOS selection switch. This switch selects the family of the IC being tested. If the MEM/PULSE switch is in the pulse position, any pulse as short as about 50 ns will be displayed on the pulse LED for about 0.3 s. If the MEM/PULSE switch is in the MEM (memory) position, any signal change at the tip (1 to 0 or 0 to 1) will activate the pulse LED. In the memory mode, the pulse LED lights continuously on any single pulse. Although not shown in the photograph in Fig. 8-25(*a*), this logic probe has a plug-in cable which is connected to power. The power cable plugs into the jack at the rear of the probe.

Another logic level measuring device used with digital ICs is the *logic monitor*. One logic monitor is shown in Fig. 8-25(*b*). The ribbon cable connects to an IC test clip something like the ones shown in Fig. 8-25(*c*). The logic levels of all 16 pins on an IC are then visible on the LEDs. The LEDs on the logic monitor light when a HIGH logic level is present. They do not light when the pin is LOW or in the undefined region between HIGH and LOW. The threshold voltage for indicating a HIGH on the unit in Fig. 8-25(*b*) is about 2.3 V for TTL and 70 percent of V_{cc} for CMOS. Notice in Fig. 8-25(*b*) that the logic monitor can check either CMOS or TTL/DTL circuits. Lower (or higher) voltage thresholds can be used to indicate HIGH levels on the logic monitor by using the variable mode and adjusting the variable threshold control shown at the lower left in Fig. 8-25(*b*).

Logic monitors are many times constructed in an enlarged IC test clip like those in Fig. 8-25(*c*). One such logic monitor is pictured in Fig. 8-25(*d*). This logic monitor is referred to as a *logic clip* by some manufacturers and technicians. The logic monitor pictured in Fig. 8-25(*d*) clips over the 16 pins of a DIP IC. This unit is designed for TTL or DTL and has a threshold voltage of about 2.0 V.

A commercial *digital logic pulser* is pictured in Fig. 8-26 (see page 170). This unit outputs a single pulse when the push button is pressed. According to the manufacturer, the pulse width is 1.5 μs with TTL selected. The pulse width

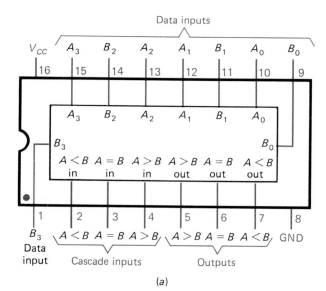

(a)

Truth Table—74HC85 Magnitude Comparator IC

COMPARING INPUTS				CASCADING INPUTS			OUTPUTS		
A_3, B_3	A_2, B_2	A_1, B_1	A_0, B_0	$A > B$	$A < B$	$A = B$	$A > B$	$A < B$	$A = B$
$A_3 > B_3$	X	X	X	X	X	X	H	L	L
$A_3 < B_3$	X	X	X	X	X	X	L	H	L
$A_3 = B_3$	$A_2 > B_2$	X	X	X	X	X	H	L	L
$A_3 = B_3$	$A_2 < B_2$	X	X	X	X	X	L	H	L
$A_3 = B_3$	$A_2 = B_2$	$A_1 > B_1$	X	X	X	X	H	L	L
$A_3 = B_3$	$A_2 = B_2$	$A_1 < B_1$	X	X	X	X	L	H	L
$A_3 = B_3$	$A_2 = B_2$	$A_1 = B_1$	$A_0 > B_0$	X	X	X	H	L	L
$A_3 = B_3$	$A_2 = B_2$	$A_1 = B_1$	$A_0 < B_0$	X	X	X	L	H	L
$A_3 = B_3$	$A_2 = B_2$	$A_1 = B_1$	$A_0 = B_0$	H	L	L	H	L	L
$A_3 = B_3$	$A_2 = B_2$	$A_1 = B_1$	$A_0 = B_0$	L	H	L	L	H	L
$A_3 = B_3$	$A_2 = B_2$	$A_1 = B_1$	$A_0 = B_0$	X	X	H	L	L	H
$A_3 = B_3$	$A_2 = B_2$	$A_1 = B_1$	$A_0 = B_0$	H	H	L	L	L	L
$A_3 = B_3$	$A_2 = B_2$	$A_1 = B_1$	$A_0 = B_0$	L	L	L	H	H	L

(b)

Fig. 8-22   CMOS magnitude comparator IC (74HC85). (*a*) **Pin diagram.** (*b*) **Truth table.**
(*Courtesy of National Semiconductor Corporation*)

is 10 μs with CMOS selected. If the push button on this digital pulser is pressed and held down for a time, a pulse train of 100 pulses per second is generated at the tip. The tip is isolated by high impedance when the push button is not pressed. This is called *tristating*. The output can have three levels: LOW, HIGH, or high impedance.

Using the digital pulser in Fig. 8-26, a single press of the pulse push button emits a flash on the pulse indicator LED. When the push button is held down for a time, the pulse LED indicator lights continuously. This indicates a pulse train. As on the logic probe, the jack on

the end of the digital pulser in Fig. 8-26 is for the removable power cable.

Beginning students and efficient technicians find an IC tester useful in making sure the ICs used in a circuit are working according to specifications. The IC tester pictured in Fig. 8-27(*a*) (see page 170) provides a quick method of verifying that a chip is satisfactory. The IC tester in Fig. 8-27(*a*) is not very sophisticated, but it is easy to use and quite inexpensive. The IC tester in Fig. 8-27(*a*) will test most TTL and CMOS ICs. The IC tester in Fig. 8-27(*a*) is indicating that the CMOS 4060 IC being tested is satisfactory.

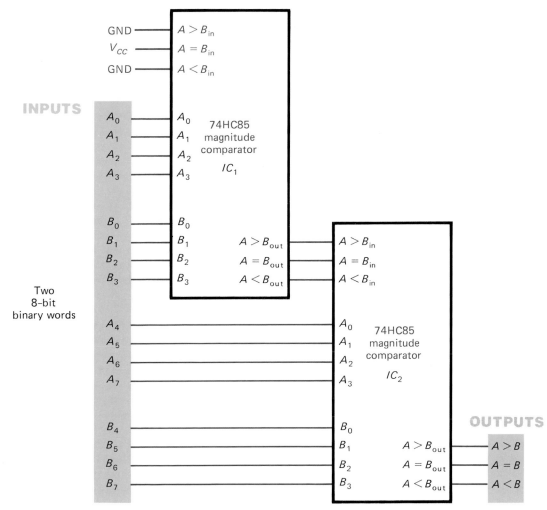

Fig. 8-23 Cascaded magnitude comparators.

A much more sophisticated IC tester is pictured in Fig. 8-27(*b*). This IC tester can test ICs both in and out of the circuit. The procedure for testing an IC out of the circuit is as follows. First, type in the number of the IC to be tested. Second, place the IC in the *zero-insertion force (ZIF)* IC socket on the front of the IC tester. Third, press the test button and the display will indicate if the IC is good or display a failure. In-circuit tests are handled through a ribbon cable running from the connector on the left to a DIP clip which clamps on the IC to be tested. The test procedure is somewhat like that for out-of-circuit testing.

As in most areas of electronics, the oscilloscope is probably the most important research, development, and troubleshooting instrument. A dual-trace oscilloscope, such as the one pictured in Fig. 8-28(*a*) (see page 170), is particularly valuable in comparing waveforms and observing important timing relationships in digital circuits. Oscilloscopes are very good for

observing repetitive or synchronous events in a digital system. However, they have difficulties when observing glitches or asynchronous events in digital systems. A *glitch* is a random signal that causes problems in a digital system. It may be caused by noise in the system or a design error. A glitch usually takes the form of a very short L-H-L or H-L-H pulse that appears when the waveform should be at a steady LOW or HIGH.

The *logic analyzer* is a specialized instrument that typically has many input channels (16 to over 40). It samples and stores multiple input signals and then displays them on a CRT (cathode-ray tube). A typical logic analyzer is pictured in Fig. 8-28(*b*) (see page 170). The advantages of the logic analyzer over the oscilloscope are its many input channels, signal sampling methods, relatively large memory or storage, sophisticated output (CRT), and ability to detect asynchronous events or glitches. These glitches may be stored for later analysis on the CRT.

Logic analyzer

Troubleshooting a counter

2-bit ripple counter

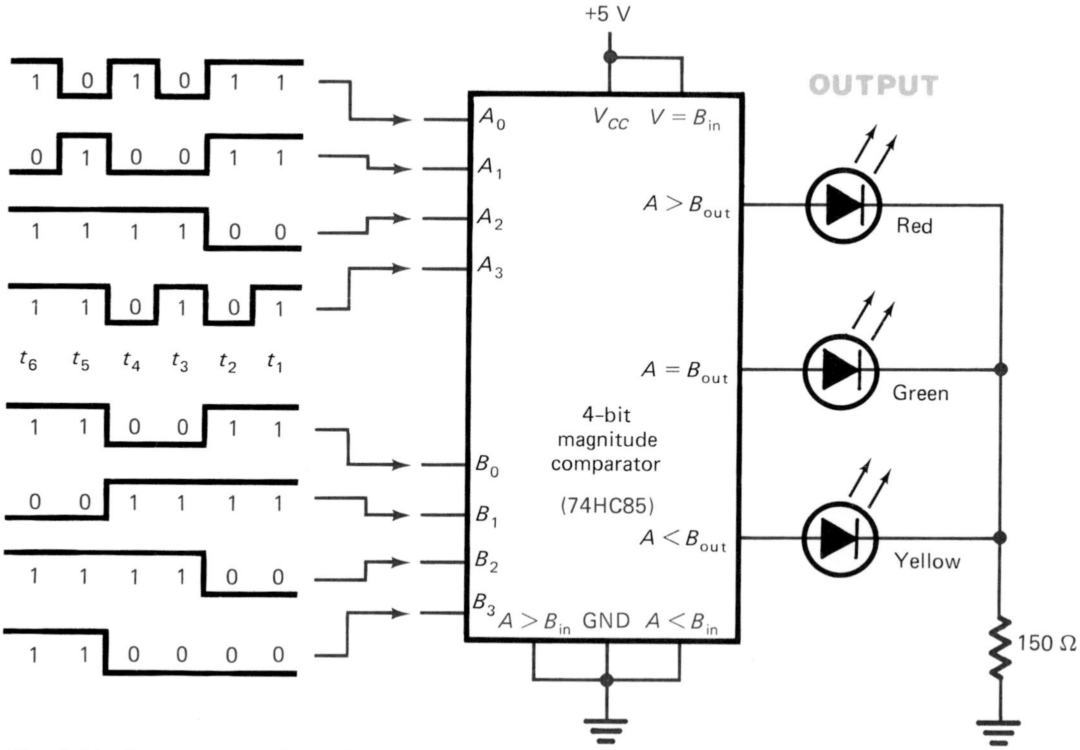

Fig. 8-24 Comparator pulse-train problem (Self-Test question 41, page 164).

Self-Test

Supply the missing word in each statement.

42. Digital circuits that use flip-flops (such as counters) are classified as ________ (combinational, sequential) logic circuits.
43. A logic ________ (monitor, probe) can check the logic level of all the pins on an IC at one time.
44. Refer to Fig. 8-26 (page 170). A single press of the pulse push button on the logic pulser sends out a ________ (stream of pulses at about 100 Hz; single pulse) to the tip.
45. An oscilloscope is a very good instrument for observing ________ (asynchronous, repetitive) digital signals.
46. The test instrument called a logic ________ (analyzer, probe) features multiple input channels, signal sampling, and signal storage for the later viewing.

8-11 TROUBLESHOOTING A COUNTER

Consider the job of troubleshooting the faulty 2-bit ripple counter shown in Fig. 8-29(*a*) (see page 171). For your convenience, a pin diagram for the 7476 IC used in this circuit is shown in Fig. 8-29(*b*). Note that all of the input and output markings in Fig. 8-29(*a*) and (*b*) are *not* the same. For instance, the asynchronous preset inputs on the logic diagram are labeled *PS*. The same inputs are labeled *PR* (for preset) by National Semiconductor Corporation. The labels on the pins may be different from manufacturer to manufacturer. However, the pins on an IC labeled as a 7476 serve the same *function* even if the labeling is different.

It is found that the faulty 2-bit counter circuit can be cleared to 00 by the reset switch at the left in Fig. 8-29(*a*). The IC seems to be operating at the correct temperature, and the technician can see no signs of trouble.

A digital logic pulser (like the one in Fig. 8-26, page 170) is used to pulse the *CLK* input on FF 1. According to the pin diagram, the tip of the digital pulser must touch pin 1 of the 7476 IC. Upon repeated single pulses, the counting sequence is 00 (reset), 01, 10, 11, 10, 11, 10, 11, and so on. The *Q* output of FF 2 seems to be "stuck HIGH"; however, the asynchronous clear (*CLR*), or reset, switch can drive it LOW.

Power is then turned off in the circuit in Fig. 8-29(*a*). A TTL logic monitor like the one in Fig. 8-25(*b*) is clipped over the pins on the 7476

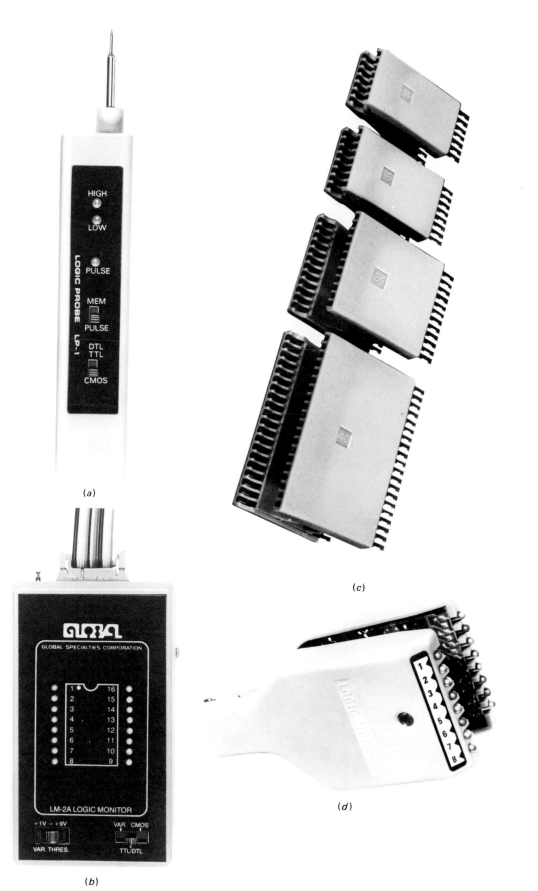

Fig. 8-25 (*a*) **Logic probe.** (*b*) **Logic monitor.** (*c*) **IC clips.** (*d*) **Logic monitor (logic clip).** (*Courtesy of Interplex Electronics, Inc.*)

Fig. 8-26 Commercial digital logic pulser for TTL or CMOS circuits. *(Courtesy of Interplex Electronics, Inc.)*

IC. Power is turned on again. The reset switch is activated. The results displayed on the logic monitor after the reset are shown in Fig. 8-29(c). Compare the logic levels shown on the logic monitor with your expectations. You must use the manufacturer's pin diagram furnished in Fig. 8-29-(b). In looking over the logic levels pin by pin, the LOW or undefined logic level at pin 7 should cause concern. This is the asynchronous preset (*PS* or *PR*) input and should be HIGH according to the logic diagram in Fig. 8-29(a). If it is LOW or in the undefined region it may cause the *Q* output of FF 2 to be in the "stuck HIGH" condition.

A logic probe [like the one in Fig. 8-25(a)] is used to check pin 7 of the 7476 IC. Both LEDs on the logic probe remain off. This means neither a LOW nor a HIGH logic level is present. Pin 7 appears to be floating in the undefined region between LOW and HIGH. The IC is in-

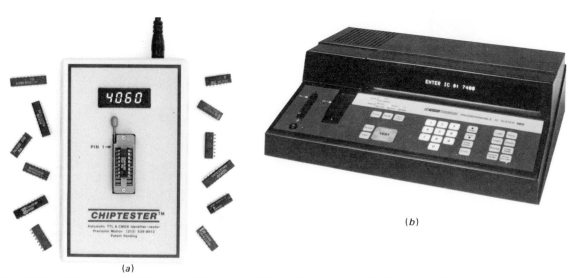

(a)

(b)

Fig. 8-27 (a) **Inexpensive IC tester.** *(Courtesy of Precision Motion)* (b) **Programmable IC tester.** *(Courtesy of Dynascan Corporation)*

(a)

(b)

Fig. 8-28 (a) **Modern dual-trace oscilloscope.** *(Courtesy of Dynascan Corporation)* (b) **Logic analyzer.** *(Courtesy of Hewlett-Packard)*

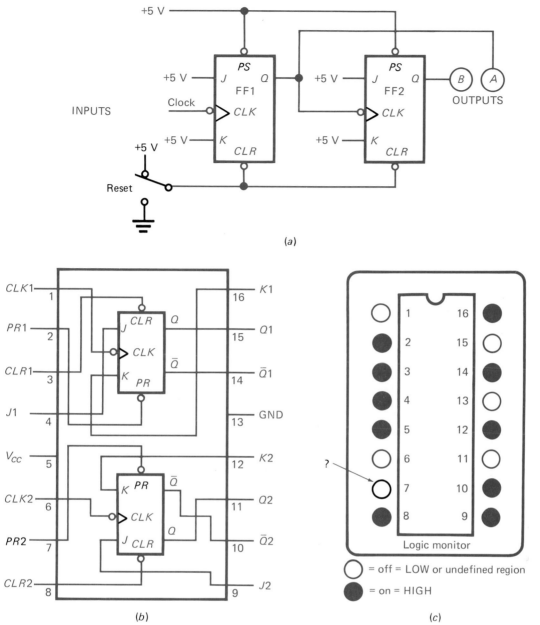

Fig. 8-29 (a) Faulty 2-bit ripple counter circuit using in troubleshooting example. (b) Pin diagram for 7476 J-K flip-flop IC. (Courtesy of National Semiconductor Corporation) (c) Logic monitor readings after momentarily resetting the faulty 2-bit counter.

terpreting this as LOW at some times and as HIGH at other times.

The IC is removed from the 16-pin DIP IC socket. It is found that pin 7 of the IC is bent under and not making contact with the IC socket. This caused it to float. The fault is illustrated in Fig. 8-30. This common fault is very hard to see when the IC is seated in the IC socket.

In this example, several tools were used in troubleshooting. First, the logic diagram and your knowledge of how it works are most important. Second, a manufacturer's pin diagram was used. Third, a digital logic pulser was used to inject single pulses. Fourth, a logic monitor

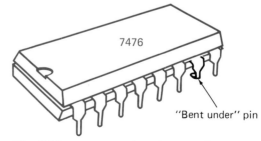

Fig. 8-30 Bent pin caused input to float.

checked the logic level at all pins of the 7476 IC. Fifth, a logic probe was used to check the suspected pin of the IC. Finally, your knowledge of the circuit and visual observation solved the problem. *Your knowledge of the circuit's normal operation and your powers of observation are probably the most important troubleshooting tools.* Logic pulsers, logic probes, logic monitors, logic analyzers, IC testers, and oscilloscopes are only aids to your powers of observation.

The example of a floating input caused by a bent-under pin is a very common problem in student-constructed circuits. It is good practice to make sure all inputs go to the proper logic level. This is true for TTL and especially true for CMOS circuits.

Self-Test

Supply the missing word or words in each statement.

47. Refer to Fig. 8-29 (see page 171). Pins 4, 9, 12, and 16 to the *J* and *K* inputs of the flip-flops should all be _____ (HIGH, LOW) in this circuit.
48. Refer to Fig. 8-29. Pins 3 and 8 to the _____ inputs of the flip-flops should follow the logic state of the reset switch.
49. Refer to Fig. 8-29. Pins 2 and 7 to the _____ inputs of the flip-flops should be _____ (HIGH, LOW) in this circuit.
50. Refer to Fig. 8-29. The fault in this circuit was located at pin _____ (number). It was _____ rather than being at a HIGH logic level.

SUMMARY

1. Flip-flops are wired together to form binary counters.
2. Counters can operate asynchronously or synchronously. Asynchronous counters are called ripple counters and are simpler to construct than synchronous counters.
3. The modulus of a counter is how many different states it goes through in its counting cycle. A mod-5 counter counts 000, 001, 010, 011, 100 (0, 1, 2, 3, 4 in decimal).
4. A 4-bit binary counter has four binary place values and counts from 0000 to 1111 (decimal 0 to 15).
5. Gates can be added to the basic flip-flops in counters to add features. Counters can be made to stop at a certain number. The modulus of a counter can be changed.
6. Counters are designed to count either up or down. Some counters have both features built into their circuitry.
7. Counters are used as frequency dividers. Counters are also widely used to count or sequence events.

8. Manufacturers produce a wide variety of self-contained IC counters. They produce detailed data sheets for each counter IC. Several TTL and CMOS counter ICs were studied in this chapter.
9. Many variations in pin labeling and logic symbols occur from manufacturer to manufacturer.
10. A magnitude comparator will compare two binary numbers and decide if $A = B$, $A > B$, or $A < B$. Magnitude comparator ICs can be cascaded to compare larger binary numbers.
11. The technician's knowledge of the circuit and powers of observation are the most important tools in troubleshooting. The logic probe, voltmeter, logic monitor, digital pulser, logic analyzer, IC tester, and oscilloscope aid the technician's observations when troubleshooting sequential logic circuits.

CHAPTER REVIEW QUESTIONS

Answer the following questions.

8-1. Draw a logic symbol diagram of a mod-8 ripple up counter. Use three J-K flip-flops. Show input *CLK* pulses and three output indicators labeled *C*, *B*, and *A* (*C* indicator is MSB).

8-2. Draw a table (similar to Fig. 8-1) showing the binary and decimal counting sequence of the mod-8 counter in question 1.

8-3. Draw a waveform diagram [similar to Fig. 8-2(b)] showing the eight CLK pulses and the outputs (Q) of FF 1, FF 2, and FF 3 of the mod-8 counter from question 1. Assume you are using negative-edge-triggered flip-flops.

8-4. Redesign the mod-8 counter in question 1 to be a mod-5 counter. Add a two-input NAND gate to your existing counter. You will need to use the CLR inputs of the J-K flip-flops.

8-5. A(n) _____ (asynchronous, synchronous) counter is the more complex circuit.

8-6. Synchronous counters have the CLK inputs connected in _____ (parallel, series).

8-7. Draw a logic symbol diagram for a 4-bit ripple down counter. Use four J-K flip-flops in this mod-16 counter. Show the input CLK pulses, PS input, and four output indicators labeled D, C, B, and A.

8-8. If the ripple down counter in question 7 is a recirculating type, what are the next three counts after 0011, 0010, and 0001?

8-9. Redesign the 4-bit counter in question 7 to count from binary 1111 to 0000 and then *stop*. Add a four-input OR gate to your existing circuit to add this self-stopping feature.

8-10. Draw a block diagram (similar to Fig. 8-12) showing how you would use two counters to get an output of 1 Hz with an input of 100 Hz. Label your diagram.

8-11. Refer to Fig. 8-13 for questions **a** to **f** on the 7493 IC counter:
  a. What is the maximum count length of this counter?
  b. This is a _____ (ripple, synchronous) counter.
  c. What must be the conditions of the reset inputs for the 7493 to count?
  d. This is a(n) _____ (down, up) counter.
  e. The 7493 IC contains _____ (number) flip-flops.
  f. What is the purpose of the NAND gate in the 7493 counter?

8-12. Refer to Fig. 8-14 for questions **a** to **f** on the 74192 counter:
  a. What is the maximum count length of this counter?
  b. This is a _____ (ripple, synchronous) counter.
  c. A logical _____ (0, 1) is needed to clear the counter to 0000.
  d. This is a(n) _____ (down, up, both up and down) counter.
  e. How could we preset the outputs of the 74192 IC to 1001?
  f. How do we get the counter to count downward?

8-13. Draw a diagram [similar to Fig. 8-15(a)] showing how you would wire the 7493 counter as a 4-bit (mod-16) ripple counter. Refer to Fig. 8-13.

8-14. Refer to Fig. 8-31 (see page 174). The 74192 counter is in the (clear, count up, load) mode during pulse *a*.

8-15. List the binary output from the 74192 counter IC after each of the eight input pulses shown in Fig. 8-31.

8-16. Refer to Fig. 8-17 for questions **a** to **e** on the 74HC393 IC counter:
  a. This is a _____ (ripple, synchronous) counter.
  b. This is a _____ (down, up, either up or down) counter.
  c. The MR pins are _____ (asynchronous, synchronous) active _____ (HIGH, LOW) inputs that clear the outputs.
  d. Each counter contains four _____ (R-S, T) flip-flops.
  e. This is a _____ (CMOS, TTL) counter.

8-17. Refer to Fig. 8-18 for questions **a** to **e** on the 74HC193 IC counter:
  a. When the MR pin is activated with a _____ (HIGH, LOW), all outputs are reset to _____ (0, 1).
  b. This is a _____ (ripple, synchronous) counter.
  c. Parallel data from the data inputs (D_0 to D_3) flow through to the outputs (Q_0 to Q_3) when the _____ input is activated with a LOW.

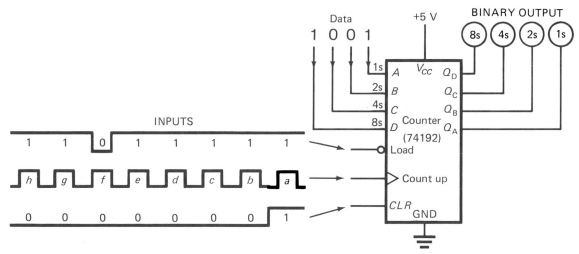

Fig. 8-31   IC counter pulse-train problem.

   d. When a clock signal enters pin CP_U, the CP_D pin must be tied to _____ (+5 V, GND).
   e. This IC counter is called *presettable* because of what operating mode?
8-18. Refer to Fig. 8-32. The 74HC193 IC is wired as a mod-_____ (number) counter in this circuit.
8-19. Refer to Fig. 8-32. List the mode of operation for the 74HC193 counter during each pulse *a* to *h*. (use answers *parallel load, count up, count down*)
8-20. Refer to Fig. 8-32. List the binary output for the 74HC193 counter IC after each pulse *a* to *h*.
8-21. Refer to Fig. 8-23. The two 74HC85 magnitude comparator ICs are said to be _____ (cascaded, subdivided) so they can compare two _____ (number)-bit binary numbers.

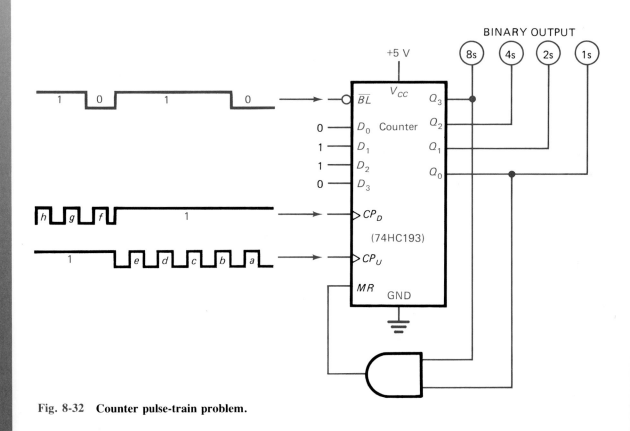

Fig. 8-32   Counter pulse-train problem.

8-22. Refer to Fig. 8-33. List the *color* of the output LED that is lit for each time period (t_1 to t_6).

8-23. While an oscilloscope is very good for observing repetitive digital signals, a logic _____ (analyzer, monitor) is better for looking at asynchronous signals.

8-24. The equipment pictured in Fig. _____ [8-25(a), 8-27(a)] is an easy-to-use instrument for testing if an IC is good or bad.

8-25. A digital _____ (IC tester, pulser) is an instrument for injecting a signal into a circuit.

8-26. List five pieces of electronic test equipment used in testing and troubleshooting sequential digital logic circuits.

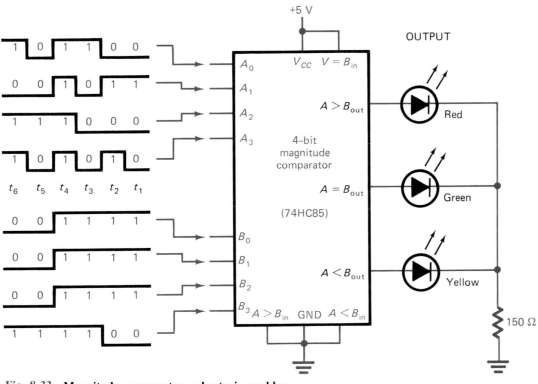

Fig. 8-33 **Magnitude comparator pulse-train problem.**

Answers to Self-Tests

1. two
2. 4
3. toggle
4. pulse $a = 00$
   pulse $b = 01$
   pulse $c = 10$
   pulse $d = 11$
   pulse $e = 00$
   pulse $f = 01$
5. ripple, decade
6. ripple, 5
7. pulse $a = 111$, then cleared to 000 just before pulse b
   pulse $b = 001$
   pulse $c = 010$
   pulse $d = 011$
   pulse $e = 100$

   pulse $f = 000$
8. synchronous
9. parallel
10. toggle
11. all the flip-flops toggle
12. toggle
13. HIGH-to-LOW
14. only FF 1 toggles
15. pulse $a = 00$
    pulse $b = 11$
    pulse $c = 10$
    pulse $d = 01$
    pulse $e = 00$
    pulse $f = 11$
16. down
17. HIGH, toggle
18. LOW, hold

19. 1000
20. 2
21. 0000 (reset)
22. four, up
23. decade, synchronous
24. 5
25. HIGH
26. point $B = 200$ Hz
    point $C = 100$ Hz
    point $D = 50$ Hz
27. 8
28. 4-bit binary
29. HIGH
30. H-to-L
31. 16, ripple
32. synchronous
33. asynchronous

34. Q_0 to Q_3
35. 0001, 0010, 0011, 0100, 0101, 0110 (1 to 6 in decimal)
36. to preset the counter to 0001 after the highest count of 0110
37. manufacturers use different standards when drawing logic symbols and labeling
38. red, too high

39. press and release switch SW_1
40. astable
41. t_1 = red
    t_2 = green
    t_3 = red
    t_4 = yellow
    t_5 = red
    t_6 = green
42. sequential

43. monitor
44. single pulse
45. repetitive
46. analyzer
47. HIGH
48. clear
49. preset (asynchronous), HIGH
50. 7, floating

CHAPTER 9

Shift Registers

*A typical example of a shift register at work is found within a calculator. As you enter each digit on the keyboard, the numbers shift to the left on the display. In other words, to enter the number 268 you must do the following. First, you press and release the 2 on the keyboard; a 2 appears at the extreme right on the display. Next, you press and release the 6 on the keyboard causing the 2 to shift one place to the left allowing for 6 to appear on the extreme right; 26 appears on the display. Finally, you press and release the 8 on the keyboard; 268 appears on the display. This example shows two important characteristics of a shift register: (1) It is a **temporary memory** and thus holds the numbers on the display (even if you release the keyboard number) and (2) it shifts the numbers to the left on the display each time you press a new digit on the keyboard. These **memory** and **shifting characteristics** make the shift register extremely valuable in most digital electronic systems. This chapter introduces you to shift registers and explains their operations.*

Shift registers are constructed by wiring flip-flops together. We mentioned in Chaps. 7 and 8 that flip-flops have a memory characteristic. This memory characteristic is put to good use in a shift register. Instead of wiring shift registers by using individual gates or flip-flops, you can buy shift registers in IC form.

Shift registers often are used to momentarily store data. Figure 9-1 shows a typical example of where shift registers might be used in a

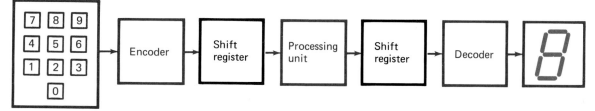

Fig. 9-1   A digital system using shift registers.

From page 177

Memory and shifting characteristics

Shift registers

On this page:

Serial in-serial out

Serial in-parallel out

Parallel in-serial out

Parallel in-parallel out

Shift register characteristics

digital system. This system could be that of a calculator. Notice the use of shift registers to hold information from the encoder for the processing unit. A shift register is also being employed for temporary storage between the processing unit and the decoder. Shift registers are also used at other locations within a digital system.

One method of describing shift register characteristics is by how data is *loaded into* and *read from* the storage units. Four categories of shift registers are illustrated in Fig. 9-2. Each storage device in Fig. 9-2 is an 8-bit register. The registers are classified as:

1. Serial in–serial out [Fig. 9-2(a)]
2. Serial in–parallel out [Fig. 9-2(b)]
3. Parallel in–serial out [Fig. 9-2(c)]
4. Parallel in–parallel out [Fig. 9-2(d)]

The diagrams in Fig. 9-2 illustrate the fundamental idea of each type of register. These classifications are often used in a manufacturer's literature.

9-1 SERIAL LOAD SHIFT REGISTERS

A basic shift register is shown in Fig. 9-3. This shift register is constructed from four D flip-flops. This register is called a *4-bit shift register* because it has four places to store data: A, B, C, D.

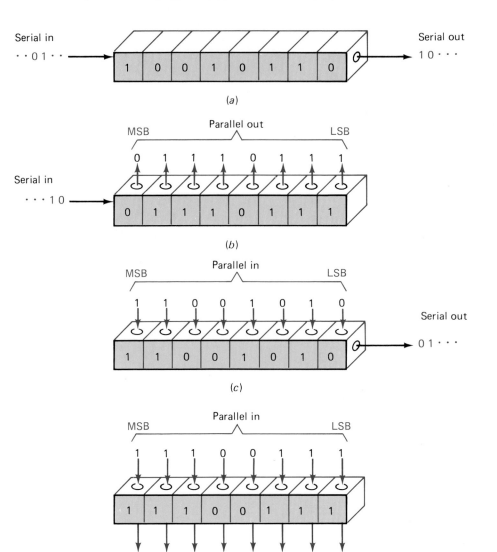

Fig. 9-2 Shift register characteristics. (*a*) Serial in–serial out. (*b*) Serial in–parallel out. (*c*) Parallel in–serial out. (*d*) Parallel in–parallel out.

With the aid of Table 9-1 on page 180 and Fig. 9-3, let us operate this shift register. First, clear (*CLR* input to 0) all the outputs (*A, B, C, D*) to 0000. (This situation is shown in line 1, Table 9-1.) The outputs remain 0000 while they await a clock pulse. Pulse the *CLK* input once; the output now shows 1000 (line 3, Table 9-1) because the 1 from the *D* input of FF *A* has has been transferred to the *Q* output on the clock pulse. Now enter 1s on the data input (clock pulses 2 and 3, Table 9-1); these 1s shift across the display to the right. Next, enter 0s on the data input (clock pulses 4 to 8, Table 9-1); you can see the 0s being shifted across the display (lines 6 to 10, Table 9-1). On clock pulse 9 (Table 9-1) enter a 1 at the data input. On pulse 10 the data input is returned to 0. Pulses 9 to 13 show the single 1 on display being shifted to the right. Line 15 shows the 1 being shifted out the right end of the shift register and being lost.

Remember that the D flip-flop is also called a *delay* flip-flop. Recall that it simply transfers the data from input *D* to output *Q after a delay of one clock pulse.*

The circuit diagramed in Fig. 9-3 is referred to as a *serial load shift register.* The term "serial load" comes from the fact that only one bit of data at a time can be entered in the register. For instance, to enter 0111 in the register, we had to go through the sequence from lines 1 through 6 in Table 9-1. It took five steps (line 2 was not needed) to serially load 0111 into the serial load shift register. To enter 0001 in this serial load shift register we need five steps, as shown in Table 9-1, lines 10 to 14. According to the classifications in Fig. 9-2, this would be a serial in–parallel out register. However, if data

were taken from only FF *D*, it becomes a serial in–serial out register.

The shift register in Fig. 9-3 could become a 5-bit shift register by just adding one more D flip-flop. Shift registers typically come in 4-, 5- and 8-bit sizes. Shift registers also can be wired using other flip-flops. J-K flip-flops and clocked R-S flip-flops are also used to wire shift registers.

Self-Test

Answer the following questions.

1. The unit shown in Fig. 9-4 on page 180 is a shift-right _____ (parallel, serial) load shift register.
2. List the contents of the register in Fig. 9-4 after each of the six clock pulses (*A* = left bit, *C* = right bit).
3. A(n) _____ (entire 3-bit group, single bit) is loaded on each clock pulse in the serial load shift register in Fig. 9-4.

9-2 PARALLEL LOAD SHIFT REGISTERS

The serial load shift register we studied in the last section has two disadvantages: it permits only one bit of information to be entered at a time, and it loses all its data out the right side when it shifts right. Figure 9-5(*a*) on page 181 illustrates a system that permits *parallel loading* of 4 bits at once. These inputs are the data inputs *A, B, C,* and *D* in Fig. 9-5. This system could also incorporate a *recirculating* feature that would put the output data back into the input so that it is not lost.

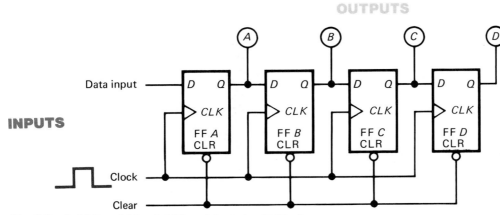

Fig. 9-3 A 4-bit serial load shift register using D flip-flops.

Table 9-1 Operation of a 4-Bit Serial Shift Register

Line Number	Clear	Data	Clock Pulse Number	FF A A	FF B B	FF C C	FF D D
	Inputs			Outputs			
1	0	0	0	0	0	0	0
2	1	1	0	0	0	0	0
3	1	1	1	1	0	0	0
4	1	1	2	1	1	0	0
5	1	1	3	1	1	1	0
6	1	0	4	0	1	1	1
7	1	0	5	0	0	1	1
8	1	0	6	0	0	0	1
9	1	0	7	0	0	0	0
10	1	0	8	0	0	0	0
11	1	1	9	1	0	0	0
12	1	0	10	0	1	0	0
13	1	0	11	0	0	1	0
14	1	0	12	0	0	0	1
15	1	0	13	0	0	0	0

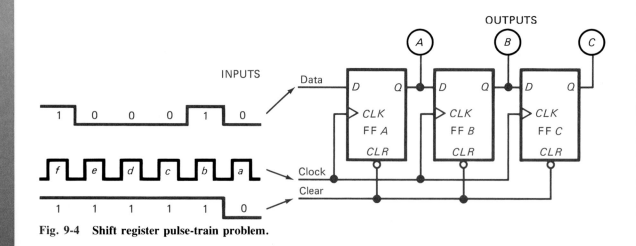

Fig. 9-4 Shift register pulse-train problem.

A wiring diagram of the *4-bit parallel load recirculating shift register* is drawn in Fig. 9-5(*b*). This shift register uses four J-K flip-flops. Notice the recirculating lines leading from the Q and $\overline{Q}$ outputs of FF D back to the J and K inputs of FF A. These feedback lines cause the data that would normally be lost out of FF D to recirculate through the shift register. The CLR input clears the outputs to 0000 when enabled by a logical 0. The parallel load data inputs A, B, C, and D are connected to the preset (PS) inputs of the flip-flops to set 1s at any output position (A,

B, C, D). If the switches attached to the parallel load data inputs are even temporarily switched to a 0, that output will be preset to a logical 1. The clock pulsing the CLK inputs of the J-K flip-flops will cause data to be shifted to the right. The data from FF D will be recirculated back to FF A.

Table 9-2, on page 182, will help you understand the operation of the parallel load shift register. As you turn on the power, the outputs may take any combination, such as the one in line 1. Line 2 shows the register being cleared with the CLR input. Line 3 shows 0100

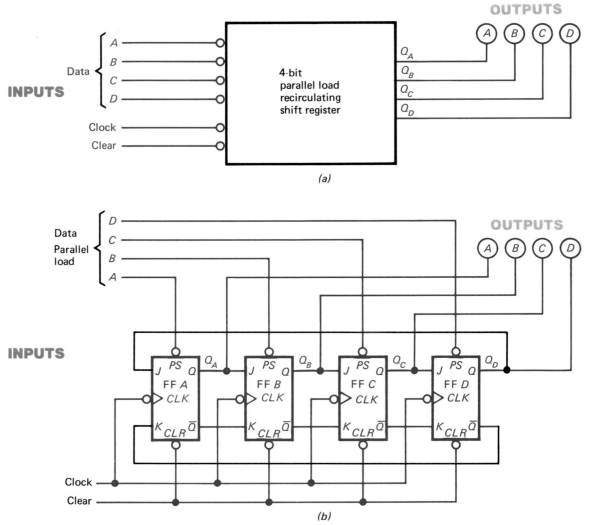

Fig. 9-5 **A 4-bit parallel load recirculating shift register.** (*a*) **Block diagram.** (*b*) **Wiring diagram.**

being loaded into the register using the parallel load data switches. Lines 4 to 8 show five clock pulses and the shifting of the data to the right. Look at lines 5 and 6: the 1 is being recirculated from the right end (FF *D*) of the register back to the left end (FF *A*). We say the 1 is being recirculated.

Line 9 shows the register being cleared again by the *CLR* input. New information (0110) is being loaded in the data inputs in line 10. Lines 11 to 15 illustrate the register being shifted five times by clock pulses. Note that it takes four clock pulsesto come back to the original data in the register (compare lines 11 and 15 or lines 4 and 8 in Table 9-2). The register in Fig. 9-5 could be classified as a parallel in–parallel out storage device.

The recirculating feature of the shift register in Fig. 9-5(*b*) can be disabled by disconnecting the two recirculating lines. The register is

then a parallel in–parallel out register. However, if only the output from FF *D* is considered, this register is a parallel in–serial out storage device.

Self-Test

Answer the following questions.

4. The unit in Fig. 9-6 is a shift-right _____ (serial, parallel) load recirculating shift register.
5. Refer to Fig. 9-6. List the mode of operation of the shift register during each of the eight clock pulses. Use as answers the terms "clear," "parallel load," and "shift-right."
6. List the contents of the register in Fig. 9-6 immediately after each of the eight clock pulses (A = left bit, C = right bit).

Table 9-2   Operation of a 4-Bit Parallel Load Recirculating Shift Register

Line Number	Clear	Parallel Load Data				Clock Pulse Number	FF A	FF B	FF C	FF D
		A	B	C	D		A	B	C	D
1	1	1	1	1	1	0	1	1	1	0
2	0	1	1	1	1	0	0	0	0	0
3	1	1	0	1	1	0	0	1	0	0
4	1	1	1	1	1	1	0	0	1	0
5	1	1	1	1	1	2	0	0	0	1
6	1	1	1	1	1	3	1	0	0	0
7	1	1	1	1	1	4	0	1	0	0
8	1	1	1	1	1	5	0	0	1	0
9	0	1	1	1	1		0	0	0	0
10	1	1	0	0	1		0	1	1	0
11	1	1	1	1	1	6	0	0	1	1
12	1	1	1	1	1	7	1	0	0	1
13	1	1	1	1	1	8	1	1	0	0
14	1	1	1	1	1	9	0	1	1	0
15	1	1	1	1	1	10	0	0	1	1

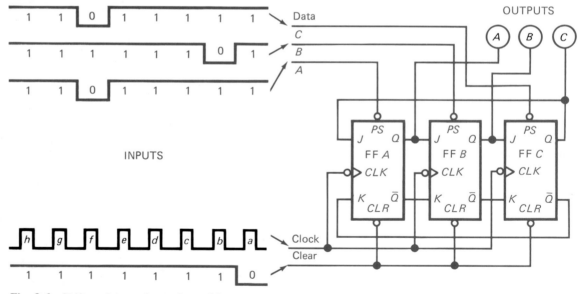

Fig. 9-6   Shift register pulse-train problem.

9-3 A UNIVERSAL SHIFT REGISTER

When reviewing data manuals you will see that manufacturers produce many shift registers in IC form. In this section one such IC shift reg-ister will be studied: the *74194 4-bit bidirectional universal shift register*.

The 74194 IC is a very adaptable shift reg-ister and has most of the features we have seen so far in one IC package. A 74194 IC register can shift right or left. It can be loaded serially or in parallel. Several 4-bit 74194 IC registers can be cascaded to make an 8-bit or longer shift register. And this register can be made to recirculate data.

The Signetics data manual contains the descriptions, diagrams, and tables shown in Fig. 9-7, on page 184. Read the description of the 74194 shift register in Fig. 9-7(a) for a good overview of what this shift register can do.

A logic diagram of the 74194 register is reproduced in Fig. 9-7(b). Because it is a 4-bit register, the circuit contains four flip-flops. Extra gating circuitry is needed for the many features of this universal shift register. The pin configuration in Fig. 9-7(c) will help you determine the labeling of each input and output. Of course, the pin diagram is also a must when actually wiring a 74194 IC.

The truth table and waveform diagrams in Fig. 9-7(d) and (e) are very helpful in determining exactly how the 74194 IC register works because they illustrate the clear, load, shift-right, shift-left, and inhibit modes of operation. As you use the 74194 universal shift register you will have occasion to look quite carefully at the truth table and waveform diagrams.

Self-Test

Answer the following questions.

7. List the five modes of operation for the 74194 universal shift register IC.
8. Refer to Fig. 9-7. If both mode control inputs ($S0$, $S1$) to the 74194 are HIGH, the unit is in the _____ mode.
9. Refer to Fig. 9-7. If both mode control inputs ($S0$, $S1$) to the 74194 IC are LOW, the unit is in the _____ mode.
10. Refer to Fig. 9-7. Shift right on the 74194 IC is accomplished when $S0$ is _____ (HIGH, LOW) and $S1$ is _____ (HIGH, LOW) and when the clock pulse goes from _____ to _____ .

9-4 USING THE 74194 IC SHIFT REGISTER

In this section we shall use the 74194 universal shift register in several ways. Figure 9-8(a) and (b) on page 185 shows the 74194 IC being used as serial load registers. A *serial load shift-*

right register is shown in Fig. 9-8(a). This register operates exactly like the serial shift register in Fig. 9-3. Table 9-1 could also be used to chart the performance of this new shift register. Notice that the mode control inputs ($S0$, $S1$) must be in the positions shown for the 74194 IC to operate in its shift-right mode. Shifting to the right is defined by the manufacturer as shifting from Q_A to Q_D. The register in Fig. 9-8(a) shifts data to the right, and as it leaves Q_D the data is lost.

The 74194 IC has been rewired slightly in Fig. 9-8(b). The shift-left serial input is used, and the mode control inputs *have been changed*. This register enters data at D (Q_D) and shifts it toward A (Q_A) with each pulse of the clock. This register is a *serial load shift-left register*.

In Fig. 9-9 on page 186, the 74194 IC is wired as a *parallel load shift-right/left register*. With a single clock pulse the data from the parallel load inputs A, B, C, and D appears on the display. The loading happens only when the mode controls ($S0$, $S1$) are set at 1, as shown. The mode control can then be changed to one of the three types of operations: shift right, shift left, or inhibit. The shift-right and shift-left serial inputs both are connected to 0 to feed in 0s to the register in the shift-right or shift-left mode of operation. With the mode control in the inhibit position ($S0 = 0$, $S1 = 0$), the data does not shift right or left but stays in position in the register. When using the 74194 IC you must remember the mode control inputs because they control the operation of the entire register. The *CLR* input clears the register to 0000 when enabled by a 0. The *CLR* input overrides all other inputs.

Two 74194 IC shift registers are connected in Fig. 9-10 to form an *8-bit parallel load shift-right register*. The *CLR* input clears the outputs to 0000 0000. The parallel load inputs A to H allow entry of all eight bits of data on a single clock pulse (mode control: $S0 = 1$, $S1 = 1$). With the mode control in the shift-right position ($S0 = 1$, $S1 = 0$), the register shifts right for each clock pulse. Notice that a recirculating line has been placed from output H (output Q_D of register 2) back to the shift-right serial input of shift register 1. Data that normally would be lost out of output H is recirculated back to position A in the register. Both inputs $S0$ and $S1$ at 0 will inhibit the shifting of data in the shift register.

74194 shift register

Modes of operation

Serial load shift-right register

Mode control inputs

Serial load shift-left register

Parallel load shift-right/left register

8-bit parallel load shift-right register

(a) DESCRIPTION

This bidirectional shift register is designed to incorporate virtually all of the features a system designer may want in a shift register. The circuit contains 45 equivalent gates and features parallel inputs, parallel outputs, right-shift and left-shift serial inputs, operating-mode-control inputs, and a direct overriding clear line. The register has four distinct modes of operation, namely:

> Parallel (broadside) load
> Shift right (in the direction Q_A toward Q_D)
> Shift left (in the direction Q_D toward Q_A)
> Inhibit clock (do nothing)

Synchronous parallel loading is accomplished by applying the four bits of data and taking both mode control inputs, $S0$ and $S1$, high. The data are loaded into the associated flip-flops and appear at the outputs after the positive transistion of the clock input. During loading, serial data flow is inhibited.

Shift right is accomplished synchronously with the rising edge of the clock pulse when $S0$ is high and $S1$ is low. Serial data for this mode is entered at the shift-right data input. When $S0$ is low and $S1$ is high, data shifts left synchronously and new data is entered at the shift-left serial input.

Clocking of the flip-flop is inhibited when both mode control inputs are low. The mode controls of the S54194/N74194 should be changed only while the clock input is high.

(b) BLOCK DIAGRAM

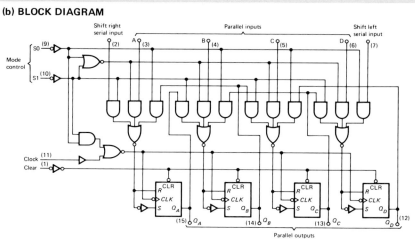

(c) PIN CONFIGURATION

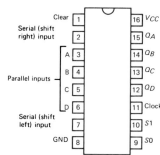

(d) TRUTH TABLE

INPUTS										OUTPUTS			
	MODE			SERIAL		PARALLEL				Q_A	Q_B	Q_C	Q_D
CLEAR	S1	S0	CLOCK	LEFT	RIGHT	A	B	C	D				
L	X	X	X	X	X	X	X	X	X	L	L	L	L
H	X	X	L	X	X	X	X	X	X	Q_{A0}	Q_{B0}	Q_{C0}	Q_{D0}
H	H	H	↑	X	X	a	b	c	d	a	b	c	d
H	L	H	↑	X	H	X	X	X	X	H	Q_{An}	Q_{Bn}	Q_{Cn}
H	L	H	↑	X	L	X	X	X	X	L	Q_{An}	Q_{Bn}	Q_{Cn}
H	H	L	↑	H	X	X	X	X	X	Q_{Bn}	Q_{Cn}	Q_{Dn}	H
H	H	L	↑	L	X	X	X	X	X	Q_{Bn}	Q_{Cn}	Q_{Dn}	L
H	L	L	X	X	X	X	X	X	X	Q_{A0}	Q_{B0}	Q_{C0}	Q_{D0}

H = high level (steady state)
L = low level (steady state)
X = irrelevant (any input, including transitions)
↑ = transition from low to high level
a,b,c,d, = the level of steady state input at inputs A,B,C, or D, respectively
Q_{A0}, Q_{B0}, Q_{C0}, Q_{D0} = the level of Q_A, Q_B, Q_C, Q_D, respectively, before the indicated steady state input conditions were established
Q_{An}, Q_{Bn}, Q_{Cn}, Q_{Dn} = the level of Q_A, Q_B, Q_C, Q_D, respectively, before the most recent ↑ transition of the clock

(e) TYPICAL CLEAR, SHIFT, AND LOAD SEQUENCES

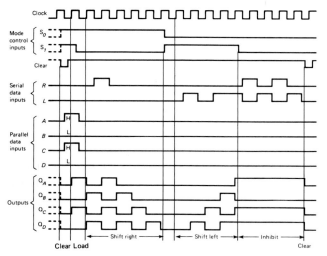

Fig. 9-7 A 4-bit TTL universal shift register (74194). (a) Description. (b) Block diagram. (c) Pin configuration. (d) Truth table. (e) Waveforms. (Courtesy of Signetics Corporation)

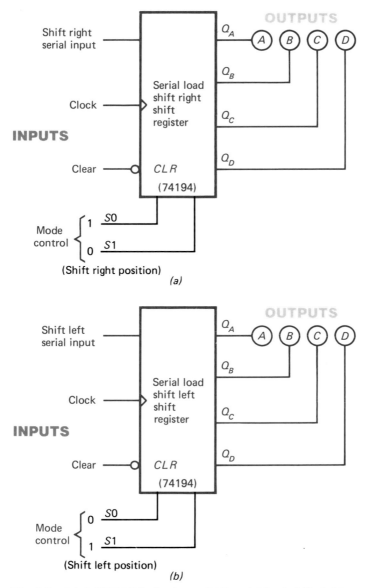

Fig. 9-8 (*a*) **A 74194 IC wired as a 4-bit serial load shift-right register.** (*b*) **A 74194 IC wired as a 4-bit serial load shift-left register.**

As you have just seen, the 74194 IC 4-bit bidirectional universal shift register is very useful. The circuits in this unit are only a few examples of how the 74194 IC can be used. Remember that all shift registers use as their basis the memory characteristic of a flip-flop. Shift registers often are used as temporary memories. Shift registers also can be used to convert serial data to parallel data or parallel data to serial data. And shift registers can be used to delay information (delay lines). Shift registers are also used in some arithmetic circuits. Microprocessors and microprocessor-based systems make extensive use

of registers similar to the ones used in this chapter.

Self-Test

Supply the missing word or words in each statement.

11. The 74194 IC is in the parallel load mode when both mode control inputs (*S0, S1*) are _____ (HIGH, LOW). The 4 bits of data at the parallel load inputs are loaded into the registers by applying _____ (number) clock pulse(s) to the *CLK* input.

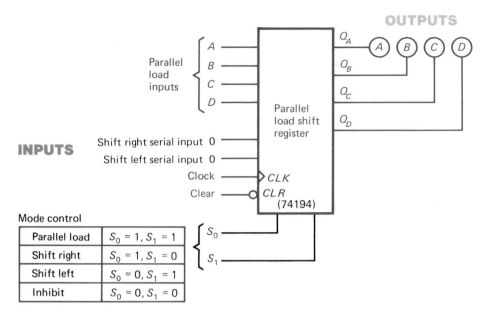

Fig. 9-9   A 74194 IC wired as a parallel load shift-right/left register.

Mode control

Parallel load	$S_0 = 1, S_1 = 1$
Shift right	$S_0 = 1, S_1 = 0$
Shift left	$S_0 = 0, S_1 = 1$
Inhibit	$S_0 = 0, S_1 = 0$

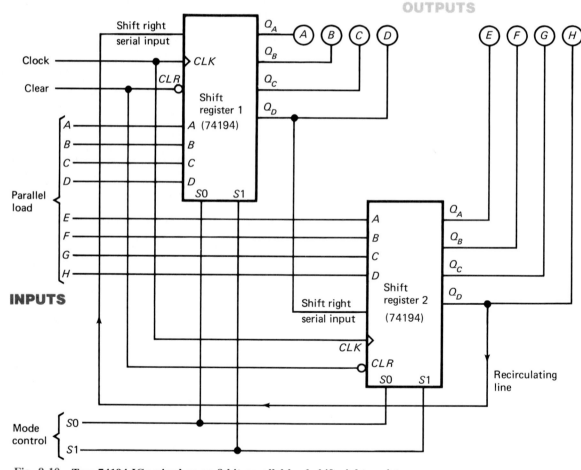

Fig. 9-10   Two 74194 ICs wired as an 8-bit parallel load shift-right register.

12. If the mode control inputs ($S0$, $S1$) to the 74194 IC are both LOW, the shift register is in the _____ mode.

13. For the 74194 IC to shift right, the mode controls are $S0 =$ _____ and $S1 =$ _____ and the serial data enters the _____ input.

14. Refer to Fig. 9-9. If $S0 = 1$, $S1 = 1$, shift left serial input = 1, and clear input = 0, then the outputs are _____ .

9-5 AN 8-BIT CMOS SHIFT REGISTER

This section will detail the operation of one of many CMOS shift registers available from manufacturers. Signetics Corporation has furnished the technical information in Fig. 9-11 (see page 188) on the 74HC164 *8-bit serial in–parallel out shift register*.

The 74HC164 CMOS IC is an 8-bit edge-triggered register with serial data entry. Parallel outputs are available from each internal D flip-flop. The detailed logic diagram in Fig. 9-11(a) shows the use of eight D flip-flops with parallel data outputs (Q_0 to Q_7).

The 74HC164 IC featured in Fig. 9-11 is described as having a serial input. Data is entered serially through one of two inputs (D_{sa} and D_{sb}). Observe on Fig. 9-11(a) that the data inputs (D_{sa} and D_{sb}) are ANDed together. The data inputs may be tied together as a single input or one may be tied HIGH using the other for data entry.

The master reset input ($\overline{MR}$) to the 74HC164 IC is shown at the lower left in Fig. 9-11(a). It is an active LOW input. The truth table in Fig. 9-11(b) shows that the MR input overrides all other inputs and clears all flip-flops to 0 when activated.

The 74HC164 IC shifts data one place to the right on each LOW-to-HIGH transition of the clock (CP) input. The clock pulse also enters data from the ANDed data inputs (D_{sa} and D_{sb}) into output Q_0 of FF 1 [see Fig. 9-11(a)].

For your reference, a pin diagram for the 74HC164 shift register IC is reproduced in Fig. 9-11(c). The helpful table in Fig. 9-11(d) describes the function of each pin on this CMOS IC.

Self-Test

Answer the following questions.

15. The 74HC164 IC's master reset pin is an active _____ (HIGH, LOW) input.
16. The clock input to the 74HC164 IC responds to a _____ (H-to-L, L-to-H) transition of the clock pulse.

17. Refer to Fig. 9-12 on page 189. List the shift register's mode of operation for each clock pulse (*a* to *f*).
18. Refer to Fig. 9-12. List the 8-bit output (Q_0 bit on left, Q_7 bit on right) after each of the six clock pulses.

9-6 USING SHIFT REGISTERS— DIGITAL ROULETTE

The roulette wheel holds great fascination for people of all ages. Variations are used in game shows and of course in gambling casinos. This section explores an electronic version of the mechanical roulette wheel. Digital roulette is a favorite project for many students.

A block diagram of a digital roulette wheel is sketched in Fig. 9-13 on page 189. This simple roulette wheel design uses only eight number markers. The number markers are LEDs in this electronic version of roulette. Only a single LED (number marker) must light at a time. A *ring counter* is a circuit that will cause the LEDs to light, one at a time, in sequence. A ring counter is simply a shift register with some added circuitry.

When turning on the power, the shift register in Fig. 9-13 must first be cleared to all zeros. Note that the system on-off switch is not represented in the block diagram. Second, when the "spin wheel" switch is pressed a *single* HIGH must be loaded into position 0 on the display lighting LED 0. The *voltage-controlled oscillator* (VCO) puts out a string of clock pulses which gradually decrease in frequency and stop. The clock pulses are directed to the ring counter (shift register) and the audio amplifier sections of the digital roulette game. Each clock pulse entering the ring counter will shift the single light around the roulette wheel. The lighting sequence should be 0, 1, 2, 3, 4, 5, 6, 7, 0, 1, and so forth, until the VCO stops emitting clock pulses. When clock pulses stop, a single LED should remain lit on the roulette wheel in some random position.

The VCO in Fig. 9-13 also sends clock pulses to the audio amplifier section. Each clock pulse is amplified to sound like the click of a roulette wheel. The frequency gradually decreases and stops, simulating a mechanical wheel coasting to a stop.

The ring counter block of the digital roulette game is detailed in Fig. 9-14(a) on page 190. Notice that the ring counter makes use of the

CMOS shift registers

74HC164 8-bit serial in-parallel out shift register

Digital roulette wheel

Ring counter

Voltage-controlled oscillator (VCO)

Audio amplifier

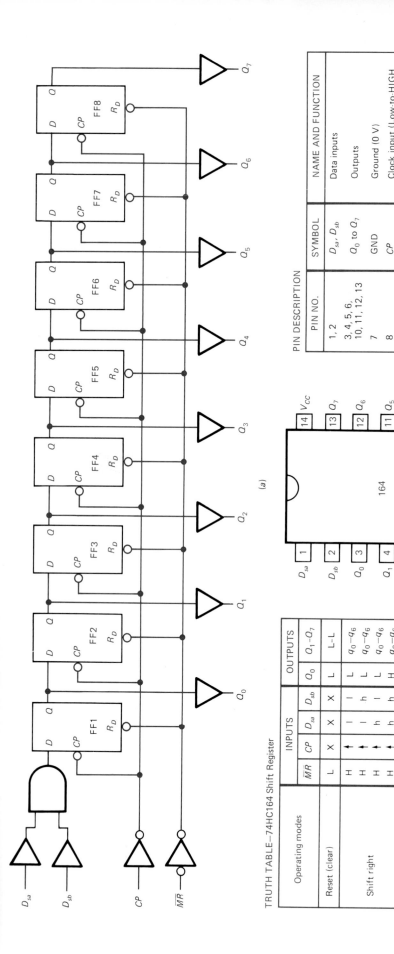

(a)

(c)

TRUTH TABLE—74HC164 Shift Register

Operating modes	INPUTS				OUTPUTS		
	$\overline{MR}$	CP	D_{sa}	D_{sb}	Q_0	Q_1–Q_7	
Reset (clear)	L	X	X	X	L	L–L	
Shift right	H	↑	l	l	L	q_0–q_6	
	H	↑	l	h	L	q_0–q_6	
	H	↑	h	l	L	q_0–q_6	
	H	↑	h	h	H	q_0–q_6	

H = HIGH voltage level
h = HIGH voltage level one set-up time prior to the LOW-
    to-HIGH clock transition
L = LOW voltage level
l = LOW voltage level one set-up time prior to the LOW-
    to-HIGH clock transition
q = lowercase letters indicate the state of the referenced
    input one set-up time prior to the LOW-to-HIGH
    clock transition
↑ = LOW-to-HIGH clock transition

(b)

PIN DESCRIPTION

PIN NO.	SYMBOL	NAME AND FUNCTION
1, 2	D_{sa}, D_{sb}	Data inputs
3, 4, 5, 6, 10, 11, 12, 13	Q_0 to Q_7	Outputs
7	GND	Ground (0 V)
8	CP	Clock input (Low-to-HIGH, edge-triggered)
9	$\overline{MR}$	Master reset input (active LOW)
14	V_{cc}	Positive supply voltage

(d)

Fig. 9-11 An 8-bit CMOS serial in–parallel out shift register (74HC164). (a) Detailed logic diagram. (b) Truth table. (c) Pin diagram. (d) Pin descriptions. (Courtesy of Signetics Corporation)

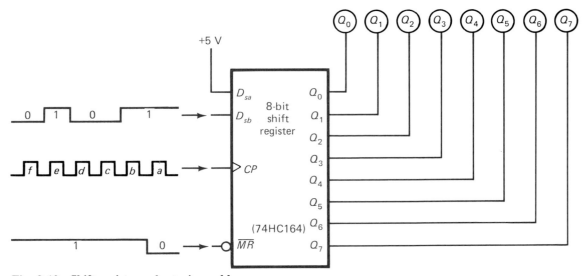

Fig. 9-12 **Shift register pulse-train problem.**

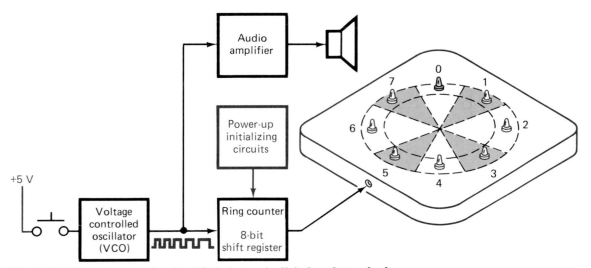

Fig. 9-13 **Block diagram for simplified electronic digital roulette wheel.**

Electronic digital roulette wheel

Recirculating line (feedback)

74HC164 8-bit shift register IC

Ring counter

74HC164 8-bit serial in-parallel out shift register IC studied earlier. When power is turned on, the circuits in the power-up initializing block clear all output to zero (all LEDs are off). Upon pressing the "spin wheel" input switch, the first pulse loads a single HIGH into the shift register. This situation is illustrated in Fig. 9-14(a). The clock pulses that follow move the single light across the display. This is illustrated in Fig. 9-14(b). Notice that on each L-to-H transition of the clock the single HIGH in the 74HC164 8-bit register shifts one position to the right. When the HIGH reaches output Q_7 [after clock pulse eight in Fig. 9-14(b)], a recirculating line (feedback) is run back to the data inputs to transfer the HIGH back to the left LED (output Q_0). In the example in Fig. 9-14(b), the switch is opened after the

twelfth pulse. This stops the light at Q_3. This is the "winning number" on the roulette wheel for this spin.

The 74HC164 8-bit shift register IC is wired as a ring counter in Fig. 9-14(a). The circuit has the two characteristics that make it a ring counter. First, it has feedback from the last flip-flop (Q_7) to the first FF (Q_0). Second, it is loaded with a given pattern of 1s and 0s and these recirculate as long as clock pulses reach the CP input of the shift register. In this case, a single 1 is loaded into the shift register and is recirculated.

In summary, the circuit in Fig. 9-14(a) is a very simple electronic roulette wheel. Pressing the spin wheel input causes the single light to be circulated through the LEDs. When the switch opens the shifting stops.

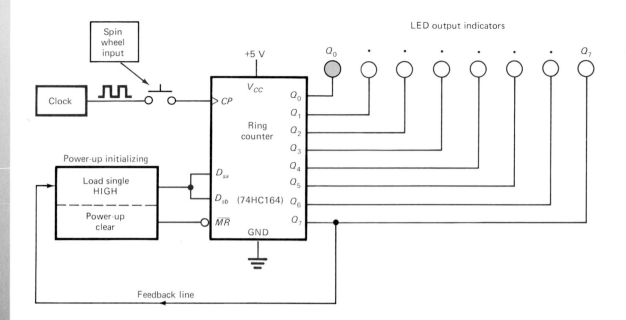

(a)

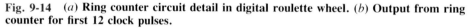

(b)

Fig. 9-14 (a) **Ring counter circuit detail in digital roulette wheel.** (b) **Output from ring counter for first 12 clock pulses.**

For added appeal, the simple digital roulette circuit in Fig. 9-14 can be changed by adding a clock that will continue to run for a time after the push button is released. Sound could also be added for a more realistic simulation. Figure 9-15 adds both features to the digital roulette wheel.

The versatile 555 timer IC is wired as a VCO in Fig. 9-15 (see next page). Pressing the spin wheel input switch turns on transistor Q_1. The 555 timer operates as a free-running MV. This square-wave output from the VCO drives both the clock input (CP) of the ring counter and the audio amplifier. Pulses from the VCO alternately turn transistor Q_2 on and off, clicking the speaker.

When the spin wheel input switch is opened the 47-μF capacitor holds a positive charge for a time which is applied to the base (B) of transistor Q_1. This keeps the transistor turned on for several seconds before the capacitor becomes discharged. As the 47-μF capacitor discharges, the voltage at the base of Q_1 becomes less and the resistance of the transistor (from emitter to collector) increases. This decreases the frequency of the oscillator. This causes the shifting light to slow down. The clicking from the speaker also decreases in frequency. This simulates the slowing of a mechanical roulette wheel.

To review, the power-up initializing circuitry block in Fig. 9-15 must first clear the shift register and then set only the first output HIGH. These two circuits have been added to the digital roulette wheel in Fig. 9-16 (see page 193).

An automatic clear circuit has been added to the roulette wheel in Fig. 9-16. It consists of the resistor-capacitor combination (R_7 and C_4). When power is turned on, the voltage at the top of the 0.01-μF capacitor starts LOW and increases quickly to a HIGH as it charges through resistor R_7. The master reset ($\overline{MR}$) input to the 74HC164 register is held LOW just long enough for the output of the shift register to be cleared to 00000000. At this point all the LEDS are off.

The circuit that loads a single 1 into the ring counter consists of the four NAND gates and two resistors (R_5 and R_6). The NAND gates are wired as an R-S latch. The two resistors (R_5 and R_6) force the output of the NAND gate (ICa) HIGH when the power is first turned on. This HIGH is applied to the data inputs (D_{sa} and D_{sb}) of the ring counter. On the very first L-to-H transition of the clock the HIGH at the data inputs is trans-ferred to output Q_0 of the 74HC164 IC. Immediately this HIGH is fed back to the input of ICd and resets the latch so that a LOW now appears at the data inputs (D_{sa} and D_{sb}). Only a single HIGH was loaded into the ring counter. Repeated clock pulses move the HIGH (light) across the display until Q_7 of the ring counter goes HIGH. This HIGH is fed back to the input of ICc setting the latch so that a 1 appears at the data inputs of the ring counter. The single HIGH has been recirculated back to Q_0.

Self-Test

Answer the following questions.

19. Refer to Fig. 9-16. Components R_4 and Q_2 form the _____ block of the digital roulette wheel circuit.
20. Refer to Fig. 9-16. The 74HC164 8-bit shift register is wired as a(n) _____ in this circuit.
21. Refer to Fig. 9-16. What components cause the 74HC164 IC to reset all outputs to 0 when the power is first turned on?
22. Refer to Fig. 9-16. Clock pulses are fed to the ring counter by the output of the 555 timer IC wired as a(n) _____ .
23. Refer to Fig. 9-16. The four NAND gates used for loading a single 1 in the ring counter are wired as a(n) _____ circuit.

9-7 TROUBLESHOOTING A SIMPLE SHIFT REGISTER

Consider the faulty serial load shift-right register drawn in Fig. 9-17 on page 194. Two 7474 D flip-flops have been wired together to form this 4-bit register.

After checking for obvious mechanical and temperature problems, the student or technician runs the following sequence of tests to observe the problem:

1. *Action:* Clear input to 0 and back to 1.
   Result: Output indicators = 0000 (not lit).
   Conclusion: Clear function operating correctly.
2. *Action:* Data input = 1.
   Single pulse to *CLK* of flip-flops from logic pulser.

Digital roulette circuit

VCO

Audio amplifier

Power-up initializing circuitry

Automatic clear circuit

Ring counter

Troubleshooting

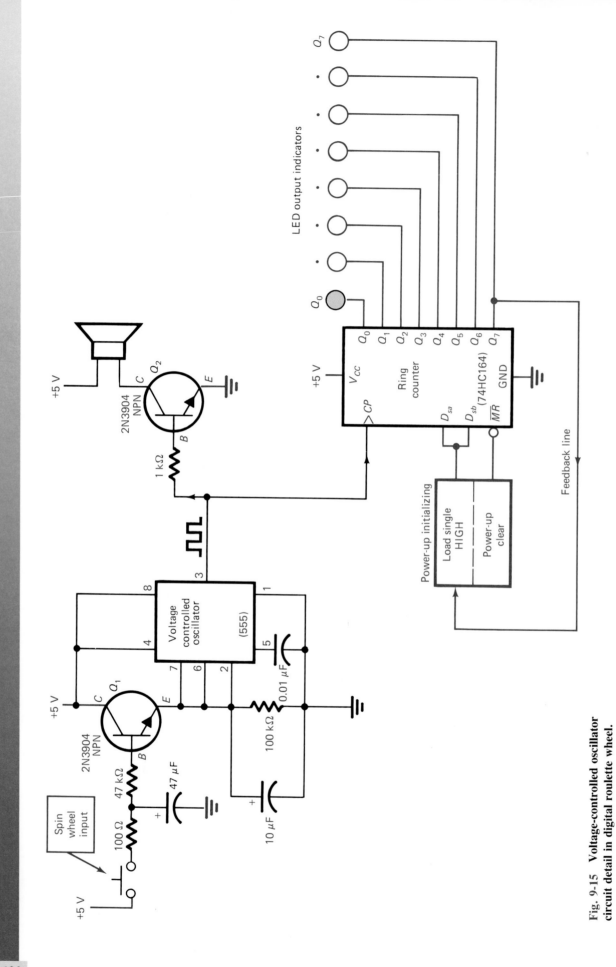

Fig. 9-15 Voltage-controlled oscillator circuit detail in digital roulette wheel.

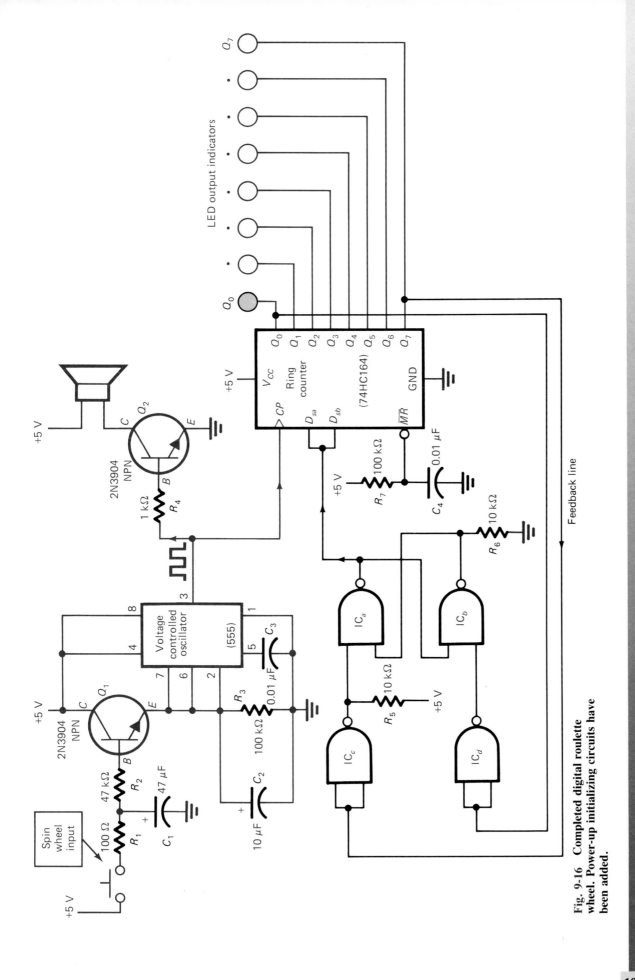

Fig. 9-16 Completed digital roulette wheel. Power-up initializing circuits have been added.

OUTPUT INDICATORS

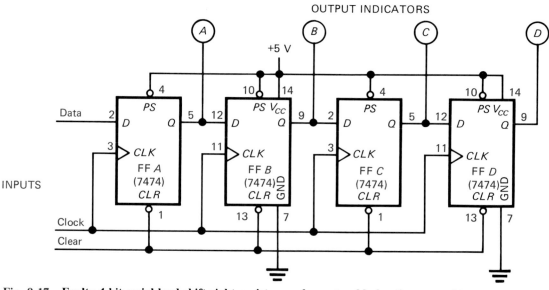

Fig. 9-17 Faulty 4-bit serial load shift-right register used as a troubleshooting example.

Result: Output indicators = 1000.
Conclusion: FF A loading 1s properly.

3. *Action:* Data input = 1.
Single pulse to *CLK* of flip-flops from logic pulser.
Result: Output indicators = 1100.
Conclusion: FF A and FF B loading 1s correctly.

4. *Action:* Data input = 1.
Single pulse to *CLK* of flip-flops from logic pulser.
Result: Output indicators = 1110.
Conclusion: FF A, FF B/FF C, and FF B loading 1s correctly.

5. *Action:* Data input = 1.
Single pulse to *CLK* of flip-flops from logic pulser.
Result: Output indicators = 1110.
Conclusion: Suspect problem near or in FF D since it did not load a HIGH properly.

6. *Action:* Logic probe at D input to FF D to see if D = 1.
Result: D = 1 on FF D.
Conclusion: HIGH data at D of FF D is correct.

7. *Action:* One pulse to *CLK* (pin 11) of FF D from logic pulser.
Result: Output indicator remains at 1110.
Conclusion: Data not being transferred from input D of FF D to output Q on a clock pulse.

8. *Action:* Logic probe to output Q of FF D (pin 9).

Result: Neither HIGH nor LOW indicator lights on logic probe.
Conclusion: Output Q (pin 9) of FF D floating between HIGH and LOW. Probably a faulty FF D in second 7474 IC.

9. *Action:* Remove and replace second 7474 IC (FF C and FF D) with exact replacement.

10. *Action:* Retest circuit, starting at step 1.
Result: All flip-flops load 1s and 0s.
Conclusion: Shift register circuit is now operating properly.

According to the sequence of tests, the *Q* output of FF D seemed to be stuck LOW while it was actually floating between LOW and HIGH. This fact made our conclusion in step 1 incorrect. This fault was caused by an open circuit within the second 7474 IC itself. Again, the technician's knowledge of how the circuit operates along with observations helped locate the fault. The logic probe and digital logic pulser aided the technician in making observations.

Sometimes the technician is not exactly sure of the appropriate logic level. In a circuit with redundant circuitry (circuits repeated over and over), the technician could go back to FF A and FF B and compare these readings with those on FF C and FF D. Digital circuits have much redundant circuitry, and at times this technique is helpful in troubleshooting.

Answer the following questions.

24. Refer to Fig. 9-17. Describe the observed problem in this circuit.

25. Refer to Fig. 9-17. What is wrong with this circuit?

26. Refer to Fig. 9-17. How can the fault in this circuit be repaired?

27. What test equipment can be used to troubleshoot this shift register circuit?

SUMMARY

1. Flip-flops are wired together to form shift registers.
2. A shift register has both a memory and a shift characteristic.
3. A serial load shift register is one that permits only one bit of data to be entered per clock pulse.
4. A parallel load shift register is one that permits all data bits to be entered at one time.
5. A register that is recirculating feeds output data back into the input.
6. Shift registers are designed to shift either left or right.
7. Manufacturers produce many adaptable universal shift registers.
8. Shift registers are widely used as temporary memories and for shifting data. They also have other uses in digital electronic systems.
9. A ring counter is a shift register that (1) has a recirculating line, and (2) is loaded with a pattern of 0s and 1s which is repeated over and over as the unit is clocked.

CHAPTER REVIEW QUESTIONS

Answer the following questions.

9-1. Draw a logic symbol diagram of a 5-bit serial load shift-right register. Use five D flip-flops. Label inputs "data," *CLK*, and *CLR*. Label outputs *A*, *B*, *C*, *D*, and *E*. The circuit will be similar to the one in Fig. 9-3.

9-2. Explain how you would clear to 00000 the 5-bit register you drew in question 1.

9-3. After clearing the 5-bit register, explain how you would enter (load) 10000 into the register you drew in question 1.

9-4. After clearing the 5-bit register, explain how you would enter (load) 00111 into the register you drew in question 1.

9-5. Refer to the register you drew in question 1. List the contents of the register after each clock pulse shown in **b** to **e** (assume data input = 0).
   a. Original output = 01001 ($A = 0$, $B = 1$, $C = 0$, $D = 0$, $E = 1$)
   b. After one clock pulse =
   c. After two clock pulses =
   d. After three clock pulses =
   e. After four clock pulses =

9-6. Refer to Table 9-2. The parallel load register using J-K flip-flops [Fig. 9-5(*b*)] needs _____ (no, one, three, four) clock pulse(s) to load data from the parallel data inputs.

9-7. Refer to Fig. 9-9. The parallel load register using the 74194 IC needs _____ (no, one, three, four) clock pulse(s) to load data from the parallel load inputs.

9-8. A _____ (serial, parallel) load shift register is the simplest circuit to wire.

9-9. A _____ (serial, parallel) load shift register is the easiest to load.

9-10. List several uses of shift registers in digital systems.

9-11. Refer to Fig. 9-7 for questions **a** to **i** on the 74194 IC shift register:
  a. How many bits of information can this register hold?
  b. List the four modes of operation for this register.
  c. What is the purpose of the mode control inputs ($S0$, $S1$)?
  d. The _____ input overrides all other inputs on this register.
  e. What type of flip-flops and how many are used in this shift register?
  f. The register shifts on the _____ (negative-, positive-) going edge of the clock pulse.
  g. What does the inhibit mode of operation mean?
  h. By definition, to shift left means to shift data from _____ to _____ (use letters).
  i. This register can be loaded _____ (serially, in parallel, either serially or in parallel).

9-12. Refer to Fig. 9-18. List the 74194 shift register's mode of operation during each of the eight clock pulses. Use as answers "clear," "inhibit," "shift right," "shift left," and "parallel load."

9-13. List the contents of the register in Fig. 9-18 after each of the eight clock pulses (A = left bit, D = right bit).

9-14. Refer to Fig. 9-11 for questions **a** to **f** on the 74HC164 shift register.
  a. How many bits of information can this register store?
  b. This shift register is a _____ (CMOS, TTL) IC.
  c. This is a _____ (parallel, serial)-load shift register.
  d. The master reset is an active _____ (HIGH, LOW) input.
  e. The register shifts data on the _____ (H-TO-L, L-TO-H) transition of the clock pulse.
  f. The register has two data inputs which are _____ (ANDed, ORed) together for loading data into FF 1.

9-15. Refer to Fig. 9-19. List the contents of the register during each of the eight clock pulses (Q_0 = left bit, Q_7 = right bit).

9-16. Refer to Fig. 9-13. The device that generates clock pulses in the digital roulette circuit is called a(n) _____ .

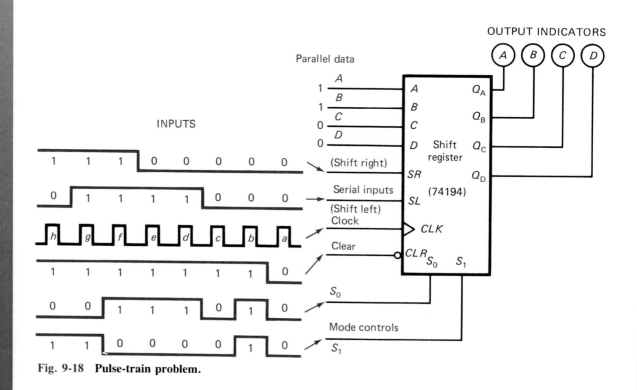

Fig. 9-18 Pulse-train problem.

9-17. Refer to Fig. 9-14(a). The 74HC164 shift register is wired as a(n) _____ in this circuit.

9-18. Refer to Fig. 9-16. The frequency of the VCO decreases as the voltage at the top of capacitor _____ (C_1, C_2, C_4) decreases.

9-19. Refer to Fig. 9-16. What is the purpose of resistor R_7 and capacitor C_4?

9-20. Refer to Fig. 9-16. Resistors R_5 and R_6 force the output of ICa _____ (HIGH, LOW) when the power is first turned on.

9-21. Refer to Fig. 9-16. If only Q_0 of the ring counter is HIGH (as shown), the R-S latch forces the output of ICa _____ (HIGH, LOW).

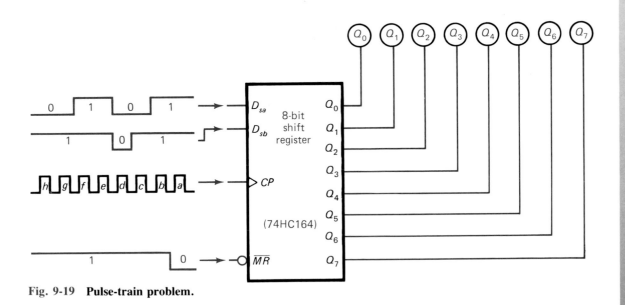

Fig. 9-19 Pulse-train problem.

Answers to Self-Tests

1. serial
2. after pulse a = 000
   after pulse b = 100
   after pulse c = 010
   after pulse d = 001
   after pulse e = 000
   after pulse f = 100
3. single bit
4. parallel
5. pulse a = clear
   pulse b = parallel load
   pulse c = shift-right
   pulse d = shift-right
   pulse e = shift-right
   pulse f = parallel load
   pulse g = shift-right
   pulse h = shift-right
6. after pulse a = 000
   after pulse b = 010
   after pulse c = 001
   after pulse d = 100
   after pulse e = 010
   after pulse f = 101

after pulse g = 110
after pulse h = 011
7. 1. clear
   2. parallel load
   3. shift-right
   4. shift-left
   5. inhibit (do nothing)
8. parallel load
9. inhibit
10. HIGH, LOW, LOW, HIGH
11. HIGH, one
12. inhibit
13. 1, 0, shift-right serial
14. 0000 (cleared)
15. LOW
16. L-to-H or LOW-to-HIGH
17. during pulse a = reset
    during pulse b = shift-right
    during pulse c = shift-right
    during pulse d = shift-right
    during pulse e = shift-right
    during pulse f = shift-right
18. during pulse a = 00000000

during pulse b = 10000000
during pulse c = 01000000
during pulse d = 00100000
during pulse e = 10010000
during pulse f = 01001000
19. audio amplifier
20. ring counter
21. R_7 and C_4
22. voltage-controlled oscillator or VCO
23. R-S latch or latch
24. will not shift a HIGH into the D position
25. Output Q (pin 9) of FF D floating; 7474 IC that contains FF C and FF D faulty
26. A new 7474 IC should be inserted, replacing FF C and FF D.
27. logic pulser, logic probe

CHAPTER 10

Arithmetic Circuits

The public's imagination has been captured by computers and modern-day calculators, probably because these machines perform human arithmetic tasks with such fantastic speed and accuracy. **This chapter deals with some logic circuits that can add and subtract. (Of course, the adding and subtracting is done in binary.) Regular logic gates will be wired together to form** adders **and** subtractors.

10-1 BINARY ADDITION

Remember that in a binary number, such as 101011, the leftmost digit is the MSB and the rightmost digit is the LSB. Also remember the place values given to the binary number is: 1s, 2s, 4s, 8s, 16s, and 32s.

You probably still recall learning your addition and subtraction tables when you were in elementary school. This is a difficult task in the decimal number system because there are so many combinations. This section deals with the simple task of adding numbers in binary. Because they have only two digits (0 and 1), the binary addition tables are simple. Figure 10-1(a) shows the binary addition tables. Just as in the case of adding with decimals, the first three problems are easy. The next problem is

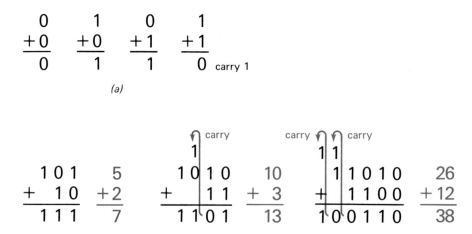

Fig. 10-1 (*a*) **Binary addition tables.** (*b*) **Sample binary addition problems.**

1 + 1. In decimal that would be 2. In binary a 2 is written 10. Therefore, in binary 1 + 1 = 0, with a carry of 1 to the next most significant place value.

Figure 10-1(*b*) shows some examples of adding numbers in binary. The problems are also shown in decimal so that you can check your understanding of binary addition. The first problem is adding binary 101 to 10, which equals 111 (decimal 7). This problem is simple using the addition tables in Fig. 10-1(*a*). The second problem in Fig. 10-1(*b*) is adding binary 1010 to 11. Here you must notice that a 1 + 1 = 0 plus a carry from the 2s place to the 4s place, as shown in the diagram. The answer to this problem is 1101 (decimal 13). In the third problem in Fig. 10-1(*b*) the binary number 11010 is added to 1100. In the figure, note two carries with the solution as 100110 (decimal 38).

Another sample addition problem is shown in Fig. 10-2(*a*). The solution looks simple until we get to the 2s column and find 1 + 1 + 1 in binary. This equals 3 in decimal, which is 11 in binary. This is one situation we left out of the first group of binary addition tables. Looking carefully at Fig. 10-2, you see that the 1 + 1 + 1 situation can arise in any column except the 1s column. So the binary addition table in Fig. 10-1(*a*) is correct for the *1s column only*. The new short-form addition table in Fig. 10-2(*b*) adds the other possible combination of 1 + 1 + 1. The addition table in Fig. 10-2(*b*), then, is for all the place values (2s, 4s, 8s, 16s, and so on) except the 1s column.

To be an intelligent worker on digital equipment you must master binary addition.

Several practice problems are provided in the first self-test.

Self-Test

Answer the following questions.

1. What is the sum of binary 1010 + 0100? (Check your answer using decimal addition.)
2. What is the sum of binary 1010 + 0111?
3. What is the sum of binary 1111 + 1001?
4. What is the sum of binary 10011 + 0111?

10-2 HALF ADDERS

The addition table in Fig. 10-1(*a*) can be thought of as a truth table. The numbers being added are on the input side of the table. In Fig. 10-3(*a*) this is the *A* and *B* input columns. The truth table needs *two* output columns, one column for the sum and one column for the carry. The sum column is labeled with the summation symbol Σ. The carry column is labeled with a C_o. The C_o stands for carry output or *carry out*. A convenient block symbol for the adder that performs the job of the truth table is shown in Fig. 10-3(*b*). This circuit is called a *half-adder* circuit. The half-adder circuit has two inputs (*A*, *B*) and two outputs (Σ, C_o).

TRUTH TABLE

INPUTS		OUTPUTS	
B	*A*	Σ	C_o
0	0	0	0
0	1	1	0
1	0	1	0
1	1	0	1
Binary digits to be added		Sum	Carry out
		XOR	AND

From page 198

Binary addition

MSB

LSB

Binary addition tables

On this page:

Half-adder

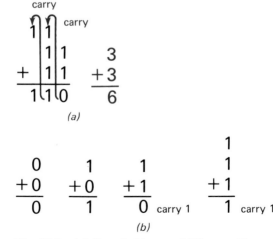

Fig. 10-2 (*a*) **Sample binary addition problem.** (*b*) **Short-form binary addition table.**

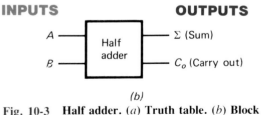

Fig. 10-3 **Half adder.** (*a*) **Truth table.** (*b*) **Block symbol.**

Take a careful look at the half-adder truth table in Fig. 10-3(a). What is the Boolean expression needed for the C_o output? The Boolean expression is $A \cdot B = C_o$. You need a two-input AND gate to take care of output C_o.

Now what is the Boolean expression for the sum (Σ) output of the half adder in Fig. 10-3(a)? The Boolean expression is $\overline{A} \cdot B + A \cdot B = \Sigma$. We could use two AND gates and one OR gate to do the job. If you look closely you will notice that this pattern is also that of an XOR gate. The simplified Boolean expression is then $A \oplus B = \Sigma$. In other words, we find that the only one 2-input XOR gate is needed to produce the sum output.

Using a two-input AND gate and a two-input XOR gate, a logic symbol diagram for a half adder is drawn in Fig. 10-4. The half-adder circuit adds only the LSB column (1s column) in a binary addition problem. A circuit called a *full adder* must be used for the 2s, 4s, 8s, and 16s, and higher places in binary addition.

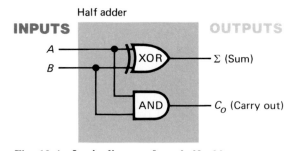

Fig. 10-4 Logic diagram for a half adder.

Self-Test

Answer the following questions.

5. Draw a block diagram of a half adder. Label inputs A and B; label outputs Σ and C_o.
6. Draw a truth table for a half adder.
7. Draw a logic symbol diagram for a half adder.

10-3 FULL ADDERS

Figure 10-2(b) is a short form of the binary addition table, with the $1 + 1 + 1$ situation shown. The truth table in Fig. 10-5(a) on page 201 shows all the possible combinations of A, B, and C_{in} (carry in). This truth table is for a full adder. Full adders are used for all binary place values except the 1s place. The full adder must be used when it is possible to have an extra *carry input*. A block diagram of a full adder is shown in

Fig. 10-5(b). The full adder has three inputs: C_{in}, A, and B. These three inputs must be added to get the Σ and C_o outputs.

One of the easiest methods of forming the combinational logic for a full adder is diagramed in Fig. 10-5(c); two half-adder circuits and an OR gate are used. The expression for this arrangement is $A \oplus B \oplus C = \Sigma$. The expression for the carry out is $A \cdot B + C_{in} \cdot (A \oplus B) = C_o$. The logic circuit in Fig. 10-6(a) (see page 202) is a full adder. This circuit is based upon the block diagram using two half adders shown in Fig. 10-5(c). Directly below this logic diagram is a logic circuit that is somewhat easier to wire. Figure 10-6(b) contains two XOR gates and three NAND gates, which makes the circuit fairly easy to wire. Notice that the circuit in Fig. 10-6(b) is exactly the same as the one in Fig. 10-6(a), except that NAND gates have been substituted for AND and OR gates.

Half and full adders are used together. For the problem in Fig. 10-2(a) we need one half adder for the 1s place and two full adders for the 2s place and the 4s place value. Half and full adders are rather simple circuits. However, many of these circuits are needed to add longer problems.

Many circuits similar to half and full adders are part of a microprocessor's *arithmetic logic unit* (ALU). These circuits are then used for adding 8-bit or even 16- or 32-bit binary numbers in a microcomputer system. The microprocessor's ALU can also subtract using the same half and full adder circuits. Later in this chapter, you will use adders to perform binary subtraction.

Self-Test

Answer the following questions.

8. Draw a block diagram of a full adder. Label inputs A, B, and C_{in}; label outputs Σ and C_o.
9. Draw a truth table for a full adder.
10. Adder circuits are widely used in the _____ section of a microprocessor.

10-4 THREE-BIT ADDERS

Half and full adders are connected to form adders that add several binary digits (bits) at one time. The system in Fig. 10-7, on page 203 adds two 3-bit numbers. The numbers being added are written as $A_2\,A_1\,A_0$ and $B_2\,B_1\,B_0$. Numbers from the 1s place value column are entered into the 1s adder, or half adder. The inputs to the 2s adder are the carry from the

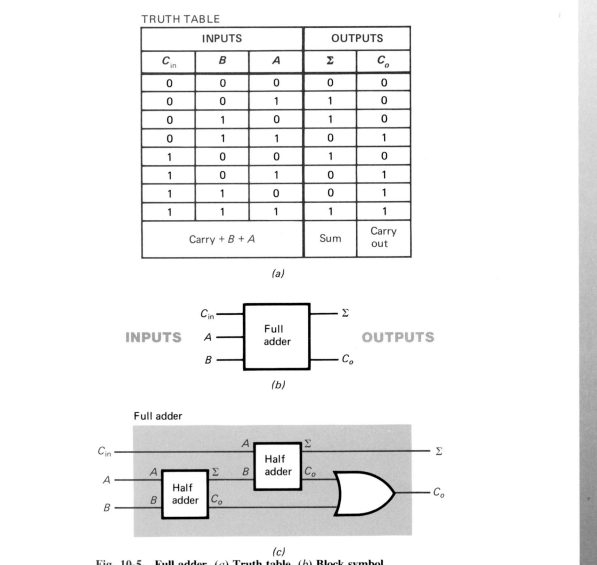

TRUTH TABLE

INPUTS			OUTPUTS	
C_{in}	B	A	Σ	C_o
0	0	0	0	0
0	0	1	1	0
0	1	0	1	0
0	1	1	0	1
1	0	0	1	0
1	0	1	0	1
1	1	0	0	1
1	1	1	1	1
Carry + B + A			Sum	Carry out

(a)

(b)

(c)

Fig. 10-5   Full adder. (*a*) **Truth table.** (*b*) **Block symbol.**
(*c*) **Constructed from half adders and an OR gate.**

half adder (C_{in}) and the new bits A_1 and B_1 from the problem. The 4s adder adds A_2 and B_2 and the carry in from the 2s adder. The total sum is shown in binary at the lower right. The output also has an 8s place value to take care of any binary number over 111 in the sum. Notice that the 4s adder's output (C_o) is connected to the 8s sum indicator.

The three-bit binary adder is organized as you would *add and carry* in hand-done arithmetic. The electronic adder in Fig. 10-7, on page 203, is just very much faster than doing the same problem by hand. Notice that multibit adders use a half adder for the 1s column only; all other bits use a full adder. This type of adder is called a *parallel adder*.

In a parallel adder, all bits are applied to the inputs at the same time. The sum appears at the output almost immediately. The parallel adder shown in Fig. 10-7 is a combinational logic circuit and typically needs various registers to latch data at the inputs and outputs.

Self-Test

Supply the missing word (or words) in each statement.

11. The unit in Fig. 10-7 uses a(n) _____ for adding the 1s column and a(n) _____ for the more significant columns.
12. Parallel adders are _____ (combinational, sequential) logic circuits.

Full adder

Parallel adder

Combinational logic circuit

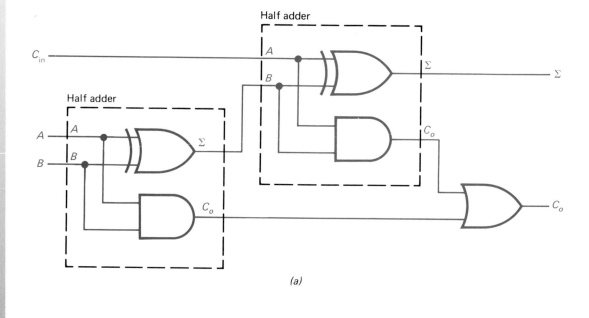

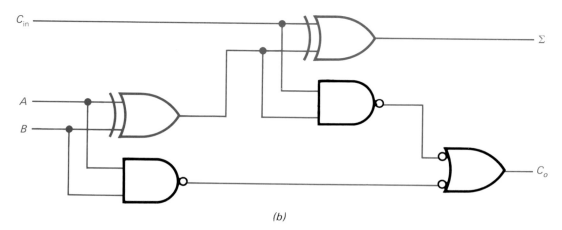

Fig. 10-6 **Full adder.** (*a*) **Logic diagram.** (*b*) **Logic diagram using XOR and NAND gates.**

10-5 BINARY SUBTRACTION

You will find that *adders* and *subtractors* are very similar. You use *half subtractors* and *full subtractors* just as you use half and full adders. Binary subtraction tables are shown in Fig. 10-8(*a*). Converting these rules to truth-table form gives the table in Fig. 10-8(*b*). On the input side, B is subtracted from A to give output Di (difference). If B is larger than A, such as in line 2, we need a *borrow*, which is shown in the column labeled B_o (borrow out).

A block diagram of a half subtractor is shown in Fig. 10-9(*a*). Inputs A and B are on the left. Outputs Di and B_o are on the right side of the diagram. Looking at the truth table in Fig. 10-8(*b*), we can determine the Boolean expressions for the half subtractor. The expression for the Di column is

$A \oplus B = Di$. This is the same as for the half adder [see Fig. 10-3(*a*)]. The Boolean expression for the B_o column is $\overline{A} \cdot B = B_o$. Combining these two expressions in a logic diagram gives the logic circuit in Fig. 10-9(*b*). This is the logic circuit for a half subtractor; notice how much it looks like the half-adder circuit in Fig. 10-4.

When you subtract several columns of binary digits, you must take into account the borrowing. Suppose you are subtracting the numbers in Fig. 10-10(*a*) (see page 204). You might keep track of the differences and borrows as shown in the figure. Look over the subtraction problem carefully, and check if you can do binary subtraction by this longhand method. (You can check yourself on the next self-test.)

A truth table that considers all the possible

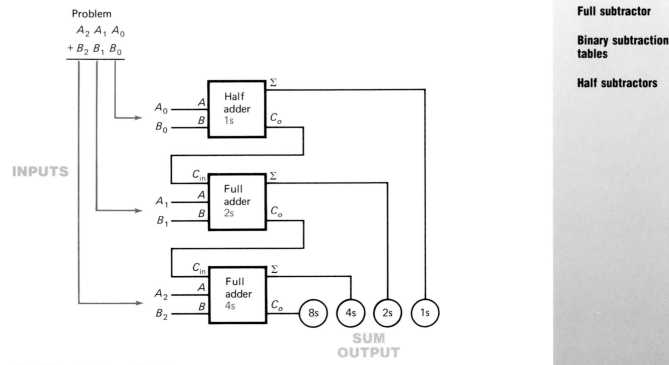

Fig. 10-7 **A 3-bit parallel adder.**

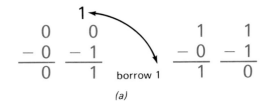

(a)

TRUTH TABLE

INPUTS		OUTPUTS	
A	*B*	*Di*	*B$_o$*
0	0	0	0
0	1	1	1
1	0	1	0
1	1	0	0
A − B		Difference	Borrow out

(b)

Fig. 10-8 (a) **Binary subtraction tables.** (b) **Truth table for the half subtractor.**

INPUTS **OUTPUTS**

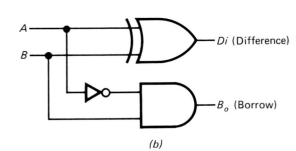

(a)

(b)

Fig. 10-9 **Half subtractor.** (a) **Block symbol.** (b) **Logic diagram.**

combinations in binary subtraction is shown in Fig. 10-10(b). For instance, line 5 of the table is the situation in the 1s column in Fig. 10-10(a). The 2s column equals line 3, the 4s column line 6, the 8s column line 3, the 16s column line 2, and the 32s column line 6 of the truth table.

A block diagram of a full subtractor is drawn in Fig. 10-11(a), on page 205. The inputs *A*, *B*, and *B$_{in}$* are on the left; the outputs *Di* and *B$_o$* are on the right. Like the full adder, the full subtractor can be wired using two half subtracters and an OR gate. Figure 10-11(b) is a full subtractor showing how half subtractors are used. A logic diagram for a full subtractor is shown in Fig. 10-11(c). This

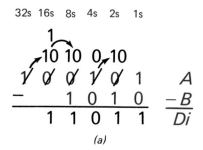

(a)

TRUTH TABLE

INPUTS			OUTPUTS	
A	B	B_{in}	Di	B_o
0	0	0	0	0
0	0	1	1	1
0	1	0	1	1
0	1	1	0	1
1	0	0	1	0
1	0	1	0	0
1	1	0	0	0
1	1	1	1	1
A $-$ B $-$ B_{in}			Difference	Borrow out

(b)

Fig. 10-10 (*a*) **Sample binary subtraction problem.** (*b*) **Truth table for a full subtractor.**

circuit performs as a full subtractor as specified in the truth table in Fig. 10-10(*b*). The AND-OR circuit on the B_o output can be converted to three NAND gates if you want. The circuit would then be similar to the full-adder circuit in Fig. 10-6(*b*).

Self-Test

Answer the following questions.

13. Do the binary subtraction problems in **a** to **f**. (Check yourself using decimal subtraction.)

    a.   11   **d.**   1010
         − 10   − 101

    b.   100   **e.** 10010
       − 10   − 11

    c.   111   **f.** 1000
       − 111   − 01

14. Draw a block diagram of a half subtractor. Label inputs A and B; outputs Di and B_o.

15. Draw a truth table for a half subtractor.
16. Draw a block diagram for a full subtractor. Label inputs A, B, and B_{in}; label outputs Di and B_o.
17. Draw a truth table for a full subtractor.

10-6 PARALLEL SUBTRACTORS

Half and full subtractors are wired together to perform as a *parallel subtractor*. You have already seen adders connected as parallel adders. An example of a parallel adder is the 3-bit adder in Fig. 10-7. A parallel subtractor is wired in a similar manner. The adder in Fig. 10-7 is considered a parallel adder because all the digits from the problem flow into the adder at the same time.

Figure 10-12 (see page 206) diagrams the wiring of a single half subtractor and three full subtractors. This forms a 4-bit parallel subtractor that can subtract binary number $B_3B_2B_1B_0$ from binary number $A_3A_2A_1A_0$. Notice that the top subtractor (half subtractor) subtracts the LSBs (1s place). The B_o output of the 1s subtractor is tied to the 2s subtractor. Each subtractor's B_o output is connected to the next more significant bit's borrow input. These borrow lines keep track of the borrows we discussed earlier.

Self-Test

Answer the following questions.

18. Refer to Fig. 10-12. This is a block diagram of a 4-bit _____ (parallel adder, parallel subtractor, serial adder, serial subtractor) circuit.
19. Refer to Fig. 10-12. The lines between subtractors (B_o to B_{in}) serve what purpose in this circuit?

10-7 USING ADDERS FOR SUBTRACTION

In earlier sections we found that there are circuits for adding or subtracting binary numbers. To simplify circuitry in a calculating machine, it is convenient to have a more universal device for calculations. With a few little tricks we can use an adder to also do subtraction.

There is a mathematical technique that helps us use an adder to do binary subtraction. The technique is outlined in Fig. 10-13 on page 206. The problem is to subtract decimal 6 from decimal 10 (binary: 1010–0110). The

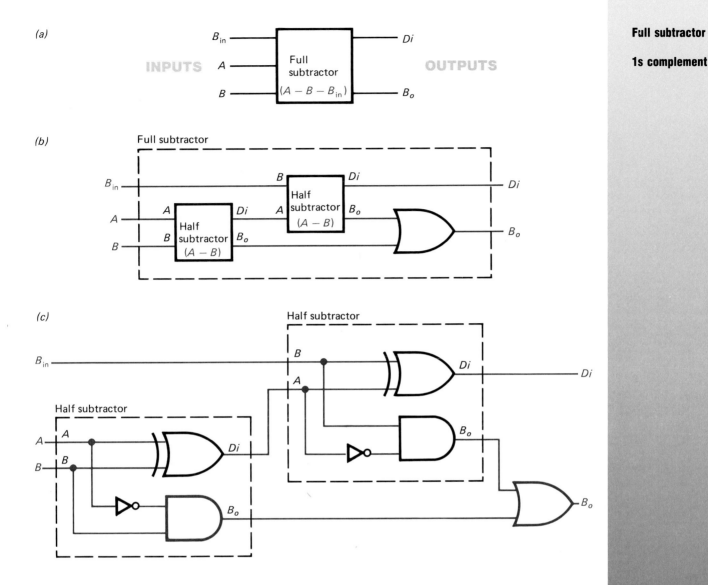

(a)

INPUTS

B_{in}

A

B

Full subtractor

$(A - B - B_{in})$

OUTPUTS

Di

B_o

(b)

Full subtractor

B_{in}

A

B

B Di

A Half subtractor B_o

$(A - B)$

A Di

Half subtractor B_o

$(A - B)$

Di

B_o

(c)

Half subtractor

B_{in}

B

A

Di

Di

B_o

Half subtractor

A

B

Di

B_o

B_o

Fig. 10-11   Full subtractor. (*a*) **Block symbol.** (*b*) **Constructed with half subtractors and an OR gate.** (*c*) **Logic diagram.**

subtracting is shown being done first in decimal, then in binary, and finally by the special technique. In the special technique the steps are first to write the *1s complement* of the number being subtracted (change all 1s to 0s and all 0s to 1s) and then add. The 1s complement of 0110 is 1001, as shown. The temporary answer to this addition is shown as 10011. Next, the last carry on the left is carried around to the 1s place (see the arrow on the diagram). This is called an *end-around carry*. When the end-around carry is added to the rest of the number, the result is the *difference* between the original binary numbers, 1010 and 0110. The answer to this problem is shown as binary 100 (decimal 4).

Using the 1s complement and end-around carry method is somewhat difficult in longhand. However, this method is quite easy to do with logic circuits. You can see that this method uses an adder to do subtraction. You should know how to do 1s complement and end-around carry subtraction. (There are a few practice problems in the next self-test.)

Now let us use adders to do binary subtraction. Figure 10-14 on page 207 shows four *full adders* (labeled FA) being used to perform binary subtraction. Pay special attention to the four inverters that complement the binary number represented by $B_3 B_2 B_1 B_0$. The inverters give the 1s complement input at the B inputs to each full adder. The adder adds the binary number

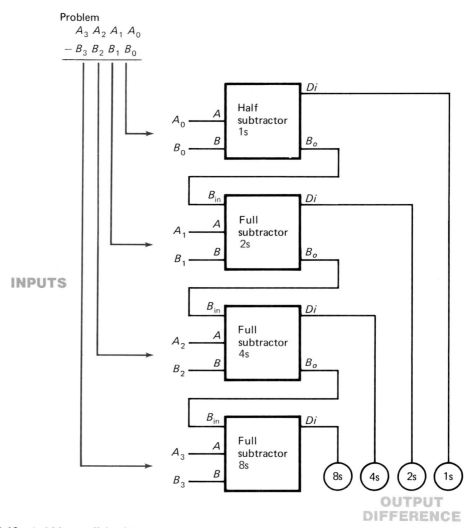

Fig. 10-12   **A 4-bit parallel subtractor.**

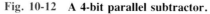

Decimal subtraction	Binary subtraction		Special technique subtraction

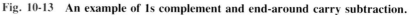

Fig. 10-13   **An example of 1s complement and end-around carry subtraction.**

represented by $A_3 A_2 A_1 A_0$ and $\bar{B}_3 \bar{B}_2 \bar{B}_1 \bar{B}_0$. The extra carry at C_o of the 8s full adder is carried back to the 1s adder by the end-around carry line shown. The indicators at the lower right show the *difference* between the binary numbers $A_3 A_2 A_1 A_0$ and $B_3 B_2 B_1 B_0$.

Computing devices may use 1s complement numbers for subtraction, but more commonly they use 2s complement numbers. Later sections cover 2s complement subtraction and subtractors.

Self-Test

Answer the following questions.

20. Do the binary subtraction problems in **a** to **f** using the 1s complement and end-around carry method.
    a. 11 − 10 =   **d.** 110 − 100 =
    b. 111 − 010 =   **e.** 1001 − 0111 =
    c. 1000 − 0111 =   **f.** 1011 − 0110 =

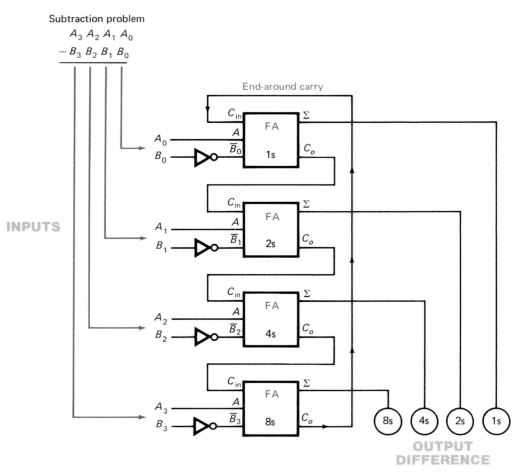

Fig. 10-14   **Using full adders and inverters to construct a 4-bit binary subtractor.**

21. The 1s complement and end-around carry method of subtraction is used when _____ (adders, half subtractors, full subtractors) are used to perform four-bit subtraction.

22. Refer to Fig. 10-14. This subtractor circuit uses inverters to form the 1s complement of the subtrahend, four _____ , and an end-around carry line.

10-8 FOUR-BIT ADDER/ SUBTRACTORS

Now that we know we can use full adders to add and subtract, let us design a system which adds and subtracts. We shall start with the subtractor system in Fig. 10-14. To make this system a *4-bit adder*, we need only bypass temporarily the four inverters and disconnect the end-around carry line. The subtractor system is redrawn in Fig. 10-15 on the next page. Four XOR gates replace the inverters. When a 0 is placed in input A of the XOR gates, the bits from the problem

pass through the XOR gate with no change (see the truth table at the lower left, Fig. 10-15). With the control at 0, the unit adds the binary number $A_3 A_2 A_1 A_0$ to $B_3 B_2 B_1 B_0$. The result (up to a sum of 1111) appears at the output indicators. The 0 on the control (add position) also disables the AND gate and does not permit the end-around carry.

To make the unit in Fig. 10-15 a *4-bit subtractor*, the control must be in the 1 position. This causes the XOR gate to act as an inverter for the B inputs to the full adders. This complementing is seen in the truth table in the lower left of Fig. 10-15. The 1 at the control also activates the AND gate so that information from the 8s full adder can take the end-around carry line back to the 1s full adder. This subtractor subtracts the binary input number $B_3 B_2 B_1 B_0$ from $A_3 A_2 A_1 A_0$. The difference appears in binary form on the output indicators. Remember that the 1s complement and end-around carry method of subtraction is used in this circuit. The XOR gates do the complementing, and the end-around carry line is labeled.

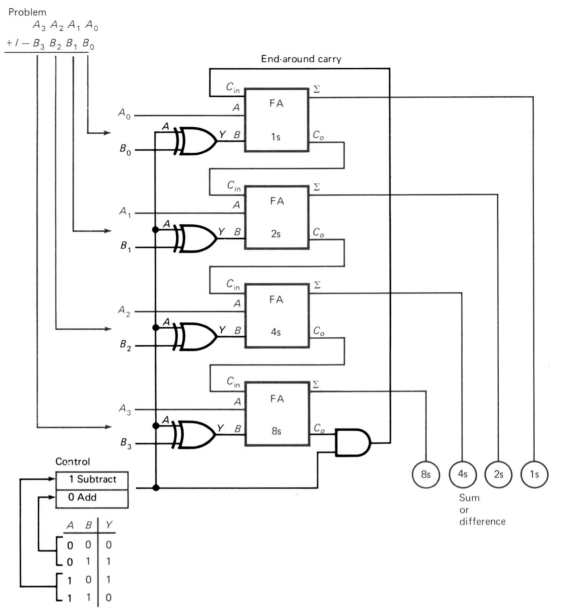

Fig. 10-15   Combination adder/subtractor circuit.

Considering that multiplication is just repeated addition and division is repeated subtraction, you can see how important the adder/subtractor circuit is in digital electronics.

Self-Test

Supply the missing word or words in each statement.

23. Refer to Fig. 10-15. When the control input of the adder/subtractor is LOW, the AND gate is _____ (disabled, enabled) and the XOR gates _____ (do, do not) complement the *B* inputs.

24. Refer to Fig. 10-15. To use this circuit for subtraction, the control input must be _____ (HIGH, LOW).

25. Refer to Fig. 10-15. When the control input of the adder/subtractor is HIGH, the AND gate is _____ (disabled, enabled) and the XOR gates _____ (do, do not) complement the *B* inputs.

10-9 IC ADDERS

IC manufacturers produce several adders. One useful arithmetic IC is the TTL *7483 4-bit binary full adder*. A block symbol for the 7483 IC adder is drawn in Fig. 10-16. The problem

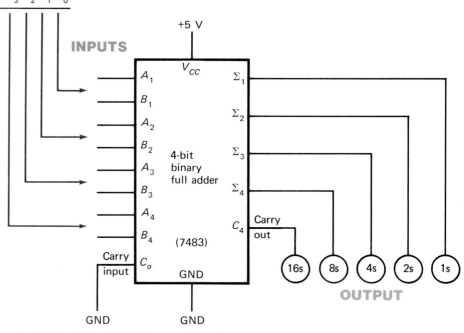

Problem
$A_3\ A_2\ A_1\ A_0$
$+\ B_3\ B_2\ B_1\ B_0$

Fig. 10-16   The 7483 4-bit binary adder IC.

TTL 7483 4-bit
binary full adder

Cascading adders

8-bit binary adder

Parallel
adder/subtractor
system

of addition of the two 4-bit binary numbers ($A_3\ A_2\ A_1\ A_0$ and $B_3\ B_2\ B_1\ B_0$) is shown being entered into the eight inputs of the 7483 IC. Notice a difference in numbering systems on the problem and the IC. For adding just two 4-bit numbers, the C_o input (end-around carry input) is held at 0. The C_o input is marked as the C_{in} input by some manufacturers. The sum outputs are shown attached to output indicators. The C_4 output is attached to the 16s output indicator. The C_4 output is marked as the C_o output by some manufacturers. This binary adder can indicate a sum as high as 11110 (decimal 30) when adding binary 1111 to 1111.

Internally the 7483 IC adder is organized very much like the unit in Fig. 10-14 (without the four inverters). The C_4 (carry out) on the 7483 IC is the same as the C_o on the 8s full adder in Fig. 10-14. The carry input labeled C_o on the 7483 IC is the same as the C_{in} on the 1s full adder in Fig. 10-14.

The 7483 IC adder can be cascaded by connecting the C_4 output of the first IC to the C_o carry input of the next 7483 IC. With two 7483 IC adders cascaded, an 8-bit binary adder is produced. The 7483 IC can also be a 4-bit subtractor using the arrangement shown in Fig. 10-14. The B inputs are inverted, or complemented, and the C_4 carry out is the end-around carry line to the C_o carry input of the 7483 IC.

The 7483 IC may also be used in the adder/subtractor circuit diagramed in Fig. 10-15. Of course, extra ICs must be used for the XOR and AND gates.

Self-Test

Supply the missing word in each statement.

26. The 7483 IC contains a 4-bit binary _____ .
27. Two 7483 ICs can be _____ to form an 8-bit parallel binary adder.
28. Refer to Fig. 10-16. To convert the 7483 IC to an adder/subtractor, several _____ gates and an AND gate have to be added to the circuit.

10-10 PARALLEL ADDER/ SUBTRACTOR SYSTEMS

In this section we shall diagram a digital electronic system with the parallel adder as its heart. Figure 10-17 on the next page is a wiring diagram of a *parallel adder/subtractor system.* This system employs parts you have already used. The inputs are shown at the left; the output is the digital readout (seven-segment display) at the right.

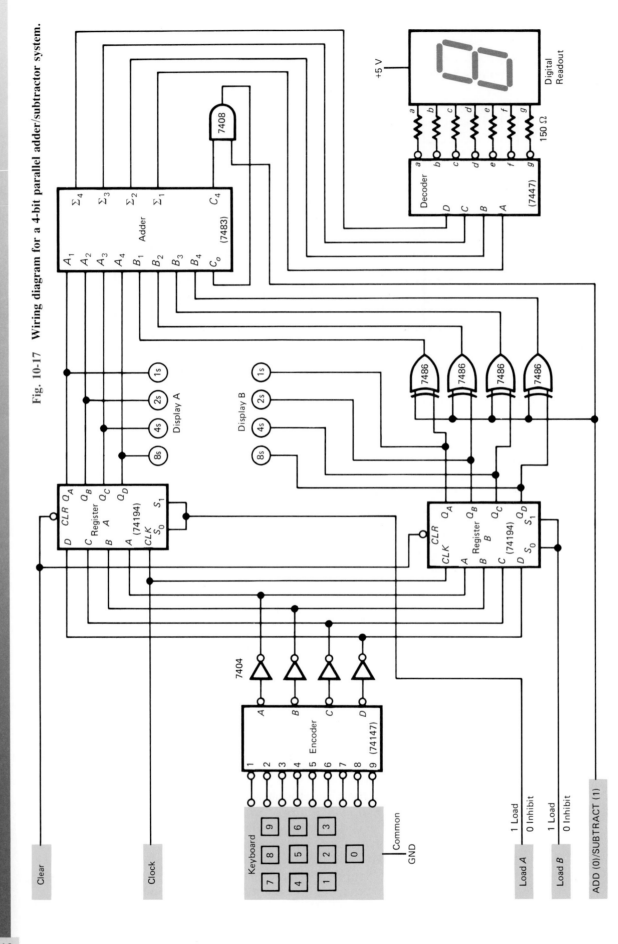

Fig. 10-17 Wiring diagram for a 4-bit parallel adder/subtractor system.

The following steps might be used to operate the parallel adder/subtractor system in Fig. 10-17. First, activate the *CLR* input control to clear both registers (*A* and *B*) to 0000. Second, set the add/subtract control to the proper mode (let us use 0 for add). Next, load each register individually with the load controls. To load register *A*, place the load *A* control at 1 (load *B* at 0). Press a keyboard number while pulsing the *CLK* input once. The binary number you loaded in register *A* now appears on display *A*. To load register *B*, place the load *B* control at 1 (load *A* at 0). Press a keyboard number while pulsing the *CLK* input once. This second binary number should appear in register *B* as indicated on display *B*. The 7483 IC 4-bit adder works immediately, and the *sum* appears on the digital display.

The subtractor subtracts the contents of register *B* from the binary number in register *A*. The procedure for subtracting with the system in Fig. 10-17 is the same as for adding *except* that the add/subtract control is set in the subtract position (1). This activates the AND and XOR gates for the 1s complement and end-around carry subtraction. The digital display gives the *difference* between register *A* and register *B*.

The 74147 encoder changes the decimal input from the keyboard to binary numbers in Fig. 10-17. The 7404 IC inverts the outputs of the 74147 encoder. From the inverters the binary is fed into the parallel load data inputs of both registers *A* and *B*. With one clock pulse the parallel data at the input of the register is transferred and stored in the registers (if S_0 and S_1 are at 1). Displays *A* and *B* indicate what binary numbers are stored in registers *A* and *B*; the binary numbers latched in the registers are applied to the inputs of the 7483 4-bit adder. The adder adds binary numbers $A_4 A_3 A_2 A_1$ and $B_4 B_3 B_2 B_1$, and the sum appears in binary at the outputs of the 7483 adder. The sum is decoded from binary to seven-segment code by the 7447 decoder. The decimal sum appears on the digital readout.

It is quite common to have an arithmetic unit serve as part of the *central processing unit* (often referred to as the CPU). A block diagram of this digital system in shown in Fig. 9-1. The processing unit in that diagram is comparable to your adder/subtractor.

74147 encoder

7404 IC

7483 4-bit adder

7447 decoder

Central processing unit

Multiplicand

Multiplier

Product

Self-Test

Supply the missing word or words in each statement.

29. Refer to Fig. 10-17. The first step in using this adder/subtractor system would be to clear registers *A* and *B* by applying a _____ (HIGH, LOW) to the clear input and then returning it to normal.
30. Refer to Fig. 10-17. The second step in using this system as a subtractor would be to load register *A* with the _____ (minuend, subtrahend) and register *B* with the _____ (minuend, subtrahend).
31. Refer to Fig. 10-17. Loading register *A* is accomplished in this system by placing the load *A* input at _____ (0, 1), pressing the number on the keyboard, and at the same time activating the _____ input.
32. Refer to Fig. 10-17. When both registers *A* and *B* are loaded, the sum/difference will appear on the digital display _____ (after one clock pulse, after four clock pulses, immediately).

10-11 BINARY MULTIPLICATION

In elementary school you learned how to multiply. You learned to lay out your multiplication problem similar to that in Fig. 10-18(*a*). You learned that the top number is called the *multiplicand*, the bottom number the *multiplier*. The solution of the problem is called the *product*. The product of 7×4, then, is 28, as shown in Fig. 10-18(*a*).

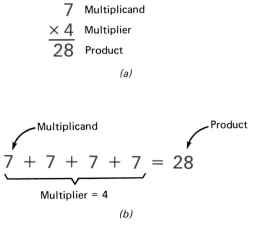

```
   7   Multiplicand
 × 4   Multiplier
  ――
  28   Product
```

(a)

Multiplicand   Product

$$7 + 7 + 7 + 7 = 28$$

Multiplier = 4

(b)

Fig. 10-18 (*a*) **Decimal multiplication problem.** (*b*) **Multiplying using the repeated addition method.**

Figure 10-18(b) shows that multiplication really is just *repeated addition*. The problem $7 \times 4 = 28$ is represented by the multiplicand (7) being added four times, because 4 is the multiplier. The product is 28.

If you want to multiply 54×14, the repeated addition system is complicated and takes a long time. The multiplicand (54) must be added 14 times to get a product of 756. Most of us were taught to multiply 54×14 in the manner shown in Fig. 10-19(a). To solve the multiplication problem 54×14, we first multiply the multiplicand, 54, by 4. This results in the first partial product (216) shown in Fig. 10-19(b). Next we multiply the multiplicand by 1. Actually the multiplicand is multiplied by a multiplier of 10, as shown in Fig. 10-19(c). The second partial product is 540. The first and second partial products (216 and 540) are then added for a final product of 756. It is normal to omit the 0 in the second partial product, as in Fig. 10-19(a).

It is important to notice the *process* in the problem in Fig. 10-19. The multiplicand is first multiplied by the LSD of the multiplier. This gives the first partial product. The second partial product is then calculated by multiplying the multiplicand by the MSD of the multiplier. The two partial products are then added, pro-

ducing the final product. This same process is used in *binary multiplication*.

Binary multiplication is much simpler than decimal multiplication. The binary system has only two digits (0 and 1), which makes the rules for multiplying simple. Figure 10-20(a) shows the rules for binary multiplication.

Rules for binary multiplication

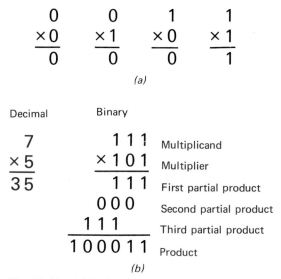

(a)

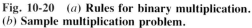

(b)

Fig. 10-20 (a) **Rules for binary multiplication.** (b) **Sample multiplication problem.**

Multiplication with binary numbers is done just as with decimal numbers. Figure 10-20(b) details a problem where binary 111 is multiplied by binary 101. First, the multiplicand (111) is multiplied by the 1s bit of the multiplier. The result is the first partial product, shown as 111 in Fig. 10-20(b). Next, the multiplicand is multiplied by the 2s bit of the multiplier. The result is the second partial product (0000). Notice that the LSB of the second partial product, 0000, is left off in Fig. 10-20(b). Third, the multiplicand is multiplied by the 4s bit of the multiplier. The result is the third partial product of 11100, shown in Fig. 10-20(b) as 111, with the two blank spaces in the 1s and 2s places. Finally, the first, second, and third partial products are added, resulting in a product of binary 100011. Notice that the same multiplication problem in decimal is shown at the left of Fig. 10-20(b) for your convenience. The binary product 100011 equals the decimal product 35.

Another binary multiplication problem is shown in Fig. 10-21. At the left the problem is in the familiar decimal form; the same problem is repeated in binary form at the right,

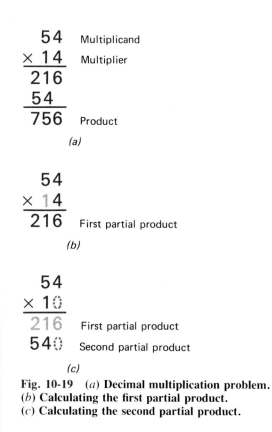

Fig. 10-19 (a) **Decimal multiplication problem.** (b) **Calculating the first partial product.** (c) **Calculating the second partial product.**

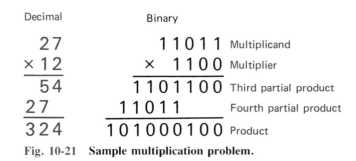

Decimal	Binary	
27	1 1 0 1 1	Multiplicand
× 1 2	× 1 1 0 0	Multiplier
54	1 1 0 1 1 0 0	Third partial product
27	1 1 0 1 1	Fourth partial product
324	1 0 1 0 0 0 1 0 0	Product

Fig. 10-21 Sample multiplication problem.

where binary 11011 is multiplied by 1100. As in decimal multiplication, the 0s in the multiplier can simply be brought down to hold the 1s and 2s places in the binary number. The binary product is shown as 101000100, which equals decimal 324.

You will gain some experiences in solving binary multiplication problems by answering the questions that follow.

Self-Test

Answer the following questions.

33. Find the product for binary 111 × 10.
34. Find the product for binary 1101 × 101.
35. Find the product for binary 1100 × 1110.

10-12 BINARY MULTIPLIERS

We can multiply numbers by repeated addition, as illustrated in Fig. 10-18(*b*). The multiplicand (7) could be added four times to obtain the product of 28. A block diagram of a circuit that performs repeated addition is shown in Fig. 10-22. The multiplicand is held in the top

register. In our example the multiplicand is a decimal 7, or a binary 111. The multiplier is held in the down counter shown on the left in Fig. 10-22. The multiplier in our example is a decimal 4, or a binary 100. The lower product register holds the product.

The repeated addition technique is shown in operation in Fig. 10-23 on the next page. This chart shows how the multiplicand (binary 111) is multiplied by the multiplier (binary 100). The product register is cleared to 00000. After one count downward, a partial product of 00111 (decimal 7) is in the product register. After the second count downward, a partial product of 01110 (decimal 14) appears in the product register. After the third count downward, a partial product of 10101 (decimal 21) appears in the product register. After the fourth downward count, the *final product* of 11100 (decimal 28) appears in the product register. The multiplication problem (7 × 4 = 28) is complete. The circuit in Fig. 10-22 has added 7 four times for a total of 28.

This type of circuit is not widely used because of the long time it takes to do the repeated addition when large numbers are multiplied. A more practical method of multiplying in digital electronic circuits is the *add and shift method* (also called the shift and add method). Figure 10-24 on the next page shows a binary multiplication problem. In this problem binary 111 is multiplied by 101 (7 × 5 in decimal). This hand-done procedure is standard except for the temporary product in line 5. Line 5 has been added to help you understand how multiplication might be done by digital circuits. Close observation of binary multiplication shows the following three important facts:

1. Partial products are always 000 if the multiplier is 0 and equal to the multiplicand if the multiplier is 1.
2. The product register needs twice as many bits as the multiplicand register.

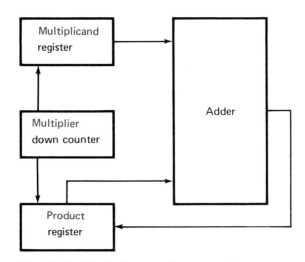

Fig. 10-22 Block diagram of a repeated addition-type multiplier system.

	Load with binary	After 1 down count	After 2 down counts	After 3 down counts	After 4 down counts
Multiplicand register	111	111	111	111	111
Multiplier counter	100	011	010	001	000
Product register	00000	00111	01110	10101	11100
	Load				Stop

Fig. 10-23   Multiplying binary 111 and 100 using the repeated addition circuit.

Line 1		1 1 1	Multiplicand
Line 2	×	1 0 1	Multiplier
Line 3		1 1 1	First partial product
Line 4		0 0 0	Second partial product
Line 5		0 1 1 1	Temporary product (line 3 + line 4)
Line 6	1 1 1		Third partial product
Line 7	1 0 0 0 1 1		Product

Fig. 10-24   Binary multiplication problem.

3. The first partial product is shifted one place to the *right* (*relative to* the second partial product) when adding.

You can observe each characteristic by looking at the sample problem in Fig. 10-24.

The important characteristics of longhand multiplication have been given. A binary multiplication circuit can be designed by using these characteristics. Figure 10-25(a) shows a circuit that does binary multiplication. Notice that the multiplicand (111) is loaded into the register at the upper left. The accumulator register is cleared to 0000. The multiplier (101) is loaded into the register at the lower right. Notice, too, that the accumulator and the multiplier are considered together. This is shown by the shading connecting the two registers.

Let us use the circuit in Fig. 10-25(a) to go through the detailed procedure for multiplying. The diagram in Fig. 10-25(b) is a step-by-step review of how binary 111 is multiplied by 101 using the add and shift method. The binary 111 is loaded into the multiplicand register. The accumulator and multiplier registers are loaded in step A in Fig. 10-25(b). Step B shows the 0000 and the 111 from the accumulator and multiplicand registers being added when the

1 is applied to the control line. This is comparable to line 3 of the multiplication problem in Fig. 10-24. Step C shifts both the accumulator and the multiplier register one place to the right. The LSB of the multiplier (1) is shifted out the right end and lost. Step D represents another *add* step. This time a 0 is applied to the control line. A 0 on the control line means *no* addition. The register contents remain the same. Step D is comparable to lines 4 and 5, Fig. 10-24. Step E shows the registers being shifted one place to the right. This time the 2s bit of the multiplier is lost as it is shifted out the right end of the register. Step F shows the 4s bit of the multiplier (1) commanding the adder to add. The accumulator contents (0001) and the multiplicand (111) are added. The result of that addition is deposited in the accumulator register (1000). This step is comparable to the left section of lines 5 to 7, Fig. 10-24. Step G is the final step in the add and shift multiplication; it shows a single shift to the right for both registers. The MSB (4s bit) of the multiplier is lost out of the right end of the register. The final product appears across both registers as 100011. Binary 111 multiplied by 101 resulted in a product of 100011 ($7 \times 5 = 35$ in decimal). The final product cal-

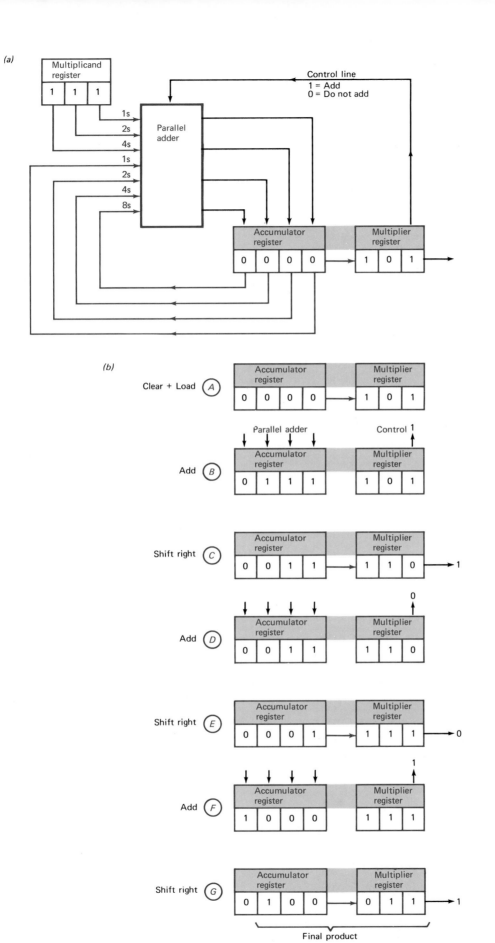

Fig. 10-25 (*a*) Diagram of an add and shift-type multiplier circuit. (*b*) Contents of the accumulator and multiplier registers in the add and shift multiplier circuit.

culated by the multiplier circuit is the same result we got in line 7, Fig. 10-24, when we multiplied by hand.

Two types of multiplier circuits have just been illustrated. The first uses repeated addition to arrive at the product. That system is shown in Fig. 10-22. The second circuit uses the add and shift method of multiplying. The add and shift system is shown in Fig. 10-25.

In many computers *the procedure*, such as add and shift method, can be programmed into the machine. Instead of permanently wiring the circuit, we simply *program*, or instruct, the computer to follow the procedure shown in Fig. 10-25(*b*). We are thus using *software* (a program) to do multiplication. This use of software cuts down on the amount of electronic circuits needed in the CPU of a computer.

Popular 8-bit microprocessors, such as the Intel 8080/8085, Motorola 6800, and the 6502/65C02 do not have circuitry in their ALUs to do multiplication. To perform binary multiplication on these processors, the programmer must write a program (a list of instructions) that multiplies numbers. Either the add and shift or the repeated addition method can be used for programming these microprocessor-based machines to do multiplication. More advanced microprocessors do have multiply instructions.

Self-Test

Answer the following questions.

36. Refer to Fig. 10-22. This circuit uses what method of binary multiplication?
37. A widely used technique for multiplying using digital circuits is the _____ method.
38. Refer to Fig. 10-25. This circuit uses what method of binary multiplication?
39. Most elementary microprocessors _____ (do, do not) have a multiply instruction.

10-13 2s COMPLEMENT NOTATION, ADDITION AND SUBTRACTION

The 2s complement method of representing numbers is widely used in microprocessors. To now, we have assumed that all numbers are positive. However, microprocessors must process both positive and negative numbers. Using 2s complement representations, the *sign as well as the magnitude* of a number can be determined.

For simplicity, assume we are using a 4-bit processor. This means that all data is transferred and processed in groups of four. The MSB is the sign bit of the number. This is shown in Fig. 10-26(*a*). A 0 sign bit means a positive number, while a 1 sign bit means a negative number.

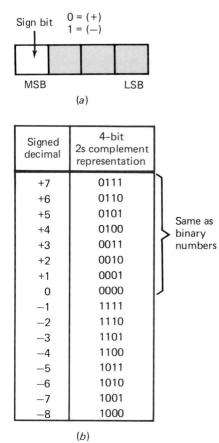

Signed decimal	4-bit 2s complement representation
+7	0111
+6	0110
+5	0101
+4	0100
+3	0011
+2	0010
+1	0001
0	0000
−1	1111
−2	1110
−3	1101
−4	1100
−5	1011
−6	1010
−7	1001
−8	1000

(*b*)

Fig. 10-26 (*a*) **MSB of 4-bit register is sign bit.** (*b*) **2s complement representation of positive and negative numbers.**

The table in Fig. 10-26(*b*) shows the *2s complement representation* for all the 4-bit positive and negative numbers from +7 to −8. The MSBs in Fig. 10-26(*b*) of the positive 2s complement numbers are 0s. All negative numbers (−1 to −8) start with a 1. Note that 2s complement representations of positive numbers are the same as binary. Therefore, +7 (decimal) = 0111 (2s complement) = 0111 (binary).

The 2s complement representation of a negative number is found by first taking the 1s complement of the number and then adding 1. An example of this process is shown in Fig. 10-27(a). The negative decimal number -4 is to be converted to its 2s complement form:

1. Convert the decimal number to its binary equivalent. In this example, convert -4_{10} to 0100_2.
2. Convert the binary number to its 1s complement by changing all 1s to 0s and all 0s to 1s. In this example, convert 0100_2 to 1011 (1s complement).
3. Add 1 to the 1s complement number, using regular binary addition. In this example, $1011 + 1 = 1100$. The answer (1100 in this example) is the 2s complement representation. Therefore, $-4_{10} = 1100$ (2s complement).

This answer can be verified by referring back to the table in Fig. 10-26(b).

To convert from 2s complement form to binary, follow the procedure shown in Fig. 10-27(b). In this example, the 2s complement number (1100) is being converted to its binary equivalent. Its equivalent decimal number can then be found from the binary.

1. Form the 1s complement of the 2s complement number by changing all 1s to 0s and

all 0s to 1s. In this example, convert 1100 to 0011.
2. Add 1 to the 1s complement number, using regular binary addition. In this example, $0011 + 1 = 0100$. The answer (0100 in this example) is in binary. Therefore, $0100_2 = 4_{10}$.

Because the MSB of the 2s complement number (1100) is a 1, the number is negative. Therefore, 2s complement 1100 equals -4_{10}.

2s complement notation is widely employed because it makes it easy to add and subtract signed numbers. Four examples of adding 2s complement numbers are shown in Fig. 10-28. Two positive numbers are added in Fig. 10-28(a). 2s complement addition looks just like adding in binary in this example. Two negative numbers (-1_{10} and -2_{10}) are added in Fig. 10-28(b). The 2s complement numbers representing -1 and -2 are given as 1111 and 1110. The MSB (overflow from 4-bit register) is discarded, leaving the 2s complement sum of 1101, or -3 in decimal. Look over examples (c) and (d) in Fig. 10-28 to see if you understand the procedure for adding signed numbers using 2s complement notation.

2s complement notation is also useful in subtracting signed numbers. Four subtraction problems are shown in Fig. 10-29. The first

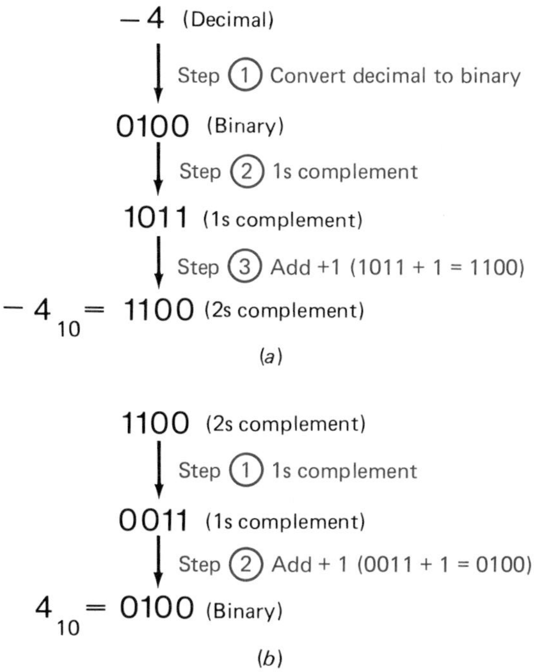

$$
\begin{array}{rl}
(+4) & 0100 \\
+\ (+3) & +\ 0011 \\
\hline
+7_{10} & 0111 \text{ (2s complement SUM)}
\end{array}
$$

(a)

$$
\begin{array}{rl}
(-1) & 1111 \\
+\ (-2) & +\ 1110 \\
\hline
-3_{10} & 1\ 1101 \text{ (2s complement SUM)}
\end{array}
$$

Discard

(b)

$$
\begin{array}{rl}
(+1) & 0001 \\
+\ (-3) & +\ 1101 \\
\hline
-2_{10} & 1110 \text{ (2s complement SUM)}
\end{array}
$$

(c)

$$
\begin{array}{rl}
(+5) & 0101 \\
+\ (-4) & +\ 1100 \\
\hline
+1_{10} & 1\ 0001 \text{ (2s complement SUM)}
\end{array}
$$

Discard

(d)

-4 (Decimal)

Step ① Convert decimal to binary

0100 (Binary)

Step ② 1s complement

1011 (1s complement)

Step ③ Add +1 (1011 + 1 = 1100)

$-4_{10} = 1100$ (2s complement)

(a)

1100 (2s complement)

Step ① 1s complement

0011 (1s complement)

Step ② Add +1 (0011 + 1 = 0100)

$4_{10} = 0100$ (Binary)

(b)

Fig. 10-27 (a) Converting signed decimal numbers to 2s complement form. (b) Converting from 2s complement form to binary numbers.

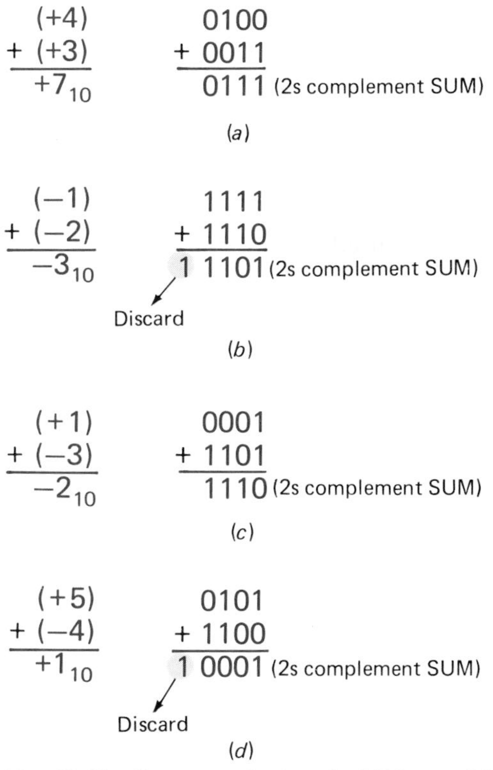

Fig. 10-28 Four sample signed addition problems using 2s complement numbers.

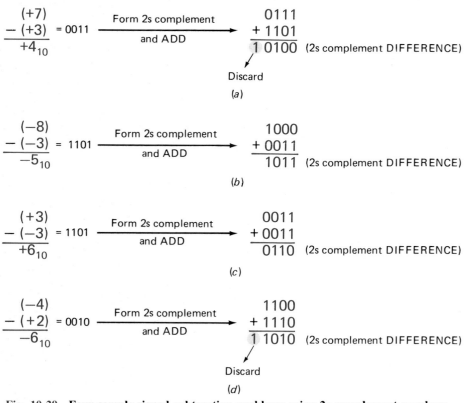

$$\begin{array}{r} (+7) \\ -\ (+3) \\ \hline +4_{10} \end{array} = 0011 \xrightarrow[\text{and ADD}]{\text{Form 2s complement}} \begin{array}{r} 0111 \\ +\ 1101 \\ \hline 1\ 0100 \end{array} \text{(2s complement DIFFERENCE)}$$

Discard

(a)

$$\begin{array}{r} (-8) \\ -\ (-3) \\ \hline -5_{10} \end{array} = 1101 \xrightarrow[\text{and ADD}]{\text{Form 2s complement}} \begin{array}{r} 1000 \\ +\ 0011 \\ \hline 1011 \end{array} \text{(2s complement DIFFERENCE)}$$

(b)

$$\begin{array}{r} (+3) \\ -\ (-3) \\ \hline +6_{10} \end{array} = 1101 \xrightarrow[\text{and ADD}]{\text{Form 2s complement}} \begin{array}{r} 0011 \\ +\ 0011 \\ \hline 0110 \end{array} \text{(2s complement DIFFERENCE)}$$

(c)

$$\begin{array}{r} (-4) \\ -\ (+2) \\ \hline -6_{10} \end{array} = 0010 \xrightarrow[\text{and ADD}]{\text{Form 2s complement}} \begin{array}{r} 1100 \\ +\ 1110 \\ \hline 1\ 1010 \end{array} \text{(2s complement DIFFERENCE)}$$

Discard

(d)

Fig. 10-29 Four sample signed subtraction problems using 2s complement numbers.

problem is $(+7) - (+3) = +4_{10}$. The subtrahend ($+3$ in this case) is converted to its 2s complement form. Next, the 2s complement of this is formed, yielding 1101. Then 0111 is *added* to 1101, yielding 1 0100. The MSB (overflow from 4-bit register) is discarded, leaving the *difference* of 0100, or $+4_{10}$. Note that an adder is used for subtraction. This is done by converting the subtrahend to its 2s complement and adding. Any carry or overflow into the fifth binary place is discarded.

Look over the sample 2s complement subtraction problems using an adder in Fig. 10-29(*b*), (*c*), and (*d*). See if you can follow the procedure in these remaining subtraction problems.

Only *4-bit* 2s complement numbers were used in the previous examples. Most commercial microprocessors use 8-, 16-, or 32-bit groupings. The procedures used with 4-bit 2s complement numbers apply to 8-bit, 16-bit, and 32-bit 2s complement numbers also. In summary, 2s complement numbers are used because they can show the sign of a number. Twos complement numbers can be used with adders to either *add* or *subtract* signed numbers.

Self-Test

Answer the following questions.

40. When microprocessors process both positive and negative numbers, _____ representations are used.
41. The 2s complement number 0111 represents _____ in binary and _____ in decimal.
42. The 2s complement number 1111 represents _____ in decimal.
43. In 2s complement representation, the MSB is the _____ bit. If the MSB is 0 the number is _____ (negative, positive), whereas if the MSB is 1, the number is _____ (negative, positive).
44. The decimal number -6 equals _____ in 2s complement representation.
45. The decimal number $+5$ equals _____ in 2s complement representation.
46. Calculate the sum of the 2s complement numbers 1110 and 1101. Give the answer in 2s complement and in decimal.
47. Calculate the sum of the 2s complement numbers 0110 and 1100. Give the answer in 2s complement and in decimal.

10-14 2s COMPLEMENT ADDER/SUBTRACTORS

A 2s complement *4-bit adder/subtractor system* is drawn in Fig. 10-30. Note the use of four full adders to handle the two 4-bit numbers. XOR gates have been added to the *B* inputs of each full adder to control the mode of operation of the unit. With the mode control at 0, the system *adds* the 2s complement numbers $A_3 A_2 A_1 A_0$ and $B_3 B_2 B_1 B_0$. The sum appears in 2s complement notation at the output indicators at the lower right. The LOW at the *A* inputs of the XOR gates permit the *B* data to flow through the gate with *no inversion*. If a HIGH enters B_0 input to the XOR gate, then a HIGH exits the gate at *Y*. The C_{in} input to the top 1s full adder is held at a 0 during the time the mode control is in the add position. In the add mode, the 2s complement adder operates just like a binary adder except that the carry out (C_o) from the 8s full adder is discarded. In Fig. 10-30, the C_o output from the 8s full adder is left disconnected.

The mode control input is placed at logical 1 for the unit to subtract 2s complement numbers. This causes the XOR gates *to invert the data at the B inputs*. The C_{in} input to the 1s full adder also receives a HIGH. The combination of the XOR gate's inversion plus adding the 1 at the C_{in} input of the 1s full adder is the same as complementing and adding 1. This is comparable to forming the 2s complement of the subtrahend (*B* number in Fig. 10-30).

The 2s complement adder/subtractor system in Fig. 10-30 looks much like the binary adder/subtractor circuit in Fig. 10-14. Remember that the system in Fig. 10-30 only uses 2s complement numbers.

Self-Test

Supply the missing word or words in each statement.

48. Refer to Fig. 10-30. The numbers to be added or subtracted in this system must be in _____ (binary, BCD, 1s complement, 2s complement).

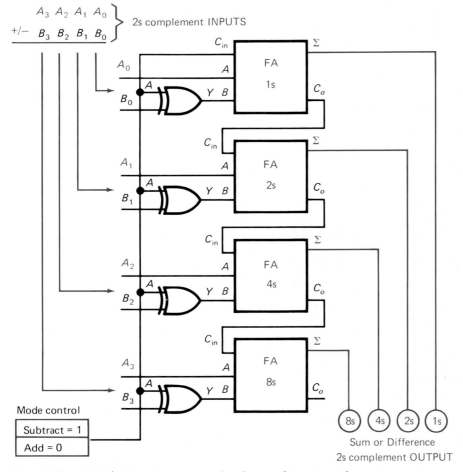

Fig. 10-30   **Adder/subtractor system using 2s complement numbers.**

INPUTS

OUTPUTS

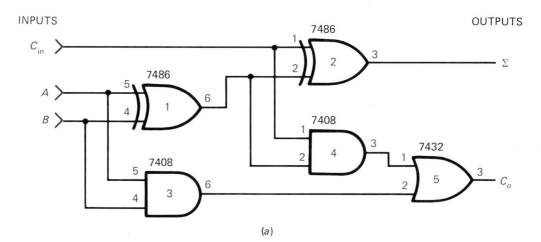

(a)

INPUTS			NORMAL OUTPUTS		ACTUAL OUTPUTS	
C_{in}	B	A	Σ	C_o	Σ	C_o
0	0	0	0	0	L	L
0	0	1	1	0	H	L
0	1	0	1	0	H	L
0	1	1	0	1	L	H
1	0	0	1	0	H	L
1	0	1	0	1	L	L
1	1	0	0	1	L	L
1	1	1	1	1	H	H

(b)

Fig. 10-31 (a) **Faulty full-adder circuit used for troubleshooting problem.** (b) **Full-adder truth table with normal and actual outputs.**

49. Refer to Fig. 10-30. The sum or difference output from this system will be in _____ (binary, BCD, 1s complement, 2s complement).

50. Refer to Fig. 10-30. This system can add or subtract _____ (signed, only unsigned) numbers.

51. Refer to Fig. 10-30. If the system is adding 0011 to 1100, the output will read _____ . This is the 2s complement representation for decimal _____ .

52. Refer to Fig. 10-30. If the system is subtracting 0010 from 0101, the output will read _____ . This is the 2s complement representation of decimal _____ .

10-15 TROUBLESHOOTING A FULL ADDER

A faulty full adder circuit is sketched in Fig. 10-31(a). The student or technician first checks the circuit visually and for signs of excessive heat. No problems are found.

The full-adder is a combinational logic cir-

cuit. For your convenience, its truth table with normal outputs is shown at the left in Fig. 10-31(b). The student or technician manipulates the full-adder inputs and with a logic probe checks the outputs (Σ and C_o). The actual logic probe outputs are shown in the right-hand columns of the truth table in Fig. 10-31(b). H stands for a HIGH logic level, while L stands for a LOW logic level. Two errors seem to appear in the C_o column in lines 6 and 7 of the truth table. These are noted in Fig. 10-31(b). A look at the truth-table results of the faulty full adder indicates no trouble in the Σ column. The Σ circuitry involves the two XOR gates labeled 1 and 2 in Fig. 10-31(a). It appears that these gates are operating properly.

The troubleshooter expects the problem to be in the OR gate or two AND gates. The bottom line of the truth table suggests that the bottom AND gate and OR gate work. The upper AND gate (labeled 4) is suspect. The technician manipulates the inputs to line 6 on the truth table ($C_{in} = 1$, $B = 0$, $A = 1$). Pins 1 and 2 of the AND gate labeled 4 should both be 1. Both inputs to gate 4 indicate a HIGH logic level when a logic probe is touched to pins 1 and 2. Output 3 of AND gate 4 is checked and remains LOW. This indicates a stuck LOW output at gate 4.

The technician carefully checks the 7408 IC and surrounding circuit board for possible short circuits to GND. None are found. Gate 4 is assumed to have a stuck LOW output, and the 7408 IC is replaced with an exact duplicate.

After replacement of the 7408 IC, the troubleshooter checks the full-adder circuit for proper operation. The circuit works according to its normal truth table. Truth tables help both technician and student with troubleshooting. Such tables define how a *normal circuit should respond*. The truth table becomes part of the technician's knowledge of the circuit. Knowledge of normal circuit operation is critical to good troubleshooting.

Self-Test

Supply the missing word or words in each statement.

53. Refer to Fig. 10-31. The fault in the _____ (combinational, sequential) logic circuit seems to be in the _____ (carry out, sum) part of the circuit.
54. Refer to Fig. 10-31. The fault in the circuit is in gate _____ (number); the output is stuck _____ (HIGH, LOW).

SUMMARY

1. Arithmetic circuits, such as adders and subtractors, are combinational logic circuits constructed with logic gates.
2. The basic addition circuit is called a half adder. Two half adders and an OR gate can be wired to form a full adder.
3. The basic subtraction circuit is called a half subtractor. Two half subtractors and an OR gate can be wired to form a full subtractor.
4. Adders (or subtractors) can be wired together to form parallel adders.
5. A 4-bit parallel adder adds two 4-bit binary numbers at one time. This adder contains a single half adder (1s place) and three full adders.
6. By using the 1s complement and end-around carry method of subtraction, an adder can be used for binary subtraction.
7. By using AND and XOR gates with a parallel adder, the add and subtract functions can be combined in one device.
8. Manufacturers produce several arithmetic ICs, such as the 7483 4-bit binary adder.
9. Adder/subtractor units are often part of the CPU of calculating machines.
10. Binary multiplication performed by digital circuits may use repeated additions or the add and shift method.
11. Microprocessors use 2s complement notation when dealing with signed numbers. Adders can be used to perform both addition and subtraction using 2s complement numbers.
12. Truth tables are a great aid in troubleshooting combinational logic circuits since they define the normal operation of the circuits.

CHAPTER REVIEW QUESTIONS

Answer the following questions.

10-1. Do binary addition problems **a** to **h** (show your work):

a. 101 + 011 =	**e.** 1000 + 1000 =
b. 110 + 101 =	**f.** 1001 + 0111 =
c. 111 + 111 =	**g.** 1010 + 0101 =
d. 1000 + 0011 =	**h.** 1100 + 0101 =

10-2. Draw a block diagram for a half adder (label two inputs and two outputs).

10-3. Draw a block diagram for a full adder (label three inputs and two outputs).

10-4. Do binary subtraction problems **a** to **h** (show your work):

a. $1100 - 0010 =$	**e.** $10000 - 0011 =$
b. $1101 - 1010 =$	**f.** $1000 \ - 0101 =$
c. $1110 - 0011 =$	**g.** $10010 - 1011 =$
d. $1111 - 0110 =$	**h.** $1001 \ - 0010 =$

10-5. Draw a block diagram of a half subtractor (label two inputs and two outputs).

10-6. Draw a block diagram of a full subtractor (label three inputs and two outputs).

10-7. Draw a block diagram of a two-bit parallel adder (use a half and a full adder).

10-8. Draw a block diagram of a three-bit parallel subtractor (use three full adders and three inverters).

10-9. Draw a block diagram of a three-bit parallel adder/subtractor (use three full adders, three XOR gates, and one AND gate).

10-10. Do binary subtraction problems **a** to **h** using the 1s complement and end-around carry method (show your work):

a. $111 \ - \ \ 101 =$	**e.** $1011 - 1010 =$
b. $1000 - 0011 =$	**f.** $1100 - 0110 =$
c. $1001 - 0010 =$	**g.** $1110 - 0100 =$
d. $1010 - 0100 =$	**h.** $1111 - 0111 =$

10-11. Draw a logic symbol diagram of a two-bit parallel adder using AND, OR, and XOR gates.

10-12. Do binary multiplication problems **a** to **h** (show your work). Check your answers using decimal multiplication.

a. $101 \ \times 011 =$	**e.** $1010 \times 011 \ =$
b. $111 \ \times 011 =$	**f.** $110 \ \times 111 \ =$
c. $1000 \times 101 =$	**g.** $1100 \times 1000 =$
d. $1001 \times 010 =$	**h.** $1010 \times 1001 =$

10-13. List two methods of doing binary multiplication with digital electronic circuits.

10-14. If the CPU of your computer has only an adder and shift registers, how can you still do binary multiplication with this machine?

10-15. Convert the following signed decimal numbers to their 4-bit 2s complement form:

a. $+1 =$	**c.** $-1 =$
b. $+7 =$	**d.** $-7 =$

10-16. Convert the following 4-bit 2s complement numbers to their signed decimal form:

a. $0101 =$	**c.** $1110 =$
b. $0011 =$	**d.** $1000 =$

10-17. Add the following 4-bit 2s complement numbers. Give each sum as a *4-bit 2s complement* number. Also give each sum as a signed decimal number.

a. $0110 + 0001 =$
b. $1101 + 1011 =$
c. $0001 + 1100 =$
d. $0100 + 1110 =$

10-18. Subtract the following 4-bit 2s complement numbers. Give each difference as a *4-bit 2s complement number*. Also give each difference as a signed decimal number.

a. $0110 - 0010 =$
b. $1001 - 1110 =$
c. $0010 - 1101 =$
d. $1101 - 0001 =$

Answers to Self-Tests

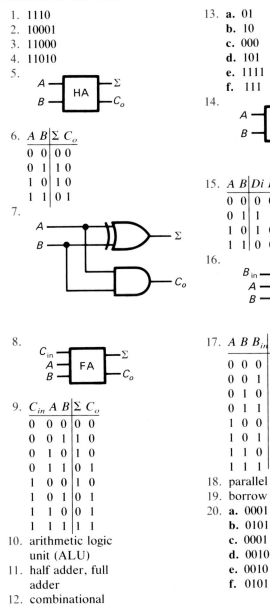

1. 1110
2. 10001
3. 11000
4. 11010
5.

6.

A	B	Σ	C_o
0	0	0	0
0	1	1	0
1	0	1	0
1	1	0	1

7.

8.

9.

C_{in}	A	B	Σ	C_o
0	0	0	0	0
0	0	1	1	0
0	1	0	1	0
0	1	1	0	1
1	0	0	1	0
1	0	1	0	1
1	1	0	0	1
1	1	1	1	1

10. arithmetic logic unit (ALU)
11. half adder, full adder
12. combinational

13. **a.** 01
  b. 10
  c. 000
  d. 101
  e. 1111
  f. 111
14.

15.

A	B	Di	B_o
0	0	0	0
0	1	1	1
1	0	1	0
1	1	0	0

16.

17.

A	B	B_{in}	Di	B_o
0	0	0	0	0
0	0	1	1	1
0	1	0	1	1
0	1	1	0	1
1	0	0	1	0
1	0	1	0	0
1	1	0	0	0
1	1	1	1	1

18. parallel subtractor
19. borrow lines
20. **a.** 0001
  b. 0101
  c. 0001
  d. 0010
  e. 0010
  f. 0101

21. adders
22. full adders
23. disabled, do not
24. HIGH
25. enabled, do
26. adder
27. cascaded
28. XOR
29. LOW
30. minuend, subtrahend
31. l, *CLK*
32. immediately
33. 1110
34. 1000001
35. 10101000
36. repeated addition
37. add and shift
38. add and shift
39. do not
40. 2s complement
41. 0111, +7
42. −1
43. sign, positive, negative
44. 1010
45. 0101
46. 1011, −5
47. 0010, +2
48. 2s complement
49. 2s complement
50. signed
51. 1111, −1
52. 0011, +3
53. combinational, carry out
54. 4, LOW

CHAPTER 11

Memories

The flip-flop forms the basic "memory cell" in many semiconductor memories. You have already used a shift register and latches to form temporary memories. Three more types of semiconductor memory devices will be investigated in this chapter. They are the RAM, the ROM, and the PROM. The chapter will also survey other types of memory or storage devices. Most calculators contain an electronic memory. A given number can be stored in this electronic memory by a simple press of the "store" key. The number can be recalled from memory by a press of the "memory recall" key. Human memory works in a similar fashion. You store information by learning, and you recall information by remembering. Most digital electronic systems use several types of memories. Electronic devices that contain a memory are almost always constructed using digital electronic circuits. Data storage is easily implemented using digital technology, but it is difficult using analog electronic circuits. A complicated digital system such as a computer uses both internal and external storage devices. The internal storage is probably in the form of semiconductor memory ICs. Some older larger computers may still use magnetic-core memory. Outside the computer, information is stored on bulk-storage memory devices which include magnetic floppy or hard disks, drums, magnetic tape or punched paper cards or tape. Besides a computer's internal storage (RAM and ROM), many temporary storage devices such as latches and registers are used within the CPU or microprocessor.

11-1 RANDOM-ACCESS MEMORY (RAM)

One type of semiconductor memory device used in digital electronics is the *random-access* *memory*. The RAM is a memory that you can "teach." After the "teaching-learning" process (called *writing*), the RAM remembers the information for a while and the RAM's stored information can be recalled, or "remembered," at any time. We say that we can *write* information (0s and 1s) into the memory and *read out*, or recall, information. The RAM is also called a *read/write memory* or a *scratch-pad memory*.

A semiconductor memory with 64 positions in which to place 0s and 1s is illustrated in Fig. 11-1. The 64 squares (mostly blank) at the right represent the 64 positions that can be filled with data. Notice that the 64 bits are organized into 16 groups called *words*. Each of the 16 words contains 4 *bits* of information. This memory is said to be organized as a 16×4 memory. That is, it contains 16 words, and each word is 4 bits long. A 64-bit memory could be organized as a 32×2 memory (32 words of 2 bits each), a 64×1 memory (64 words of 1 bit each), or an 8×8 memory (8 words of 8 bits each).

Address	Bit D	Bit C	Bit B	Bit A
Word 0				
Word 1				
Word 2				
Word 3	0	1	1	0
Word 4				
Word 5				
Word 6				
Word 7				
Word 8				
Word 9				
Word 10				
Word 11				
Word 12				
Word 13				
Word 14				
Word 15				

Fig. 11-1   Organization of a 64-bit memory.

The memory in Fig. 11-1 looks very much like a truth table on a scratch pad. On the table after word 3 we have written the contents of word 3 (0110). We say we have stored, or *written*, a word into the memory; this is the "write" operation. To see what is in the memory at word location 3, just read from the table in Fig. 11-1; this is the "read" operation. The write operation is the process of putting new information into the memory. The read operation is the process of taking information out of the memory. The read operation is also re-

ferred to as the *sense* operation because it senses, or reads, the contents of the memory.

You could write any combination of 0s and 1s in the table in Fig. 11-1 rather like writing on a scratch pad. You could then read any word(s) from the memory, as from a scratch pad. Notice that the information in the memory remains even after it is read. Now it should be obvious why this memory is sometimes called a 64-bit scratch-pad memory. The memory has a place for 64 bits of information, and the memory can be written into or read from very much like a scratch pad.

The memory in Fig. 11-1 is called a random access memory because you can go directly to word 3 or word 15 and read its contents. In other words, you have access to any bit (or word) at any instant. You merely skip down to its word location and read that word. A location in the memory, such as word 3, is referred to as the storage location or *address*. In the case of Fig. 11-1, the address of word 3 is 0011_2 (3_{10}). However, the data stored at this address is 0110.

The RAM cannot be used for permanent memory because it loses its data when the power to the IC is turned off. The RAM is considered a *volatile* memory because of this loss of data. Volatile memories are thus used for the *temporary* storage of data. However, some memories are permanent; they do not "forget" or lose their data when the power goes off. Such permanent memories are called *nonvolatile* storage devices.

RAMs are used where only a temporary memory is needed. RAMs are used for calculator memories, buffer memories, cache memories, and microcomputer user memories.

Self-Test

Supply the missing word or words in each statement.

1. The letters "RAM" stand for _____ .
2. Copying information into a storage location is called _____ into memory.
3. Copying information from a storage location is called _____ from memory.
4. A RAM is also called a _____ or scratch-pad memory.
5. Refer to Fig. 11-1. This 64-bit unit is organized as a _____ memory.
6. A disadvantage of the RAM is that it is _____ ; it loses its data when the power is turned _____ (off, on).

From page 224:

RAM

ROM

PROM

Semiconductor memory ICs

Magnetic-core memory

Bulk-storage memory latches

Registers

Random-access memory

On this page:

Read/write memory

Memory organization

Address

Write operation

Read operation

Volatile memory

Nonvolatile storage devices

11-2 STATIC RAM ICS

The *7489 read/write TTL RAM* is a 64-bit data storage unit in IC form. Figure 11-2(*a*) is a logic symbol for the 7489 RAM. The memory cells are arranged like the layout of the table in Fig. 11-1. The memory can hold 16 words; each word in the 7489 IC is 4-bits wide. The 7489 RAM is said to be organized as a 16×4-bit memory. A pin diagram for the 7489 IC is given in Fig. 11-2(*b*).

A simplified truth table for the 7489 RAM is shown in Fig. 11-2(*c*). The *memory enable* ($\overline{ME}$) input is used to "turn on" or "select" the RAM for either reading or writing. The top line in the truth table shows both the $\overline{ME}$ and the *write enable* ($\overline{WE}$) inputs LOW. The 4 bits at the data inputs ($D1$ to $D4$) are stored in the memory location selected by the address inputs ($A3$ to $A0$). The RAM was in the *write mode*.

Let us write data into the 7489 memory chip. Suppose we want to write 0110 into word 3 lo-

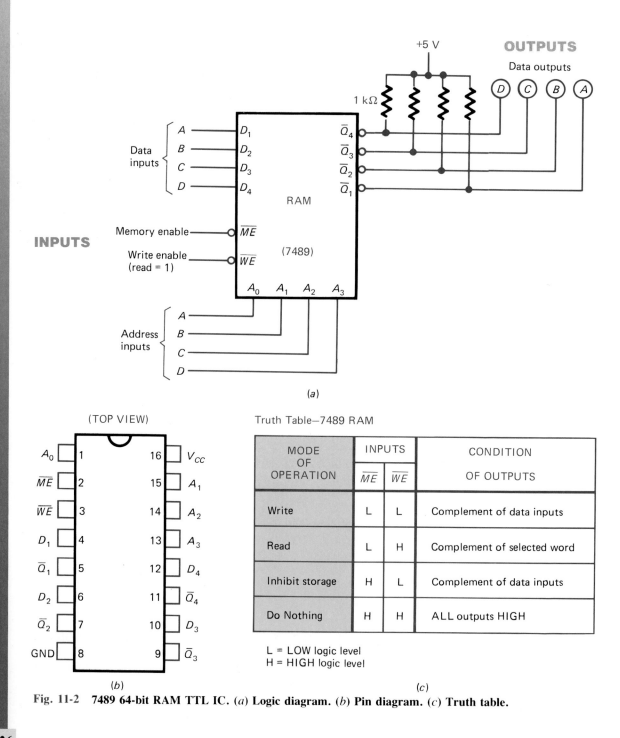

(*a*)

(TOP VIEW)

Truth Table—7489 RAM

MODE OF OPERATION	INPUTS		CONDITION
	$\overline{ME}$	$\overline{WE}$	OF OUTPUTS
Write	L	L	Complement of data inputs
Read	L	H	Complement of selected word
Inhibit storage	H	L	Complement of data inputs
Do Nothing	H	H	ALL outputs HIGH

L = LOW logic level
H = HIGH logic level

(*b*)

(*c*)

Fig. 11-2 7489 64-bit RAM TTL IC. (*a*) **Logic diagram.** (*b*) **Pin diagram.** (*c*) **Truth table.**

cation, as shown in Fig. 11-1. The address for word 3 is $A3 = 0$, $A2 = 0$, $A1 = 1$, and $A0 = 1$. Word 3 is located in the memory by placing a binary 0011 on the *address inputs* of the 7489 RAM [see Fig. 11-2(*a*)]. Next, place the correct input data at the *data inputs*. To enter 0110, place a 0 at input *A*, a 1 at input *B*, a 1 at input *C*, and a 0 at input *D*. Next, place a LOW at the write enable ($\overline{WE}$) input. Last, place a LOW at the memory enable ($\overline{ME}$) input. Data is written into the memory in the storage location called word 3.

Now let us *read*, or *sense*, what is in the memory. If we were to read out the data stored at word 3, we first set the address inputs to binary 0011 (decimal 3). The write enable ($\overline{WE}$) input should be in the read position, or HIGH according to the truth table in Fig. 11-2(*c*). The memory enable ($\overline{ME}$) input should be LOW. The data outputs will indicate 1001. This output is the *complement* of the actual memory contents, which is 0110. Inverters could be attached to the outputs of the 7489 IC to make the output data the same as that in the memory. This illustrates the use of the *read mode* on the 7489 RAM.

The last two lines in the truth table in Fig. 11-2(*c*) inhibit both the read and write processes. When both $\overline{ME}$ and $\overline{WE}$ inputs are HIGH, all outputs go HIGH. When the $\overline{ME}$ input is HIGH and the $\overline{WE}$ input is LOW, the outputs are the complement of the inputs but no reading or writing is taking place.

The 7489 RAM has open-collector outputs. This is suggested by the use of pull-up resistors on the outputs in the diagram in Fig. 11-2(*a*). The 7489 RAM is referred to as an *82S25* or *3102A 64-bit bipolar scratch-pad memory IC* by Signetics Corporation. A close relative of the 7489 is the 74189 64-bit RAM with the same configuration and pins except its outputs are of the tristate type instead of the open-collector type. A *tristate* output has three levels: LOW, HIGH, or high impedance.

You will find that although different manufacturers use various labels for the inputs and outputs on this IC, all 7489 ICs have the inputs and outputs shown in Fig. 11-2. IC manufacturers usually include even very small memories like the 7489 RAM in separate data manuals that cover semiconductor memories.

The 7489 RAM is just one example of a semiconductor memory in IC form. Many other semiconductor memories are available from IC manufacturers. Semiconductor memories are relatively new, but they are being widely used in many applications. They have the advantage of being inexpensive, small, reliable, and able to operate at high speeds. Microprocessor-based digital equipment makes extensive use of semiconductor read/write RAMs in IC form.

Semiconductor RAM ICs are subdivided by manufacturers into static and dynamic types. The *static RAM* stores data in a flip-flop-like element. It is called a static RAM because it holds its 0 or 1 as long as the IC has power. The *dynamic RAM* IC stores its logic state as an electric charge in an MOS device. The stored charge leaks off after a very short time and must be refreshed many times per second. Refreshing the logic elements of a dynamic RAM requires rather extensive refresh circuitry. The dynamic RAM logic element is simpler and therefore takes up less space on the silicon chip. Dynamic RAMs come in larger sizes than static RAMs. The newer dynamic RAMs have the refresh circuitry on the chip. Because of their ease of use, static RAMs will be used in this chapter.

A popular MOS memory IC is the 2114 static RAM. The 2114 RAM will store 4096 bits which are organized into 1024 words of 4 bits each. A logic diagram of the 2114 RAM is sketched in Fig. 11-3(*a*) on page 228. The 2114 RAM has 10 address lines which can access 1024 (2^{10}) words. It has *chip select* ($\overline{CS}$) and $\overline{WE}$ control inputs. The $\overline{CS}$ input is similar to the $\overline{ME}$ input on the 7489 RAM. The four input/output (I/O_1, I/O_2, I/O_3, I/O_4) pins serve as inputs when the RAM is in the write mode and outputs when the IC is in the read mode. The 2114 RAM is powered by a +5-V power supply.

A block diagram of the 2114 RAM is illustrated in Fig. 11-3(*b*). Especially note the three-state buffers used to isolate the input/output (I/O) pins from a computer data bus. Note that the address lines are also buffered. The 2114 RAM comes in 18-pin DIP IC form.

An important characteristic of a RAM is its access time. The *access time* is the time it takes to locate and output (or input) a piece of data. The access time of the 7489 TTL RAM is about 33 ns. The access time of the 2114 MOS RAM ranges between 100 and 250 ns depending on what version of the chip you purchase. The TTL RAM is said to be faster than the 2114 memory chip because of its shorter access time.

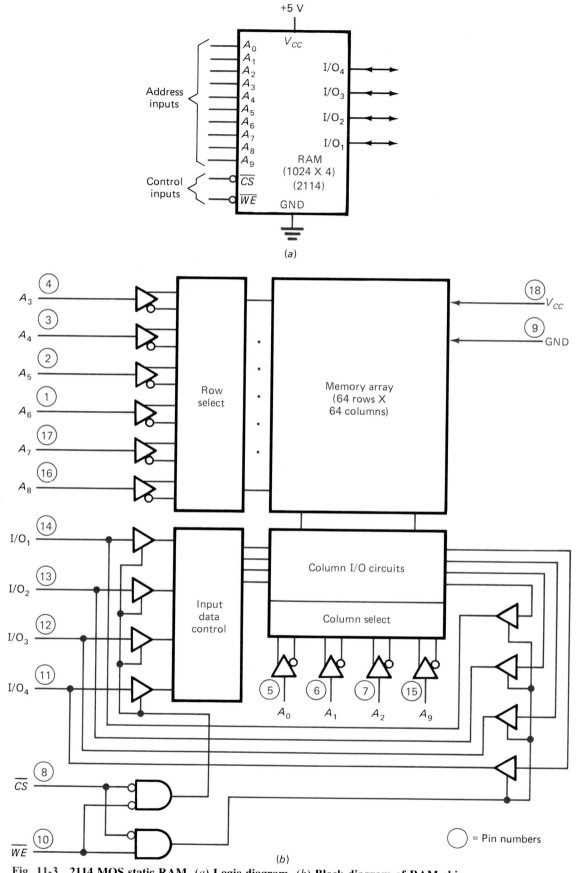

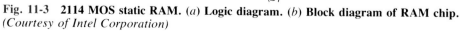

Fig. 11-3 2114 MOS static RAM. (*a*) **Logic diagram.** (*b*) **Block diagram of RAM chip.**
(*Courtesy of Intel Corporation*)

Self-Test

Supply the missing word in each statement.

7. The 7489 IC is a 64-bit _____ .
8. The 7489 memory IC can hold _____ words, each word being _____ bits wide.
9. Refer to Fig. 11-2. If the address inputs = 1111, write enable = 0, memory enable = 0, and data inputs = 0011, then the 7489 IC is in the _____ (read, write) mode. Input data 0011 is being _____ (read from, written into) memory location _____ (decimal number).
10. A _____ (dynamic, static) RAM must be refreshed many times per second.
11. The 2114 RAM IC will store _____ bits of data, and each of the 1024 words is _____ bits wide.
12. Both the 7489 and 2114 memory chips are _____ (dynamic, static) RAMs.
13. Refer to Fig. 11-3. When both control inputs ($\overline{CS}$ and $\overline{WE}$) are LOW, the four I/O pins of the 2114 RAM serve as _____ (inputs, outputs).

11-3 USING A RAM

We need some practice in using the 7489 read/write RAM. Let us *program* it with some usable information. To program the memory is to write in the information we want in each memory cell.

Probably you cannot remember how to count from 0 to 15 in the Gray code, and so let us take the Gray code and program it into the 7489 RAM. The RAM will remember the Gray code for us, and we can then use the RAM to convert from binary numbers to Gray code numbers.

Table 11-1 shows the Gray code numbers from 0 to 15. For convenience binary numbers are also included in Table 11-1. The 64 logical 1s and 0s in the Gray code number column of the table must be written into the 64-bit RAM. The 7489 IC is perfect for this job because it contains 16 words; each word is 4 bits long. This is the same pattern we have in the Gray code column of Table 11-1. The decimal number in the table will be the word number (see Fig. 11-1). The binary number is the number applied to the address input of the 7489 RAM (see Fig. 11-2). The Gray code number is applied to the data inputs of the RAM [see Fig. 11-2(a)]. When the $\overline{ME}$ and $\overline{WE}$ inputs are ac-

tivated, the Gray code is written into the 7489 RAM. The RAM remembers this code as long as the power is not turned off.

Table 11-1 Gray Code

Decimal Number	Binary Number	Gray Code Number
0	0000	0000
1	0001	0001
2	0010	0011
3	0011	0010
4	0100	0110
5	0101	0111
6	0110	0101
7	0111	0100
8	1000	1100
9	1001	1101
10	1010	1111
11	1011	1110
12	1100	1010
13	1101	1011
14	1110	1001
15	1111	1000

After the 7489 RAM is programmed with the Gray code, it is a *code converter*. Figure 11-4(a) on the next page shows the basic system. Notice that we input a binary number. The code converter reads out the equivalent Gray code number. The system is a *binary-to-Gray-code converter*.

How do you convert binary 0111 (decimal 7) to the Gray code? Figure 11-4(b) shows the binary number 0111 being applied to the address inputs of the 7489 RAM. The $\overline{ME}$ input is at 0. The $\overline{WE}$ input is in the read position (logical 1). The 7489 IC then reads out the stored word 7 in inverted form. The four inverters complement the output of the RAM. The result is the correct Gray code output. The Gray code output for binary 0111 is shown as 0100 in Fig. 11-4(b). You may input any binary number from 0000 to 1111 and get the correct Gray code output.

The binary-to-Gray-code converter in Fig. 11-4 works fine. It demonstrates how you can program and use the 7489 RAM. It is not very practical, however, because the RAM is a volatile memory. If the power is turned off for even an instant, the storage unit loses all its

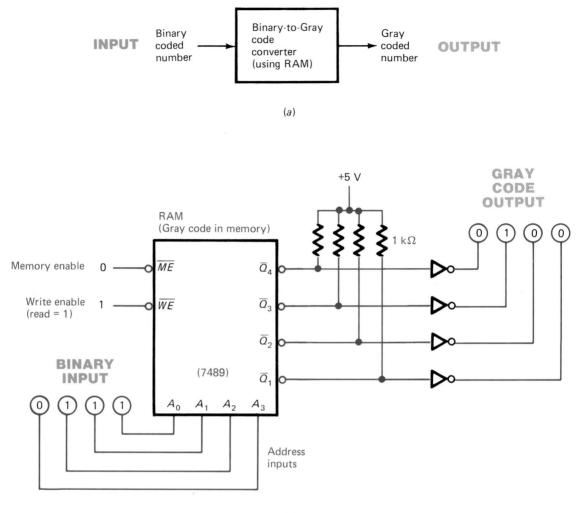

Fig. 11-4   **Binary-to-Gray-code converter. (*a*) System diagram. (*b*) Wiring diagram using RAM.**

memory and "forgets" the Gray code. We say the memory has been *erased*. You then have to again program, or teach, the Gray code to the 7489 RAM.

16. If power to the 7489 IC in Fig. 11-4 is turned off for an instant, the RAM will _____ (lose its program and have to be reprogrammed, still hold the Gray code in its memory cells).

Self-Test

Supply the missing word or words in each statement.

14. Refer to Fig. 11-4. The RAM is programmed as a _____ -code converter in this example.
15. Refer to Fig. 11-4. If the address inputs = 1000, $\overline{WE} = 1$, and $\overline{ME} = 0$, then the output at the displays on the right will be _____ . This is the _____ code equivalent of binary _____ .

11-4 READ-ONLY MEMORY (ROM)

Many digital devices including microcomputers must store some information permanently. This is typically stored in a *read-only memory* or *ROM*. The ROM is programmed by the manufacturer to the user's specifications. Smaller ROMs may be used to solve combinational logic problems like decoding.

ROMs are classified as nonvolatile memories because they do not lose their data when power is turned off. The read-only memory is also referred to as the *mask-programmed ROM*. The ROM is used in only high-volume production applications because of the expensive initial setup costs. Programmable read-only memories (PROMs) are used for lower-volume applications where a permanent memory is required.

The primitive diode ROM circuit in Fig. 11-5 on page 232 can perform the task of translating from binary to Gray code. The Gray code along with decimal and binary equivalents is listed in Table 11-1.

If the rotary switch in Fig. 11-5(*a*) has selected the decimal 6 position, what will the ROM output indicators display? The outputs (*D*, *C*, *B*, *A*) will indicate LHLH or 0101. The *D* and *B* outputs are connected directly to ground through the resistors and read LOW. The *C* and *A* outputs are connected to +5 V through two forward-biased diodes and the output voltage will read about +2 to +3 V, which is a logical HIGH. Notice that the pattern of diodes in the diode ROM matrix in Fig. 11-5(*a*) is similar to the pattern of 1s in the Gray code column in Table 11-1. Each new position of the rotary switch will give the correct Gray code output. In a memory, such as the ROM in Fig. 11-5, each position of the rotary switch is referred to as an *address*.

A refinement in the diode ROM is shown in Fig. 11-5(*b*). The diode ROM circuit in Fig. 11-5(*b*) uses a 1-of-10 decoder (7442 TTL IC) and inverters for row selection. This example shows a binary input of 0101 (decimal 5). This activates output 5 of the 7442 with a LOW. This drives the inverter, which outputs a HIGH. The HIGH forward biases the three diodes connected to the row 5 line. The outputs would be LHHH or 0111. This is the Gray code equivalent for binary 0101 according to Table 11-1.

The diode ROM suffers many disadvantages. Their logic levels are marginal. The diode ROM also suffers in that it has very limited drive capability. The diode ROMs do not have input and output buffering needed when working with systems that contain data and address busses.

Practical ROMs are available from many manufacturers. These can range from very small bipolar TTL units to quite large capacity CMOS or NMOS ROMs. Most commercial ROMs can be purchased in DIP form. As examples, a very small capacity unit might be the TTL 74S370 2048-bit ROM organized as a 512-word by 4-bit memory. A larger-capacity unit might be CMOS TMS47C512 524, 288-bit ROM organized as a 65,536-word by 8-bit

memory. The 65,536 × 8 unit has an access time of from 200 to 350 ns depending on the version you purchase. Even some personal computers have ROMs of this large capacity.

As an example of a commercial product, the Texas Instruments TMS4764 ROM will be featured. The TMS4764 is an 8192-word by 8-bit ROM. Its 8192 × 8 organization makes it useful in microprocessor-based systems which typically store data in 8-bit groups, or bytes.

A pin diagram for the TMS4764 ROM is reproduced in Fig. 11-6(*a*) on page 233. The ROM is housed in a 24-pin DIP. The names and functions of the pins are given in the chart in Fig. 11-6(*b*). Notice that a total of 13 address lines (A0 to A12) are needed to address the 8192 (2^{13}) memory locations. A0 is the LSB and A12 is the MSB of the word address. The access time of the TMS4764 ROM varies from 150 to 250 ns depending on the version of the chip you purchase. Permanently stored data is output via the pins labeled Q1 through Q8. Q1 is considered the LSB while Q8 is the MSB. The output pins (Q1 to Q8) are enabled by pin 20. Pin 20 may be programmed by the manufacturer to be an active HIGH or active LOW $\overline{CS}$ or $\overline{CE}$ input. When the three-state outputs are disabled, they are in a high-impedance state which means they may be connected directly to a data bus in a microcomputer system.

Read-only memories are used to store permanent data and programs. Computer system programs, look-up tables, decoders, and character generators are but a few uses of the ROM; ROMs can also be used for solving combinational logic problems. General-purpose microcomputers allocate a larger proportion of their internal memory to RAM. However, dedicated computers allocate more addresses to ROM and usually contain only small amounts of RAM. About 500 different ROMs were available in one recent listing.

A computer program is typically referred to as *software*. However, when a computer program is stored in a ROM it is called *firmware* because of the difficulty of making changes.

Self-Test

Supply the missing word or words in each statement.

17. The letters "ROM" stand for _____ .
18. Read-only memories never forget data and are called _____ memories.

Mask-programmed ROM

Diode ROM

Address

1-of-10 decoder

TTL 745370 2048-bit ROM

CMOS TMS47C512 524, 288-bit ROM

TMS4764 ROM

Data bus

Microcomputer system

Software

Firmware

Diode ROM disadvantages

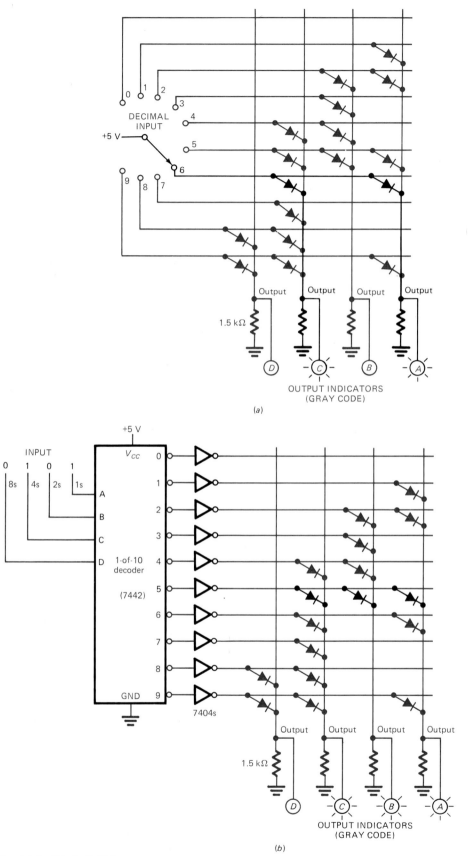

Fig. 11-5 Diode ROMs. (*a*) Primitive diode ROM programmed with Gray code.
(*b*) Diode ROM with input decoding (programmed with Gray code).

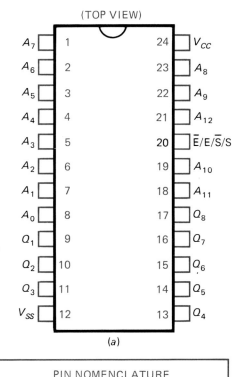

(a)

PIN NOMENCLATURE	
A_0–A_{12}	Address inputs
$\overline{E}$/E/$\overline{S}$/S	Chip Enable/Power Down or Chip Select
Q_1–Q_8	Data out
V_{CC}	5-V supply
V_{SS}	Ground

(b)

Fig. 11-6 TMS4764 ROM IC. (a) Pin diagram. (b) Pin nomenclature. *(Courtesy of Texas Instruments Incorporated)*

19. The term _____ is used to describe microcomputer programs that are permanently held in ROM.
20. Read-only memories are programmed by the _____ (manufacturer, computer operator) to your specifications.

11-5 USING A ROM

Suppose you have to design a device that will give the decimal counting sequence shown in Table 11-2: 1, 117, 22, 6, 114, 44, 140, 17, 0, 14, 162, 146, 134, 64, 160, 177, and then back to 1. These number are to read out on seven-segment displays and must appear in the order shown.

Knowing you will use digital circuits, you convert the decimal numbers to BCD numbers. This is shown in Table 11-2. You find you have

Table 11-2 Counting Sequence Problem

Decimal Readout			Binary-Coded Decimal Number		
100s	10s	1s	100s	10s	1s
		1	0	000	001
1	1	7	1	001	111
	2	2	0	010	010
		6	0	000	110
1	1	4	1	001	100
	4	4	0	100	100
1	4	0	1	100	000
	1	7	0	001	111
		0	0	000	000
	1	4	0	001	100
1	6	2	1	110	010
1	4	6	1	100	110
1	3	4	1	011	100
	6	4	0	110	100
1	6	0	1	110	000
1	7	7	1	111	111

16 rows and 7 columns of logical 0s and 1s. This section forms a truth table. As you look at the truth table, the problem seems quite complicated to solve with logic gates or data selectors. You decide to try a ROM. You think of the inside of a memory as a truth table. The BCD section of Table 11-2 reminds you that a memory organized as a 16 × 7 storage unit will do the job. This 16 × 7 ROM will have 16 words for the 16 rows on the truth table. Each word will contain seven bits of data for seven columns on the truth table. This will take a 112-bit ROM.

A 112-bit ROM is shown in Fig. 11-7 on the next page. Notice that it has four address inputs to select one of the 16 possible words stored in the ROM. The 16 different addresses are shown in the left columns of Table 11-3 (see page 235). Suppose the address inputs are binary 0000. Then the first line in Table 11-3 shows that the stored word is 0 000 001 (*a* to *g*). After decoding in Fig. 11-7, this stored word reads out on the digital displays as a decimal 1 (100s = 0, 10s = 0, 1s = 1).

Let us consider another example. Apply binary 0001 to the address inputs of the ROM in Fig. 11-7. The second row on Table 11-3 shows us that the stored word is 1 001 111 (*a* to *g*). When decoded this word reads out on the digital display as decimal 117 (100s = 1, 10s = 1, 1s = 7). Remember that the 0s and 1s in the

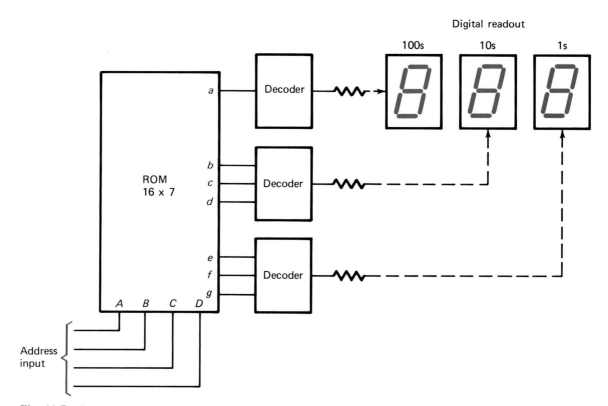

Fig. 11-7   **System diagram for the counting sequence problem using a ROM.**

center section of Table 11-3 are *permanently* stored in the ROM. When the address at the left appears at the address input of the ROM, a row of 0s and 1s or word appears at the outputs.

You have solved the difficult counting sequence problem. Figure 11-7 diagrams the basic system to be used. The information in Table 11-3 shows the addressing and programming of the 112-bit ROM and the decoded BCD as a decimal readout. You would give the information in Table 11-3 to a manufacturer, who would custom-make as many ROMs as you need with the correct pattern of 0s and 1s.

It is quite expensive to have just a few ROMs custom-programmed by a manufacturer. You probably would not use the ROM if you did not have need for many of these memory units. Remember that this problem also could have been solved by a combinational logic circuit using logic gates.

Semiconductor memories usually come in 2^n sizes or 64-, 256-, 1024-, 4096-, 8192-bit and larger units. A 112-bit memory is an unusual size. The 112-bit memory was used in the example because its truth table in Table 11-3 is exactly the truth table of the 744 IC. You used the 7447 IC as BCD-to-seven-segment de-

coder in Chap. 9. You will want to use the 7447 IC as a ROM in the laboratory.

Read-only memories are used for encoders, code converters, look-up tables, microprograms, character generators, and microcomputer system firmware.

Self-Test

Supply the missing word or words in each statement.

21. Refer to Fig. 11-7. If power is turned off and then back on, the counting sequence programmed into the ROM will ___ (be lost from, remain in) memory.
22. Refer to Table 11-3 and Fig. 11-7. If the ROM address input = 1111, the digital readout will be _____ .
23. A ROM is programmed by the _____ (manufacturer, user).

11-6 PROGRAMMABLE READ-ONLY MEMORY (PROM)

Mask-programmable ROMs are programmed by the manufacturer using photographic masks

Table 11-3 Counting Sequence Problem

Inputs				ROM Outputs							Decimal Readout		
Address or Word Location				100s	10s			1s					
				1s	4s	2s	1s	4s	2s	1s			
D	C	B	A	a	b	c	d	e	f	g	100s	10s	1s
0	0	0	0	0	0	0	0	0	0	1			1
0	0	0	1	1	0	0	1	1	1	1	1	1	7
0	0	1	0	0	0	1	0	0	1	0		2	2
0	0	1	1	0	0	0	0	1	1	0			6
0	1	0	0	1	0	0	1	1	0	0	1	1	4
0	1	0	1	0	1	0	0	1	0	0		4	4
0	1	1	0	1	1	0	0	0	0	0	1	4	0
0	1	1	1	0	0	0	1	1	1	1		1	7
1	0	0	0	0	0	0	0	0	0	0			0
1	0	0	1	0	0	0	1	1	0	0		1	4
1	0	1	0	1	1	1	0	0	1	0	1	6	2
1	0	1	1	1	1	0	0	1	1	0	1	4	6
1	1	0	0	1	0	1	1	1	0	0	1	3	4
1	1	0	1	0	1	1	0	1	0	0		6	4
1	1	1	0	1	1	1	0	0	0	0	1	6	0
1	1	1	1	1	1	1	1	1	1	1	1	7	7

to expose the silicon die. *Mask-programmable ROMs* have long development times and the initial costs are high. Mask-programmable ROMs are usually simply called ROMs.

Field-programmable ROMs (PROMs) are also available. They shorten development time and many times lower costs. It is also much easier to correct program errors and update products when PROMs can be programmed (burned) by the local developer. The regular PROM can only be programmed once like a ROM but its advantage is that it can be made in limited quantities and can be programmed in the local lab or shop.

The *EPROM (erasable programmable read-only memory)* is a variation of the PROM. The EPROM is programmed or burned in the local lab using a *PROM burner*. If an EPROM needs to be reprogrammed, a special window on the top of the IC is used. Ultraviolet (UV) light is directed at the chip under the window of the EPROM for about an hour. The UV light erases the EPROM by setting all the memory cells to a logical 1. The EPROM can then be reprogrammed. A 24-pin EPROM DIP IC is shown in Fig. 11-8. The actual EPROM chip is visible through the window on top of the IC.

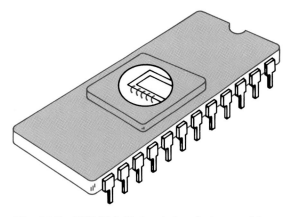

Fig. 11-8 EPROM. Note window in top used to erase EPROM with ultraviolet light.

The EEPROM is another variation of a programmable read-only memory. The *EEPROM* is an *electrically erasable PROM* also referred to as a E^2PROM. Because EEPROMs can be erased electrically, it is possible to erase and reprogram the PROMs without removing them from the circuit.

The basic idea of a PROM is illustrated in Fig. 11-9. This simplified 16-bit (4 × 4) PROM is similar to the diode ROM studied in the

Mask-programmable ROMs

Field-programmable ROMs (PROMs)

EPROM (erasable programmable read-only memory)

PROM burner

EEPROM

Electrically erasable PROM

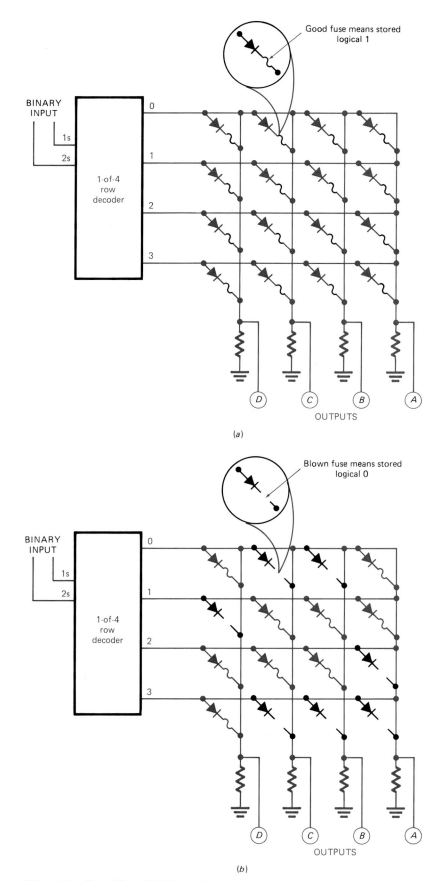

Fig. 11-9 Simplified PROM (*a*) PROM before programming. All fuses good (all 1s). (*b*) PROM after programming. Seven fuses blown (seven 0s programmed).

previous section. In Fig. 11-9(*a*), each memory cell contains a diode and a good fuse. This indicates that all of the memory cells are storing a logical 1. This is how the PROM might look before programming.

The PROM in Fig. 11-9(*b*) has been programmed with seven 0s. To program or *burn* the PROM, tiny fuses must be blown as shown in Fig. 11-9(*b*). A blown fuse in this case disconnects the diode and means a logical 0 is permanently stored in this memory cell. Because of the permanent nature of burning a PROM, the unit cannot be reprogrammed. A PROM of the type illustrated in Fig. 11-9 can only be programmed once.

A popular EPROM family is the 27XXX series. These are available from many manufacturers. A short summary of some models in the 27XXX series is shown in Table 11-4. Notice that they are all organized with byte-wide (8-bit-wide) outputs, making them compatible with most microprocessor-based systems. Many versions of each of these basic numbers are available such as low-power CMOS units, EPROMs with different access times, and even pin-compatible PROMs, EEPROMs, and ROMs.

Table 11-4 27XXX Series EPROMs

EPROM 27XXX	Organization	Number of Bits
2708	1024 × 8	8192
2716	2048 × 8	16384
2732	4096 × 8	32768
2764	8192 × 8	65536
27128	16384 × 8	131072
27256	32768 × 8	262144
27512	65536 × 8	524288

A sample IC from the 27XXX series EPROM family is illustrated in Fig. 11-10 on the next page. The pin diagram in Fig. 11-10(*a*) represents the *2732A 32K (4K × 8) ultraviolet-erasable PROM*. The 2732 EPROM has 12 address pins (A_0 to A_{11}) which can access 4096 (2^{12}) byte-wide words in the memory. The 2732 EPROM uses a 5-V power supply and can be erased using UV light. The $\overline{CE}$ input is like the chip select *CS* inputs on some other memory chips. The $\overline{CE}$ input is activated with a LOW. The $\overline{OE}/V_{PP}$ pin serves a dual purpose. It has one purpose during reading and another during writing. Under normal use the EPROM is being read. A LOW at the output enable ($\overline{OE}$) pin during a memory read activates the outputs driving the data bus of the computer system. The eight output pins are labeled O_0 to O_7 on the 2732 EPROM. A block diagram is drawn in Fig. 11-10(*c*) to show the organization of the 2732 EPROM chip.

When the 2732 EPROM is erased, all memory cells are returned to logical 1. Data is introduced by changing selected memory cells to 0s. The 2732 is in the *programming mode* (writing into the EPROM) when the dual-purpose $\overline{OE}/V_{PP}$ input is at 21 V. During programming (writing), the input data is applied to the data output pins (O_0 to O_7). The word to be programmed into the EPROM is addressed using the 12 address lines. A very short (less than 55 ms) TTL level LOW pulse is then applied to the $\overline{CE}$ input to complete the write process.

Erasing and programming an EPROM is handled by special equipment called PROM burners. After erasing and reprogramming it is common to protect the EPROM window (see Fig. 11-8) with an opaque sticker. The sticker over the EPROM window protects the chip from UV light from fluorescent lights and sunlight. The EPROM can be erased by direct sunlight in about one week or room-level fluorescent lighting in about three years.

The RAM has the disadvantage that it is volatile. When power is turned off, all data is lost from a RAM. The PROM has the disadvantage that temporary data cannot be stored in its memory cells. A newer product called *nonvolatile RAM (NVRAM)* has become available. The NVRAM has some of the advantages of both the RAM and the PROM.

A block diagram of a nonvolatile RAM is shown in Fig. 11-11 (see page 239). The NVRAM in Fig. 11-11 combines a 256 × 8 static RAM (SRAM) with an identical size EEPROM. Nonvolatile data can be stored in the EEPROM, while independent data is accessed simultaneously in the static RAM. Data can be transferred back and forth between static RAM and EEPROM by store and recall operations. A nonvolatile store signal transfers data from the static RAM to the nonvolatile EEPROM, where it is safely stored when power is turned off. When power is turned on, the data in the EEPROM is automatically recalled into the SRAM. Data stored in the nonvolatile EEPROM can be recalled an unlimited number of times.

Burn the PROM

Byte-wide

2732A 32K (4K × 8) ultraviolet-erasable PROM

Nonvolatile RAM (NVRAM)

Static RAM (SRAM)

EEPROM

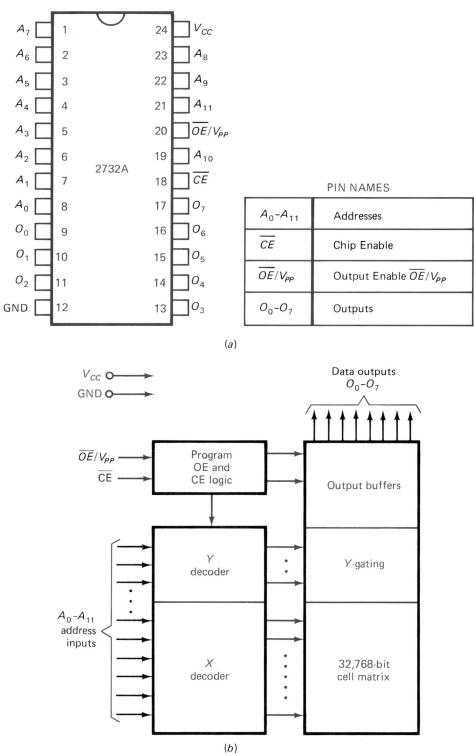

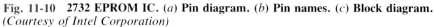

Fig. 11-10 2732 EPROM IC. (*a*) **Pin diagram.** (*b*) **Pin names.** (*c*) **Block diagram.** (*Courtesy of Intel Corporation*)

The NVRAMs are currently only available in limited sizes. The unit shown in Fig. 11-11 represents a 2K (256 × 8) NVRAM; NVRAMs are sometimes also referred to as *shadow RAMs*.

Self-Test

Supply the missing word or words in each statement.

24. The letters PROM stand for _____ .

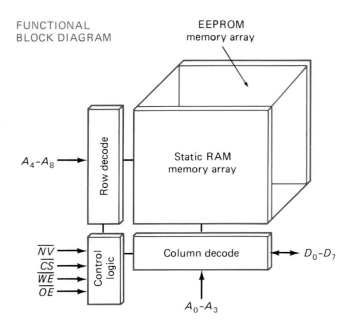

FUNCTIONAL
BLOCK DIAGRAM

EEPROM
memory array

Static RAM
memory array

Row decode

A_4-A_8

Control
logic

$\overline{NV}$
$\overline{CS}$
$\overline{WE}$
$\overline{OE}$

Column decode

D_0-D_7

A_0-A_3

PIN NAMES

A_0-A_8	Address inputs		$\overline{WE}$	Write Enable
D_0-D_7	Data I/O		$\overline{OE}$	Output Enable
$\overline{CS}$	Chip Select		NC	No connection
$\overline{NV}$	Non-Volatile Enable		V_{CC}	+ 5 volts ± 10%

Fig. 11-11 Block diagram and pin names on a typical NVRAM.

25. The letters EPROM stand for _____ .
26. The letters EEPROM stand for _____ .
27. The letters NVRAM stand for _____ .
28. Erasing EPROMs can be done _____ or by shining _____ light through a special window in the top of the IC.
29. See Table 11-4. The 27512 EPROM can store a total of _____ bits of data organized as _____ words, each 8 bits wide.

11-7 MAGNETIC-CORE MEMORY

Digital computers need large amounts of internal and external data storage. Inside the computer, semiconductor and *magnetic-core memories* are used. The magnetic-core memory is a time-proven method of storing data in the central memory of a computer. Magnetic-core memory is gradually becoming obsolete.

The magnetic-core memory is based upon the characteristics of the tiny *ferrite core*. A ferrite core is a small doughnut-shaped piece of ferromagnetic material, such as iron. The core is baked and compressed into a ceramic-like doughnut. Figure 11-12(*a*) on the next page shows a highly magnified view of a ferrite core. A typical core might measure about 1/16 in. across.

The ferrite core is used as a small magnet. Figure 11-12(*b*) shows a *write wire* threaded through the ferrite core. When current passes through the write wire in a given direction, the magnetic flux travels in a counterclockwise (ccw) direction. The magnetic flux direction is shown by an arrow on the core. We have defined the situation in Fig. 11-12(*b*) as a logical 1. In other words, ccw movement of the magnetic flux in the core means it contains a logical 1.

Figure 11-12(*c*) shows the current being reversed. With the −*I* pulse we find that the magnetic flux in the core reverses. The magnetic flux is now traveling in a clockwise (cw) direction in the core. The situation in Fig. 11-12(*c*) is defined as a logical 0; that is, a logical 0 is stored in the core when the flux

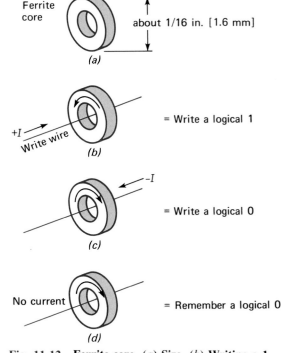

Fig. 11-12 Ferrite core. (*a*) Size. (*b*) Writing a 1. (*c*) Writing a 0. (*d*) Remembering a 0.

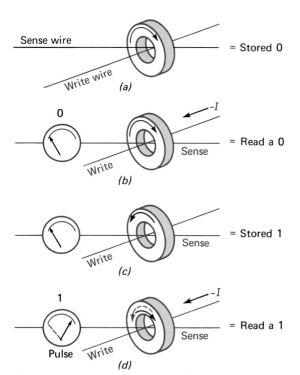

Fig. 11-13 Ferrite core. (*a*) Adding a sense wire. (*b*) Reading a 0. (*c*) Stored 1. (*d*) Reading a 1.

travels in a cw direction. With no current flow in the write wire the ferrite core still is a magnet. Depending upon which direction the core is magnetized, it remembers a logical 0 or a logical 1. Figure 11-12(*d*) shows the ferrite core with no current flowing in the write wire. The core still has magnetic flux moving in a cw direction. We say the core is storing a logical 0.

In Fig. 11-12 we saw how we write data into a magnetic core and how it remembers by staying magnetized. Like a flip-flop circuit, the ferrite core will always be either a logical 0 or a logical 1. You have seen that the magnetic-core memory is a nonvolatile type of memory. It remembers even when the power is turned off.

The *read* process requires another wire. The added wire is called a *sense* wire, as shown in Fig. 11-13(*a*). To *read* the contents of the ferrite core we apply a −*I* pulse to the core, as shown in Fig. 11-13(*b*). Assuming the core is a logical 0, there will be *no change* in magnetic flux in the core. With no change in magnetic flux we have no current, or 0 A, *induced* in the sense wire. We read that the core contains a 0.

In Fig. 11-13(*c*) we assume that the core contains a logical 1. This is shown with the ccw

arrow on the core. To read the contents of the core we apply a −*I* pulse in Fig. 11-13(*d*). The magnetic flux changes direction from ccw to cw, as shown by the arrow. When the magnetic flux *changes direction,* a pulse is *induced* in the sense wire. The pulse in the sense wire tells us a logical 1 was stored in the core. Notice that the 1 which was stored in the ferrite core was destroyed by the read process. The core must be restored to the 1 state or the data will be lost.

When more than one ferrite core is used we need more than one write line. Figure 11-14 shows four ferrite cores, with two write wires passing through each. If we want to write a 1 in core 4, we pass one-half the amount of current needed through lines Y_2 and X_2. The two half currents add and write a 1 into core 4. Core 1 in Fig. 11-14 receives no current. Cores 2 and 3 receive only half current, and therefore nothing is written into these memory cells. A memory wired like the one in Fig. 11-14 is called a *coincident-current memory*. If we want to write a 1 into core 1, we put + ½ *I* in lines Y_1 and X_2. A −*I* pulse through all the lines resets all the cores to 0. Normally many more ferrite cores are wired together, as in Fig. 11-14 to form a memory plane.

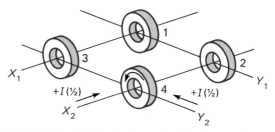

Fig. 11-14 Writing into a coincident-current core memory.

Next we shall look at the read operation in a coincident-current memory. Figure 11-15 shows four ferrite cores wired with the write wire (X_1, X_2, Y_1, Y_2). A third wire, called the *sense wire*, is added to the matrix. The sense wire is shown in Fig. 11-15. To read the contents of core 4, a $-\frac{1}{2} I$ pulse is applied to the X_2 and Y_2 lines. Core 4 is reset to 0 by the addition of these two currents. As it changes magnetic state (from ccw to cw), core 4 *induces* a pulse in the sense line, which tells us we have a 1 in core 4. Core 1 receives no current. Cores 2 and 3 receive only half the current needed to reset them to 0. Cores 2 and 3 do not change magnetic state. Only the core that was addressed (core 4) can change state and trigger the sense wire.

A group of ferrite cores wired somewhat like the one in Fig. 11-16 (on the next page) form a *plane*. One plane contains many cores. Figure 11-16 shows how 64 cores are arranged in a plane. In this arrangement four planes are stacked on top of one another. This forms a 64×4-bit memory. This 256-bit magnetic core memory has 64 words, each word 4 bits long. Figure 11-16 shows how cores X_6Y_5 on all four planes are located. This is word X_6Y_5. The dots on each plane represent the cores being addressed by the X and Y select lines. The lines shown are the write lines we mentioned earlier. Notice that these write (X and Y select) lines are wired in series from plane to plane.

Sense lines from each board (or plane) are brought out for a parallel readout of the four planes. The sense outputs are represented on the right. The sense wires are threaded through the ferrite cores parallel to the X select lines.

A fourth wire, called the *inhibit wire*, is threaded through each ferrite core. The inhibit wires run through the cores parallel to the Y select lines. The inhibit leads for each plane are represented on the left side of Fig. 11-16.

Suppose we are writing a 1 into the word X_6Y_5 location, as shown in Fig. 11-16. This means that binary 1111 is written into the four cores. What if we want to write 1011 ($D = 1$, $C = 0, B = 1$, $A = 1$) instead? We simply activate the C plane inhibit line; it cancels the command from the select (write) lines. The inhibit lines can be thought of as a means of writing a 0 in a word.

The 256-bit magnetic-core memory shown in Fig. 11-16 is a nonvolatile memory. This memory still needs decoding for selecting the correct address. This memory destroys its contents when reading. Schemes for writing contents back into the cores after the reading process is complete are available. You can imagine that a core memory of this type is difficult and expensive to manufacture.

Magnetic-core memories were commonly used as internal central storage in large computers. Magnetic-core memories are being replaced by less expensive types, such as semiconductor memories. Magnetic-core memories may well be used in applications where high temperatures and radiation are a problem.

Self-Test

Supply the missing word in each statement.

30. Magnetic-core memory is based on the characteristics of the tiny _____ .
31. The magnetic-core memory is _____ (nonvolatile, volatile).
32. The expensive magnetic-core memories are being replaced by _____ memories.

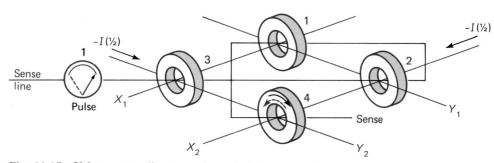

Fig. 11-15 Using a sense line to read a coincident-current core memory.

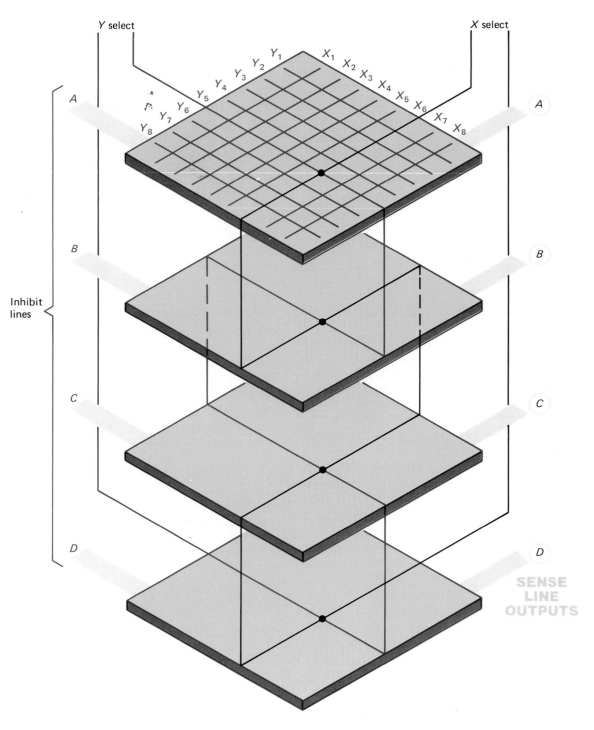

Fig. 11-16   A 64 × 4 coincident-current core memory.

11-8 COMPUTER BULK STORAGE DEVICES

Semiconductor memories are used for internal storage on most modern computers. Magnetic-core memory serves the same purpose in older units. The computer's internal storage is also called *primary storage*. It is not possible to store all data inside the computer itself. For instance, it is not necessary nor desirable to store last month's payroll information inside the computer after the checks are printed and cashed. Thus most data is stored outside the computer. External storage is also called *secondary storage*. Several methods are used to store information for immediate and future use by a computer. External storage devices can be classified as either *mechanical* or *magnetic*.

Mechanical bulk storage devices are the punched paper card and punched or perforated paper tape. The punched card was developed before 1900 by Herman Hollerith who adapted them for use in the 1890 United States census. These cards commonly have holes punched in them to represent alphanumeric data. The code used is called the *Hollerith card code*. A typical Hollerith punched card is made of heavy paper and measures about 3.25 by 7.5 inches. A common punched card can hold 80 characters. Punched paper cards are less popular now than in the past.

Perforated paper tape is another method of mechanically storing data. The paper tape is a narrow strip of paper with holes punched across the tape at places selected according to a code. The paper tape can be stored on reels.

Common magnetic bulk storage devices are the magnetic tape, the magnetic disk, and the magnetic drum. Each device operates much like a common tape recorder. Information is recorded (stored) on the magnetic material. Information can also be read from the magnetic material. Magnetic tape has been widely used for many years as a secondary storage medium. It is still very popular for backing up data because it is quite inexpensive. The main disadvantage of magnetic tapes is that they are sequential-access devices. That is, to find information on the tape you must search through the tape sequentially which makes the access time long.

Magnetic disks have become particularly popular in recent years. Magnetic disks are random-access devices, which means that any data can be accessed easily and in a short time. Magnetic disks are manufactured in both rigid and floppy (flexible) disk form. The floppy disk is an extremely popular form of secondary storage used by most microcomputers. The rigid or hard disk is more expensive than the floppy disk. The hard disk is very popular on large computers and is becoming popular on the more expensive microcomputers.

Self-Test

Answer the following questions.

33. External computer bulk storage devices can be classified as either _____ or _____ .

34. List several computer bulk storage devices.

11-9 MICROCOMPUTER MEMORY

A simplified microcomputer system is sketched in Fig. 11-17 (see next page). The keyboard is the input device, while the monitor or television receiver is the output device. The CPU controls the operation of the microcomputer system and processes data. The *internal memory* on a typical microcomputer system is composed of two types of semiconductor memory. The volatile memory is shown as RAM, or read/write memory. The nonvolatile semiconductor memory is shown in Fig. 11-17 as ROM and NVRAM. The nonvolatile RAM is commonly *RAM with battery backup* in current microcomputer systems.

The *external memory* devices in a microcomputer system store large amounts of data. Programs and data are commonly stored on magnetic disks. Two types of disk storage units are shown in Fig. 11-17. The *floppy disk* is the most widely used bulk storage device. The disk drive is the unit that reads and writes on the floppy disk. A typical floppy disk can store from about 150K bytes to about 1.2M bytes of data. The floppy disk can be removed from the disk drive and stored or transported to another location. The *hard disk* is permanently housed in a hard disk drive. The drive spins the rigid disk at high speeds while the precision read/write heads float just above the surface for reading and writing data and programs on the magnetic coating. Hard disks hold vast quantities of data. Typical small hard disks used with microcomputers have storage capacities of 5M to 40M bytes. Hard disks have a faster access time than floppy disks. The access time for a hard disk is about 25 to 70 ms, while the access time for a floppy disk is about 100 ms. Hard disks are more expensive than floppy disk drive systems. Many microcomputer systems now use a combination of a hard disk and floppy disk drive. This gives the system the advantage of the fast access time and large storage capacity of the hard disk and the portability of the floppy disk.

Strictly speaking, each device shown in Fig. 11-17 contains smaller memory devices. The keyboard, video monitor, disk drives, and CPU all contain memory devices such as latches, shift registers, and buffer memories.

The internal semiconductor memory (ROM and RAM) usually comes in DIP ICs and is mounted on PC boards, as depicted in Fig. 11-17. It is common for ROM and RAM PC boards to be filled with memory and other chips.

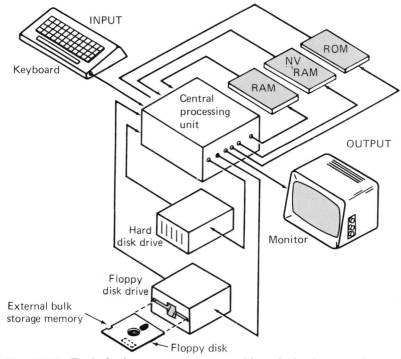

Fig. 11-17 Typical microcomputer system with typical primary and secondary storage.

Floppy disks, or *diskettes,* come in several sizes. The floppy disks commonly used with microcomputers are either 3.5 or 5.25 in. across, although an older 8-in. version is also used. A diagram of a typical floppy disk is shown in Fig. 11-18(*a*). The thin, circular, plastic floppy disk is enclosed in a plastic jacket. The plastic disk is coated with a magnetic material similar to that on audio tape. Several holes are cut in both sides of the plastic jacket. These are shown in Fig. 11-18(*a*). The round center hole in the jacket provides access to the center hub area of the disk. The hub of the disk drive clamps on this area to spin the disk at 300 r/min. The larger hole in the jacket near the bottom of the disk in Fig. 11-18(*a*) exposes part of the disk. The read/write head of the disk drive touches the spinning disk at this point to either store data on the disk or retrieve data from it. The small round hole cut in the jacket and disk is used as an index hole by some microcomputers.

Magnetic disks hold an advantage over tape for microcomputer bulk storage. Disks are random access, while tapes are sequential-access storage devices. The disk drive can jump to any location on the disk, whereas you must sequence through much data on tape to locate a certain program. The access time for disks is much shorter than the average access time for magnetic tape.

Floppy disks are organized in *tracks* and *sec-*

tors. Figure 11-18(*b*) shows how one microcomputer manufacturer formats the mini-floppy disk. The disk is organized into 35 circular tracks numbered from 00 to 34. Figure 11-18(*b*) shows only tracks 00 and 34. Each track is divided into 16 sectors. The 16 sectors are shown in Fig. 11-18(*c*). Each sector has 35 short tracks. This is illustrated near the bottom in Fig. 11-18(*c*). Using this manufacturer's format, each short track holds 256 8-bit words, or 256 *bytes.* A byte is an 8-bit group.

With the formatting shown in Fig. 11-18(*c*), a mini-floppy disk can hold about 140,000 8-bit words, or bytes, of data. This is about 1,000,000 bits of data on a single 5.25-in. mini-floppy disk. It should be noted that there is no standard way to format mini-floppy disks.

The mini-floppy disk is a random-access bulk storage memory device, which is widely used with home, school, and office microcomputers. Large amounts of data can be permanently stored on the magnetic material on a mini-floppy disk. Care must be taken when handling floppy disks. Do not touch the magnetic disk itself and do not write on the plastic jacket. Magnetic fields and high temperatures can also harm stored data on floppy disks. Because of the danger of surface abrasion, keep disks in a clean area.

Hard disks are commonly made of aluminum and covered on one or both sides with a thin coating of high-quality magnetic material. They

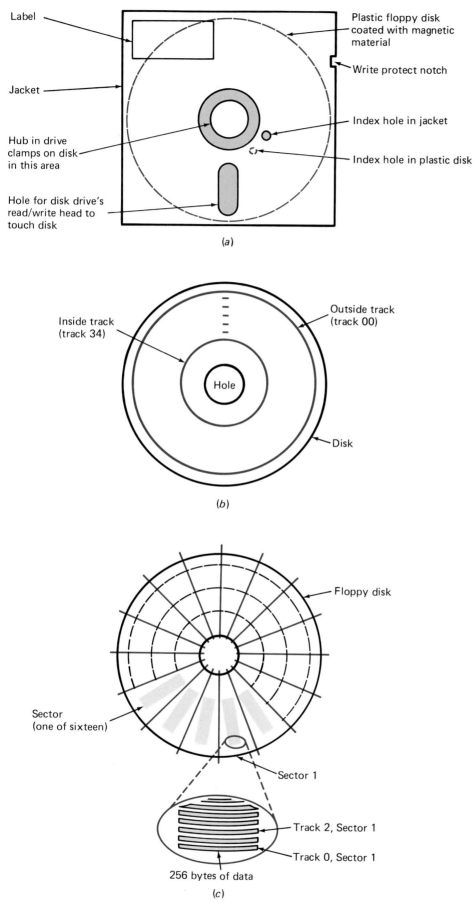

Fig. 11-18 (*a*) **Floppy disk.** (*b*) **Showing location of invisible tracks on floppy disk. (Do not remove disk from protective jacket as in this illustration.)** (*c*) **Sectoring a floppy disk.**

are formatted using tracks and sectors like floppy disks. Hard disks can have many more tracks due to the precision of the read/write head mechanism and the fixed disk. The read/write head floats just above the magnetic surface of a hard disk while it contacts the surface of a floppy disk. Hard disks are also called *fixed disks* or the drives are sometimes referred to as *Winchesters*. Hard disks cost from two to five times as much as a floppy disk drive but can store much more data.

Various kinds of storage devices used with microcomputers are compared on a cost/performance basis in Fig. 11-19. The cost of memory is usually stated in *cents per bit*. Performance is usually stated in access time. *Access time* is the time it takes to locate a piece of data in the memory. A graph of cost versus access time for several types of internal and bulk storage memories is shown in Fig. 11-19. The highest-performance (lowest access time) devices are semiconductor RAMs. Those RAMs using bipolar technology are faster than those using MOS technology. However, the cost per bit for bipolar RAMs is higher than for MOS RAMs. Microprocessors used in microcomputers are not fast processors; therefore, the MOS type RAM is used in these systems. Both static and dynamic MOS RAMs are used in microcomputer systems.

Microcomputer ROM devices have a slower access time (lower performance) than RAMs and also a slightly lower average cost per bit. The access time of ROMs is still fast enough to make this type of unit practical in most microprocessors for *main*, or internal, memory.

The *average* access time for magnetic disks and tapes is also shown in Fig. 11-19. Note that the disk is faster but somewhat more costly than the tape bulk storage device.

The internal memory capacity of a microcomputer is given as 1K, 4K, 48K, 64K, 128K, 256K, 512K, or 640K. To have 1K of RAM means to have 1024 (2^{10}) bytes of memory. Each 1K of memory actually contains 1024 bytes instead of the exactly 1000 bytes the "K" would imply. Thus a microcomputer with 64K of memory has $1024 \times 64 = 65,536$ 8-bit storage locations.

The cost of semiconductor memory chips has decreased in recent years. For this reason, most new microcomputer models have much larger internal memories. A typical newer model microcomputer might have a minimum of about 256K of memory.

Self-Test

Supply the missing word(s) or number(s) in each statement.

35. Most microcomputer internal memory takes the form of _____ and _____ ICs which are mounted on PC boards.
36. Both _____ disk and _____ disk drives are becoming common bulk storage devices used on higher-cost microcomputer systems.
37. Magnetic disks have an advantage over magnetic tapes in that they are _____ -access bulk storage devices.
38. A microcomputer with 256K of memory has 262,144 _____ (bits, bytes) of internal memory. (Note: $1024 \times 256 = 262,144$)

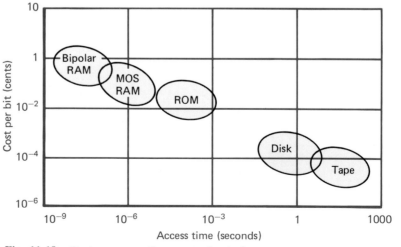

Fig. 11-19   **Cost-versus-performance chart of memory.**

SUMMARY

1. Common semiconductor memories are RAMs, ROMs, PROMs, and NVRAMs.
2. A RAM is considered a read/write random-access memory device.
3. A ROM is considered a permanent storage unit that has the read-only characteristic.
4. A PROM operates just like a ROM. PROMs are one-time write devices. PROMs come in many varieties, generally known as EPROM, or EROM, and EAROM. These "E" prefixed PROMs can be erased electrically or by shining ultraviolet light through a special transparent "window" on the top of the IC.
5. The write process stores information in the memory. The read, or sense, process detects the contents of the memory.
6. A RAM can be reprogrammed easily but is a volatile memory.
7. A 64 × 4 memory holds 64 words each 4 bits long. It holds a total of 256 bits of data.
8. A NVRAM is a nonvolatile RAM that contains a RAM and an EEPROM. Nonvolatile RAM memory is currently implemented in several new microcomputers using RAMs with a battery backup.
9. The magnetic-core memory is based upon the magnetic characteristics of the ferrite core for storing 0s and 1s.
10. Ferrite cores are organized into planes, and the planes are stacked. Read/write, sense, and inhibit lines are threaded through tiny cores in a magnetic-core coincident-current memory.
11. Computer external storage methods are magnetic tapes, disks, and drums; punched cards; and perforated tape.
12. Microcomputers typically use both RAM and ROM for internal main memory. Both floppy and hard disks are popular bulk storage devices for home, school, and small-business microcomputers.
13. One measure of memory performance is access time. Short access time means superior memory performance.
14. In microcomputers, 1K of memory means 1024 bytes. A byte is an 8-bit word.

CHAPTER REVIEW QUESTIONS

Answer the following questions.

11-1. Press the store key on a calculator. This activates the _____ (read, write) process in the memory section.

11-2. Press the recall key on a calculator. This activates the _____ (read, write) process in the memory section.

11-3. The following abbreviations stand for what?
   a. RAM   d. EPROM
   b. ROM   e. EAROM
   c. PROM   f. NVRAM

11-4. A _____ (RAM, ROM) is a volatile memory.

11-5. A _____ (RAM, ROM) has both the read and write capability.

11-6. A _____ (RAM, ROM) is a permanent memory.

11-7. A _____ (RAM, PROM) is a nonvolatile memory.

11-8. A _____ (RAM, ROM) has a read/write input control.

11-9. A _____ (RAM, ROM) has data inputs.

11-10. A RAM such as the 7489 IC is sometimes also called by what two other names?

11-11. A 32 × 8 memory can hold _____ words. Each word is _____ bits long.

11-12. Draw a diagram of how a 32 × 8 memory looks in table form. The diagram will be similar to Fig. 11-1.

11-13. Draw a logic diagram of a ROM organized as a 32 × 8 storage unit. Label address inputs E, D, C, B, and A. Label outputs = DO_1, DO_2, DO_3, DO_4, DO_5, DO_6, DO_7, and DO_8.

11-14. How would you program a ROM such as the one in questions 12 and 13?

11-15. List at least three advantages of semiconductor memories.

11-16. A _____ (RAM, ROM) can be erased easily.

11-17. List at least three uses of ROMs.

11-18. What two methods are used for erasing an EPROM?

11-19. Memory or data storage is much easier to implement using _____ (analog, digital) electronic circuitry.

11-20. The 2114 IC is a _____ (dynamic, static) RAM.

11-21. The 2114 IC (see Fig. 11-3) is organized as a 1024 × 4 RAM. The chip must have _____ address lines to access all of the 1024 memory locations.

11-22. The access time of the TTL 7489 RAM is _____ (faster, slower) than that of the MOS 2114 RAM.

11-23. Refer to Fig. 11-5(*b*). If the input to the decoder is binary 0010, the output from the ROM will be _____ in Gray code.

11-24. Computer programs that are permanently held in ROM are called _____ .

11-25. Refer to Table 11-4. Which 27XXX series EPROM could be used to implement a 16K ROM in a microcomputer?

11-26. A _____ (dynamic RAM, NVRAM) is also sometimes called a "shadow RAM."

11-27. The ferrite core is the memory cell in a _____ -type memory.

11-28. The write wire through a ferrite core is used for _____ and reading.

11-29. The sense wire through a ferrite core is used during the _____ (read, write) process.

11-30. In a ferrite core, if magnetic flux travels in a cw direction for a logical 0, then it will travel _____ for a logical 1.

11-31. A _____ (magnetic core, RAM) unit has a nonvolatile memory.

11-32. A coincident-current core memory uses four wires threaded through the cores. Name the wires.

11-33. List at least four common types of computer bulk (external) storage.

11-34. List two *mechanical* methods used for the bulk storage of data outside the computer.

11-35. Magnetic disks have much _____ (faster, slower) access time than magnetic tapes.

11-36. Magnetic disks are manufactured in both _____ and floppy (flexible) disk form.

11-37. A modern microcomputer probably uses ROMs, _____ (bipolar, MOS) RAMs, and battery backup NVRAMs to implement its _____ (primary, secondary) memory.

11-38. Why are many high-performance microcomputer systems equipped with both floppy and hard disk drives?

11-39. A mini-floppy disk is a circular plastic sheet coated with a _____ (conductive, magnetic) materials. The disk is housed in a plastic _____ (box, jacket) for protection.

11-40. List three precautions you must take when handling floppy disks.

11-41. Short access time in a memory device is a measure of _____ (good, poor) performance.

11-42. A _____ (floppy disk, RAM) has the shorter access time.

11-43. A microcomputer with 4K of memory has 4096 _____ (bits, bytes) of storage.

Answers to Self-Tests

1. random-access memory
2. writing
3. reading
4. read/write
5. 16 × 4 bit
6. volatile, off
7. RAM
8. 16, 4
9. write, written into, 15
10. dynamic
11. 4096
12. static
13. inputs
14. binary-to-Gray
15. 1100, Gray, 1000
16. lose its program and have to be reprogrammed
17. read-only memory
18. nonvolatile
19. firmware
20. manufacturer
21. remain in
22. 177
23. manufacturer
24. programmable read-only memory
25. erasable programmable read-only memory
26. electrically erasable programmable read-only memory
27. nonvolatile random-access memory
28. electrically, ultraviolet
29. 524,288, 65,536
30. ferrite core
31. nonvolatile
32. semiconductor
33. mechanical, magnetic
34. 1. punched paper tape
    2. punched cards
    3. magnetic tapes
    4. magnetic disk (hard and floppy)
    5. magnetic drum
35. RAM, ROM
36. floppy, hard (Winchester)
37. random
38. bytes

CHAPTER 12

Digital Systems

Most digital devices we use every day are **digital systems,** *such as hand-held calculators, digital wrist-watches, or even digital computers.* **Calculators, digital clocks, and computers are assemblies of subsystems. Typical subsystems might be adder/subtractors, counters, shift registers, RAMs, ROMs, encoders, decoders, data selectors, clocks, and display decoder/drivers. You have already used most of these subsystems. This chapter discusses various digital systems and how they transmit data. A digital system is formed by the proper assembly of digital subsystems.**

12-1 ELEMENTS OF A SYSTEM

Most mechanical, chemical, fluid, and electrical systems have certain features in common. Systems have an *input* and an *output* for their product, power, or information. Systems also act on the product, power, or information; this is called *processing*. The entire system is organized and its operation directed by a *control* function. The *transmission* function transmits products, power, or information. More complicated systems also contain a *storage* function. Figure 12-1 illustrates the overall organization of a system. Look carefully and you can see that this diagram is general enough to apply to nearly any system, whether it is transportation, fluid, school, or electronic. The transmission from device to device is shown by the colored lines and arrows. Notice that the data or whatever is being transferred

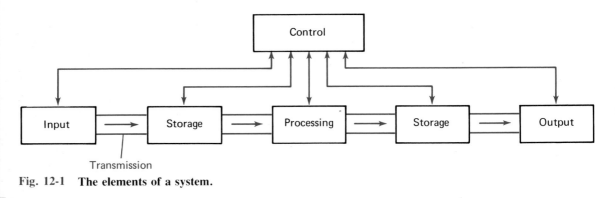

Fig. 12-1 **The elements of a system.**

always moves in one direction. It is common to use double arrows on the control lines to show that the control unit is directing the operation of the system as well as receiving feedback from the system.

The general system shown in Fig. 12-1 will help explain several digital systems in this chapter. In a digital system we shall be dealing only with transmitting data (usually numbers).

Self-Test

Answer the following questions.

1. There is a two-way path from the _____ section of a system to all other parts.
2. The keyboard of a microcomputer is classified as what part of the system?

12-2 A DIGITAL SYSTEM ON AN IC

We have learned that all digital systems can be wired from individual AND and OR gates and inverters. We have also learned that manufacturers produce subsystems on a single IC (counters, registers, and so on). We shall find that manufacturers have gone even a step further: some ICs contain nearly an entire digital system.

Texas Instruments defines the least complex digital ICs as *small-scale integrations* (SSI). An SSI contains circuit complexity up to about 10 equivalent gates or circuitry of similar complexity. Small-scale ICs include the gate and flip-flop ICs you have used.

A *medium-scale integration* (MSI) has the complexity of from 12 to 100 equivalent gates. ICs that are classified as MSIs belong to the subsystem group. Typical examples are adders, registers, code converters, counters, data selector/multiplexers, RAMs and ROMs. Most of the ICs you have studied and used to this time have been SSIs or MSIs.

A *large-scale integration* (LSI) has the complexity of more than 100 equivalent gates. A major subsystem or an entire digital system is fabricated in a single IC. Examples are digital clock ICs and calculator ICs.

A *very large-scale integration* (VLSI) has the complexity of about 1000 or more gates or circuitry of similar complexity. Many memory chips and microprocessors fit into this category. Very large-scale ICs are considered digital systems on a *chip*. The term "chip" refers to the single silicon wafer (perhaps ¼ in. square) that contains all the electronic circuitry in an IC. Various manufacturers define the terms "SSI," "MSI," "LSI," and "VLSI" differently. At this time neither JEDEC (Joint Electron Device Engineering Council) nor IEEE (Institute of Electrical and Electronic Engineers) has defined VLSI.

Self-Test

Supply the missing word in each statement.

3. A medium-scale integration is an IC that contains the equivalent of _____ gates.
4. A VLSI is an IC that contains the equivalent of over _____ gates.

12-3 THE CALCULATOR

The pocket calculator in nearly everyone's pocket or desk is a very complicated digital system. Knowing this, it is disappointing to take apart a modern miniature calculator. You will find a battery, the tiny readout displays, a few wires from the keyboard, and a circuit board with an IC attached. That single IC is most of the digital system we call a calculator and contains an LSI chip that performs the task of hundreds or thousands of logic gates. The single IC performs the storage, processing, and control functions of the calculating system. The keyboard is the input, and the displays are the output of the calculator system.

What happens inside the calculator chip when you press a number or add two numbers? The diagram in Fig. 12-2 on the next page will help us figure out how a calculator works. Figure 12-2 shows three components: the keyboard, the seven-segment displays, and the power supply. These parts are the only functional ones *not* contained in the single LSI IC in most small calculators. The keyboard is obviously the input device. The keyboard contains simple, normally open switches. The decimal display is the output. The readout unit in Fig. 12-2 contains only six 7-segment displays. The power supply is a battery in most inexpensive hand-held calculators. Many modern calculators use solar cells as their power supply, CMOS ICs, and LCD displays.

The calculator chip (the IC) is divided into several functional subsystems, as shown in Fig. 12-2.

From page 250:

Subsystems

Elements of a system

Input

Output

Processing

Control

Transmission

Storage

On this page:

Small-scale integration (SSI)

Medium-scale integration (MSI)

Large-scale integration (LSI)

Very large-scale integration (VLSI)

Chip

Calculator chip

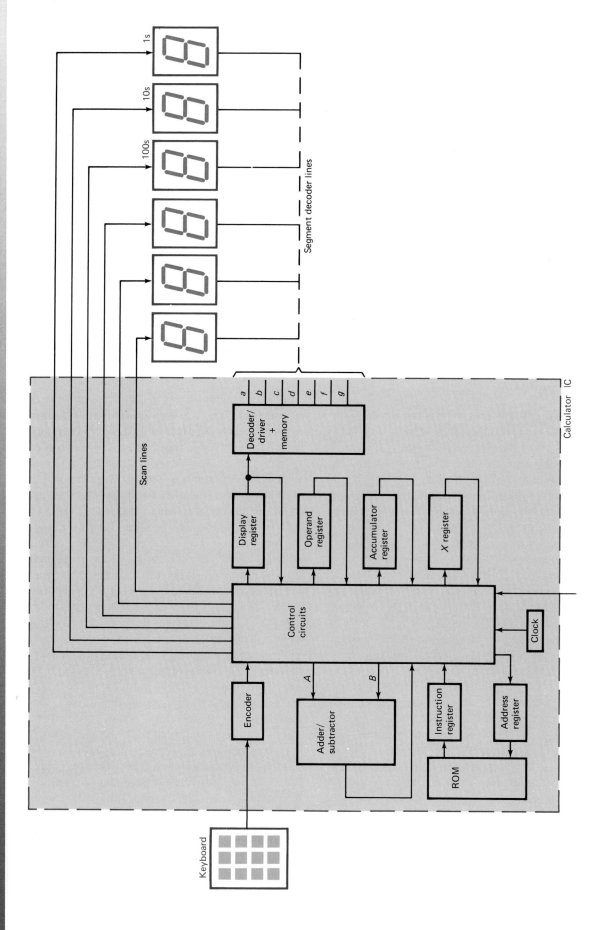

Fig. 12-2 Internal organization of a calculator.

The organization shown is only one of several ways to get a calculator to operate. The heart of the system is the adder/subtractor subsystem, which operates very much like the 4-bit adders you studied. The clock subsystem pulses all parts of the system at a constant frequency. The clock frequency is fairly high, ranging from 25 to 500 kHz. When the calculator is turned on, the clock runs constantly and the circuits "idle" until a command comes from the keyboard.

Suppose we add 2 + 3 with this calculator. As we press the 2 on the keyboard, the encoder translates the 2 to a BCD 0010. The 0010 is directed to the display register by the control circuitry and is stored in the display register. This information is also applied to the seven-segment decoder, and lines a, b, d, e, and g are activated. The first (1s display) seven-segment display shows a 2 when the scan line pulses that unit briefly. This scanning continues at a high frequency, and the display appears to be lit continuously even though it is being turned on and off many times per second. Next, we press + on the keyboard. This operation is transferred to and stored in code form in an extra register (X register). Now we press the 3 on the keyboard. The encoder translates the 3 to a BCD 0011. The 0011 is transferred to the display register by the controller and is passed to the display decoder/driver, which also places a 3 on the display. Meanwhile, the controller has moved the 0010 (decimal 2) to the operand register. Now we press the = key. The controller checks the X register to see what to do. The X register says to add the BCD numbers in the operand and display registers. The controller applies the contents of the display and operand registers to the adder inputs. The results of the addition collect in the accumulator register. The result of the addition is a BCD 0101. The controller routes the answer to the display register, shown on the readout as a 5.

For longer and more complex numbers containing decimal points, the controller follows directions in the instruction register. For complicated problems the unit may cycle through hundreds of steps as programmed into the ROM. Amazingly, however, even hundreds of operations take less than $\frac{1}{10}$ s.

The registers in Fig. 12-2 are rather large units compared to the ones you have used in the laboratory. The ROM also has a large capacity (many thousands of bits). The calculator in Fig. 12-2 is but one example of how a calculator can operate. Each commercial unit operates in its own unique way. This discussion serves only to point out that many of the subsystems you have already used are found in a complex digital system like the calculator.

Only the original IC designers need to know the organization of the subsystems in Fig. 12-2. This organization is sometimes referred to as the *architecture* of the calculator. Notice that all the elements of a system are present in this electronic calculator.

You may be asked to troubleshoot and repair a calculator. Most inexpensive hand-held calculators are designed as throw-away items. However, at least one manufacturer does reconditioning of its line of calculators.

Modern calculator designs use a keyboard, an LSI chip, perhaps a circuit board, a display module, and a power supply. Without calculator wiring diagrams, you still can check some obvious trouble spots in the calculator. Carefully check the battery or power supply. Either replace the battery or use a load test on the cell. Also look carefully at the battery connectors for signs of broken or loose connections. With a VOM, check the voltage as near to the IC as possible if you suspect a loss of power.

A second trouble spot in calculators is the keyboard. Because they are usually very inexpensive and mechanical in nature, keyboards cause many problems. Many keyboards are sealed and cannot be accessed by the technician. However, some can at least be cleaned out with compressed air and checked for broken connections. Besides cleaning keyboards and replacing batteries, check for obvious loose or broken wires or connections. Except on expensive calculators or in reconditioning laboratories, troubleshooting usually does not extend to the display module or LSI chip.

Self-Test

Answer the following questions.

5. A single _____ chip performs the storage, processing, and control functions in a modern calculator.
6. The display is the _____ device in a calculator system.
7. The keyboard is the _____ device in a calculator system.
8. List two typical problem areas to look at when troubleshooting an inexpensive calculator.

12-4 THE COMPUTER

The most complex digital systems are *computers*. Most digital computers can be divided into the five functional sections shown in Fig. 12-3. The input device may be a keyboard, mouse, joystick, graphics tablet, card reader, magnetic tape unit, or telephone line. This equipment lets us pass information from *person to machine*. The input device must *encode* human language into the binary language of the computer.

The memory section is the storage area for data and program. This storage can be supplemented by storage outside the processing unit. Much of the memory in the CPU traditionally has been magnet-core memory, but now semiconductor memories are being used in the CPU.

The arithmetic unit is what most people think of as being inside a computer. The arithmetic unit adds, subtracts, compares, and does other logic functions. Notice that a two-way path exists between the memory and arithmetic sections. In other words, data can be sent to the arithmetic section for action and the results sent back to storage in the memory. The arithmetic unit is sometimes referred to as the ALU.

The control section is the nervous system of the computer. It directs all other sections to operate in the proper order and tells the input when and where to place information in the memory. It directs the memory to route information to the arithmetic section and tells the arithmetic section to add. It routes the answer back to the memory and to the output device. It tells the output device when to operate. This is only a sampling of what the control section can do.

The output section is the link between the *machine and a person*. It can communicate to humans through a printer. It can put out information on a CRT display. Output information can also be placed on bulk storage devices, such as punched cards, magnetic tape, or disk. The output section must *decode* the language of the computer into human language

The entire center section in Fig. 12-3 is often called the CPU. The arithmetic and memory sections and most of the control section are frequently housed in a single cabinet. Devices located outside the CPU are often called *peripheral devices*.

The block diagram of the computer in Fig. 12-3 could well be the diagram for a calculator. Up to this point the basic systems operate the same. The basic difference between the calculator and computer is *size* and the use of a *stored program* in the computer. Figure 12-4 shows that two types of information are put into the computer. One is the program (instructions) telling the control unit how to proceed in solving the problem. This program, which has to be carefully written by a programmer, is stored in the central memory while the problem is being solved. The second type of information fed to the computer is *data*, to be acted on by the computer. Data is the facts and figures needed to solve the problem. Notice that the program information is placed in storage in the memory and used only by the control unit. The data information, however, is directed to various positions within the computer and is processed by the ALU. The data need never go to the control unit. The auxiliary memory is extra memory that may be needed to store the vast amount of data in some com-

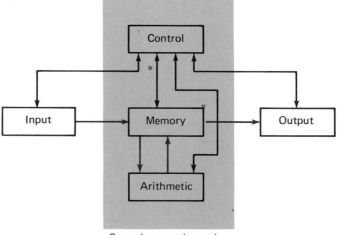

Central processing unit

Fig. 12-3 Sections of a digital computer.

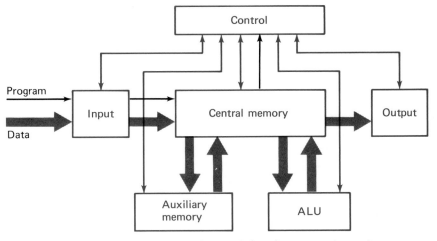

Fig. 12-4   **Flow of program instructions and data in a computer system.**

Microprocessor
(μP)

Microcomputer

MPU

Stored-program
digital computer

Address bus

Data bus

plex problems. It may not be in the CPU. Data may be stored in peripheral devices.

In summary, the computer is organized into five basic functional sections: input, memory, control, ALU, and output. Information fed into the CPU is either program instructions or data to be acted upon. The computer's stored program and size make it different from the calculator.

Computers, the most complex of digital systems, are not covered in depth in this section. There are entire books about the organization and architecture of computers. Remember, however, that all the circuits in the digital computer are constructed from logic gates, flip-flops, and subsystems such as the ones you have studied.

Self-Test

Answer the following questions.

9. Devices located outside the computer's CPU are often called _____ devices.
10. The basic difference between the calculator and the computer is _____ and the use of a _____ .
11. List the two types of information fed into a digital computer.

12-5 THE MICROCOMPUTER

Computers have been in general use since the 1950s. Formerly, digital computers were large, expensive machines used by governments or large businesses. The size and shape of the digital computer have changed in the past decade as a result of a new device called the micro-

processor. The *microprocessor* (MPU, for "microprocessing unit") is an IC that contains much of the processing capabilities of a larger computer. The MPU is a small but extremely complex VLSI device that is *programmable*. The MPU IC forms the heart of a microcomputer. The *microcomputer* is a *stored-program digital computer* which is smaller, slower, and less expensive than its larger cousins.

The organization of a typical smaller microcomputer system is diagramed in Fig. 12-5 on the next page. This microcomputer contains all the five basic sections of a computer: the *input* unit, the *control* and *arithmetic* units contained within the MPU, the *memory* units, and the *output* unit.

The MPU controls all the units of the system using the control lines shown at the left in Fig. 12-5. Besides the control lines, the *address bus* (16 parallel conductors) selects a certain memory location, input port, or output port. The *data bus* (eight parallel conductors) on the right in Fig. 12-5 is a *two-way path* for transferring data into and out of the MPU. It is important to note that the MPU can send data to memory or an output port or receive data from memory or an input port.

The microcomputer's ROM commonly contains a program. A *program* is a list of specially coded instructions that tell the MPU *exactly* what to do. The ROM in Fig. 12-5 is the place where the program resides in this example. In actual practice, the ROM contains a start-up, or initializing, program and perhaps other programs. Program listings can also be loaded into RAM from auxiliary memory. These are user programs.

The RAM area in Fig. 12-5 is identified in

**Microcomputer
system**

Peripheral devices

Keyboard

Mouse

Joystick

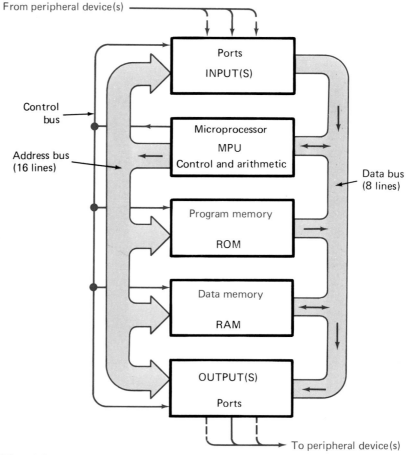

From peripheral device(s)

Ports
INPUT(S)

Control
bus

Microprocessor
MPU
Control and arithmetic

Address bus
(16 lines)

Data bus
(8 lines)

Program memory
ROM

Data memory
RAM

OUTPUT(S)
Ports

To peripheral device(s)

Fig. 12-5   Block diagram of a microcomputer system.

this example as the data memory. Data used in the program resides in this memory.

The CPU and memory sections of the microcomputer are not very useful by themselves. The CPU must be interfaced with *peripheral devices* for input, output, and storage. Typical peripheral devices used for input, output, and storage on modern microcomputers are diagrammed in Fig. 12-6. The keyboard, mouse, and joystick are probably the most common input devices connected to most microcomputers. Several other input devices con-

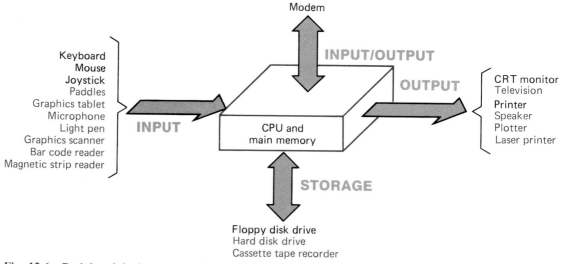

Modem

Keyboard
Mouse
Joystick
Paddles
Graphics tablet
Microphone
Light pen
Graphics scanner
Bar code reader
Magnetic strip reader

INPUT/OUTPUT

OUTPUT

INPUT

CPU and
main memory

CRT monitor
Television
Printer
Speaker
Plotter
Laser printer

STORAGE

Floppy disk drive
Hard disk drive
Cassette tape recorder

Fig. 12-6   Peripheral devices commonly attached to the CPU of a microcomputer.

nected to microcomputers are shown at the left in Fig. 12-6.

The floppy disk drive is currently the most popular secondary storage device connected to most microcomputers. Other secondary storage devices interfaced with microcomputers are the hard disk and to a lesser extent the cassette tape recorder. The CRT monitor and printer are the most common output peripheral devices used with typical microcomputers. Other output devices include TVs, speakers and sound systems, plotters, and laser printers.

A *modem* (modulator-demodulator) is the peripheral device that enables the microcomputer to transmit and receive data over telephone lines. Notice that the modem is classified as an *input/output* peripheral device in Fig. 12-6. It is an output device when transmitting data and an input device when receiving data.

Self-Test

Supply the missing word in each statement.

12. Refer to Fig. 12-5. The address bus is a one-way path, whereas the _____ bus is a two-way pathway for information.

13. Refer to Fig. 12-5. The ROM typically holds _____ (data, programs).
14. Refer to Fig. 12-5. The exact memory location, or input/output port, is selected by the MPU's output on the _____ bus.
15. A _____ is an input/output peripheral device that enables the microcomputer to send and receive data over telephone lines.
16. Refer to Fig. 12-6. The _____ is probably the most popular peripheral output device used with low-cost microcomputers.
17. Refer to Fig. 12-6. The _____ is a hand-operated input device that has a ball on the bottom and switch on the top. It is used to control the direct movement of the cursor on the CRT screen.

12-6 MICROCOMPUTER OPERATION

As an example of microcomputer operation, refer to Fig. 12-7. In this example the following things are to happen:

1. Press the "A" key on the keyboard.
2. Store the letter "A" in memory.
3. Print the letter "A" on the screen of the CRT monitor.

Floppy disk drive

CRT monitor

Printer

Modem

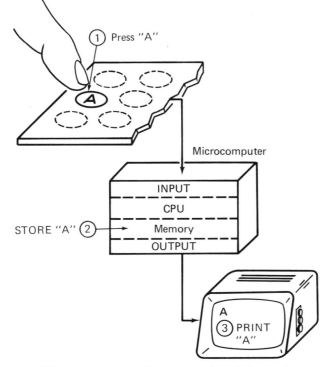

Fig. 12-7 An example of a common input-store-output microcomputer operation.

The input-store-output procedure outlined in Fig. 12-7 is a typical microcomputer system operation. The electronic hardware used in a system like that in Fig. 12-7 is quite complicated. However, the transfer of data within the system will help explain the use of several different units within the microcomputer.

The more detailed diagram in Fig. 12-8 will aid understanding of the typical microcomputer input-store-output procedure. First, look carefully at the *contents* section of the program memory in Fig. 12-8. Note that instructions have already been loaded into the first six memory locations. From Fig. 12-8, it is determined that the instructions currently listed in the program memory are:

1. Input data from input port 1.

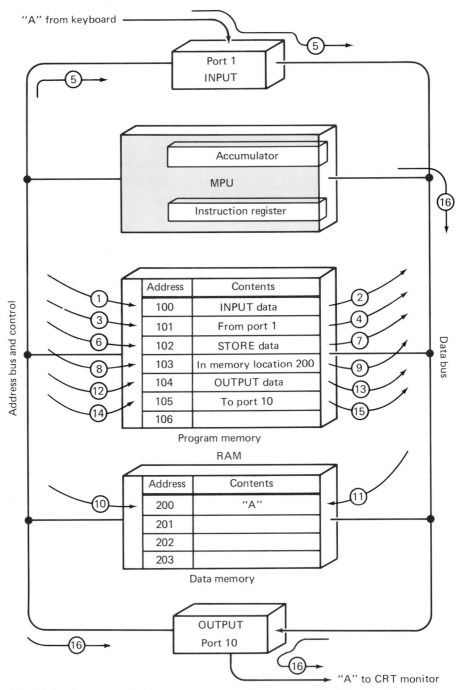

Fig. 12-8 Sequence of microcomputer operations in executing the input-store-output program.

2. Store data from port 1 in data memory location 200.
3. Output data to output port 10.

Note that there are only three instructions in the above program. It appears that there are six instructions in the program memory in Fig. 12-8. The reason for this is that instructions are usually broken into parts. The first part of instruction 1 above is to input data. The second part tells where the data comes from (from port 1). The first, *action* part of the instruction is called the *operation* and the second part the *operand*. The operation and operand are located in separate memory locations in the program memory in Fig. 12-8. For the first instruction in Fig. 12-8, program memory location 100 holds the input operation while memory location 101 holds the operand (port 1) telling where information will be input from.

Two new sections are identified inside the MPU in Fig. 12-8. These two sections are called registers. These special registers are the *accumulator* and the *instruction register*.

The sequence of events happening within the microcomputer in the input-store-output "A" example is outlined in Fig. 12-8. The flow of instructions and data can be followed by keying on the circled numbers in the diagram. Remember that the *MPU is the center of all data transfers and operations*. Refer to Fig. 12-8 for all steps below.

1. The MPU sends out address 100 on the address bus. A control line *enables* the read input on the program memory IC. This step is symbolized in Fig. 12-8 by the encircled number 1.
2. The program memory sends the first instruction (input data) on the data bus, and the MPU receives this coded message. The instruction is transferred to a special memory location within the MPU called the instruction register. The MPU *decodes*, or interprets, the instruction and determines that it needs the operand to the input data instruction.
3. The MPU sends out address 101 on the address bus. The control line enables the read input of the program memory.
4. The program memory places the operand (from port 1) on the data bus. The operand was located at address 101 in program memory. This coded message (the address for port 1) is received off the data bus and transferred to the instruction register. The

MPU now decodes the entire instruction (input data from port 1).

5. The MPU uses the address bus and control lines to the input unit to cause port 1 to open. The coded form for "A" is transferred to and stored in the accumulator of the MPU.

It is important to note that the MPU always follows a *fetch-decode-execute sequence*. It first fetches the instruction from program memory. Second, the MPU decodes the instruction. Third, the MPU executes the instruction. Try to notice this fetch-decode-execute sequence in the next two instructions. Continue with the program listed in the program memory in Fig. 12-8.

6. The MPU addresses location 102 on the address bus. The MPU uses the control lines to enable the read input on the program memory.
7. The code for the store data instruction is sent on the data bus and received by the MPU, where it is transferred to the instruction register.
8. The MPU decodes the store data instruction and determines that it needs the operand. The MPU addresses the next memory location (103) and enables the program memory read input.
9. The code for "in memory location 200" is placed on the data bus by the program memory. The MPU accepts this operand and stores it in the instruction register. The entire "store data in memory location 200" has been fetched from memory and decoded.
10. The execute process now starts. The MPU sends out address 200 on the address bus and enables the *write* input of the data memory.
11. The MPU sends the information stored in the accumulator on the data bus to data memory. The "A" is received off the data bus and is written into location 200 in data memory. The second instruction has been executed. This store process does not destroy the contents of the accumulator. The accumulator still also contains the coded form of "A".
12. The MPU must fetch the next instruction. It addresses location 104 and enables the read input of the program memory.
13. The code for the output data instruction is sent to the MPU on the data bus. The MPU receives the instruction and transfers it to the instruction register. The MPU de-

MPU-based systems

Microcomputer address decoding

Address decoder

Chip select line

Three-state buffers

High-impedance state

codes the instruction and determines that it needs an operand.

14. The MPU places address 105 on the address bus and enables the read input of the program memory.
15. The program memory sends the code for the operand (to port 10) to the MPU via the data bus. The MPU receives this code in the instruction register.
16. The MPU decodes the entire instruction "output data to port 10." The MPU activates port 10, using the address bus and control lines to the output unit. The MPU sends the code for "A" (still stored in the accumulator) on the data bus. The "A" is transmitted out of port 10 to the CRT monitor.

Most MPU-based systems transfer information in a fashion similar to the one detailed in Fig. 12-8. The greatest variations are probably in the input and output sections. Several more steps may be required to get the input and output sections to operate properly.

It is important to notice that the MPU is the center of and controls all operations. The MPU follows the fetch-decode-execute sequence. The actual operations of the MPU system, however, are dictated by the instructions listed in program memory. Instructions are usually performed in sequence (100, 101, 102, and so on).

All three instructions in the example would be fetched, decoded, and executed in 0.0001 s or so by most small microcomputers. The advantage of MPU-based systems is their fast operation and flexibility. They are flexible because they can be reprogrammed to perform many tasks.

Microcomputers are complex digital systems containing an MPU IC (or set of ICs), some memory, and inputs and outputs. The MPU chip itself is a complex, highly integrated subsystem that can process instructions at a high rate of speed. It is expected that microcomputers will be a growth industry for many years to come. The last two sections gave only a brief overview of the basic operation and organization of a microcomputer.

Self-Test

Supply the missing word or words in each statement.

18. The action part of a microcomputer instruction is called the _____ . The second part of the instruction is called the _____ .
19. Refer to Fig. 12-8. Program memory location _____ holds the operation part of the first instruction, whereas location _____ holds the operand part of the instruction.
20. Refer to Fig. 12-8. In this microcomputer, the _____ is the center of all data transfers and operations.
21. The microcomputer's MPU always follows a fetch-_____ sequence when running.
22. Program instructions are usually performed in _____ (random, sequential) order in a microcomputer.

12-7 MICROCOMPUTER ADDRESS DECODING

Consider the simple 4-bit MPU-based system shown in Fig. 12-9(a). This system uses only eight conductors in the address bus and four conductors in the data bus. The RAMs are tiny, 64-bit (16×4) units. These RAMs are like the 7489 RAMs you studied in the last chapter.

Two problems become apparent when working with a system like the one shown in Fig. 12-9(a). First, how does the MPU select which RAM to read data from when it sends the same 4-bit address to each? Second, how can several devices send data over a common data bus if, generally, outputs of logic devices cannot be tied together? The solutions to both these problems are shown in Fig. 12-9(b).

The *address decoder* shown in Fig. 12-9(b) decodes which RAM is to be used and sends the enabling signal over the chip select line. *Only one chip select line is activated at a time.* The address decoder block consists of familiar combinational logic gates. RAM 0 is selected when the address is 0 through 15. However, RAM 1 is selected when the address is 16 through 31.

The *three-state buffers* shown in Fig. 12-9(b) disconnect the RAM outputs from the data bus when the memory is not sending data. Only one device is allowed to send on the shared data bus at a given time. For this reason, the chip select line is also used to control, or turn on, the three-state buffers. When the three-state buffers are in the turned off mode, it is said that the buffer outputs are in their *high-impedance state* and are effectively discon-

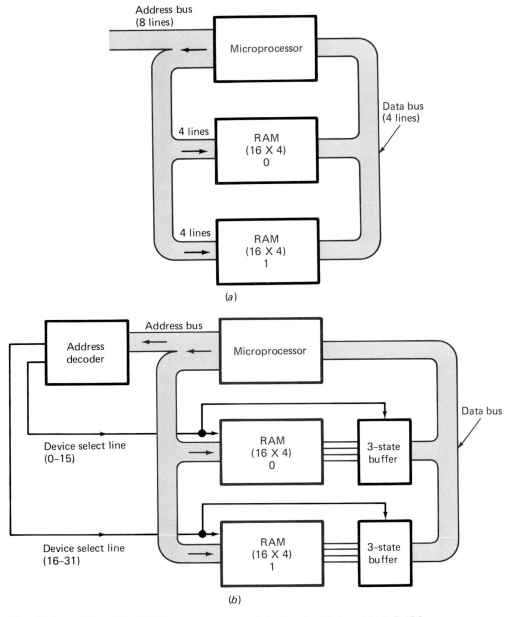

Fig. 12-9 (*a*) **Simplified 4-bit microprocessor interfaced with two 64-bit RAMs.**
(*b*) **Address decoder and three-state buffers added to 4-bit microprocessor-based system.**

nected from the four data lines at the inputs of the buffers.

The logic circuits used in a simple address decoder are shown in Fig. 12-10 on the next page. In this example, only when the four address lines (A_7 to A_4) are all 0 is the output of the bottom four-input OR gate LOW. When address lines A_7 to A_4 are 0000, then RAM 0 is enabled with a LOW at its memory enable ($\overline{ME}$) input.

When the four address lines going into the address decoder in Fig. 12-10 are 0001 ($A_7 = 0$, $A_6 = 0$, $A_5 = 0$, $A_4 = 1$), the top OR gate is activated. The 0001 causes the top OR gate in

the address decoder to generate a LOW output, which activates the bottom device-select line. This enables the bottom RAM (RAM 1).

The address decoder in Fig. 12-10 decodes only the four most significant address lines to generate the correct $\overline{ME}$ logic level. The RAMs internally decode the four least significant address lines (A_0 to A_3) to locate the exact 4-bit word in RAM.

The MPU-based system in Figs. 12-9 and 12-10 uses eight address lines. This means that the MPU can generate 256 (2^8) unique addresses. In the system in Figs. 12-9 and 12-10, the first 16 addresses are used by RAM 0 while

Address decoder

Hexadecimal
notation

Memory map

74125 quad
three-state buffer
TTL IC

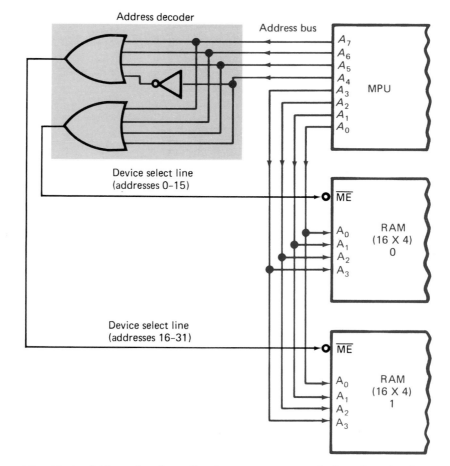

Fig. 12-10 **Address decoder gating to generate correct device select signals.**

the next 16 addresses are used by RAM 1. It is customary to draw a *memory map* of an MPU-based system. The memory map of our sample system is drawn in Fig. 12-11. This shows that the first 16 (0F in hexadecimal) addresses are used by RAM 0. These addresses range from 0 to 15 (00 to 0F in hexadecimal). The second 16 addresses are used by RAM 1. These addresses range from 16 to 31 (10 to 1F in hexadecimal). The third through sixteenth groups of addresses are not used in this very tiny system. It is customary to use hexadecimal notation in specifying addresses in an MPU-based system.

In Fig. 12-9(b), two blocks are labeled three-state buffers. The symbol for a buffer is drawn in Fig. 12-12(a). It has a data input (A) and noninverted output (Y). When the control input (C) is deactivated with a 1, output Y goes to its high-impedance (high-Z) state and is effectively disconnected from the input.

A commercial version of the three-state buffer is shown in Fig. 12-12(b). This is the pin diagram for the *74125 quad three-state buffer*

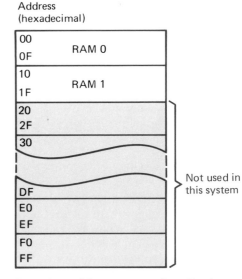

Fig. 12-11 **Memory map of small microprocessor-based system using two 16 × 4 RAMs.**

TTL IC. The truth table for the 74125 IC is shown in Fig. 12-12(c).

In summary, an address decoder is used to

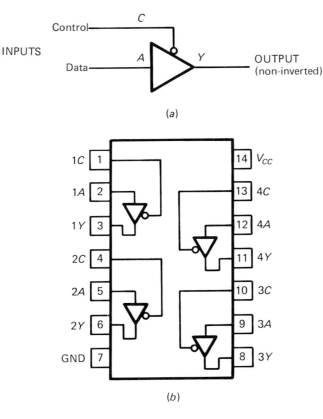

Control — C

INPUTS

Data — A Y OUTPUT (non-inverted)

(a)

1C	1		14	V_{CC}
1A	2		13	4C
1Y	3		12	4A
2C	4		11	4Y
2A	5		10	3C
2Y	6		9	3A
GND	7		8	3Y

(b)

TRUTH TABLE

INPUTS		OUTPUT
C	A	Y
L	L	L
L	H	H
H	X	(Z)

L   = LOW voltage level
H   = HIGH voltage level
X   = Don't care
(Z) = High impedance (off)

(c)

Fig. 12-12 (*a*) **Logic symbol for a three-state buffer.** (*b*) **Pin diagram for commercial 74125 quad three-state buffer IC.** *(Courtesy of Signetics Corporation)* (*c*) **Truth table for 74125 three-state buffer IC.** *(Courtesy of Signetics Corporation)*

select *which* device will be connected to the data bus in an MPU-based system. Address decoders are usually constructed of combinational logic circuits (simple gating circuits).

To permit many devices to use a common data bus, three-state buffers are used. A three-state buffer has a control input that, when disabled, places the output in the high-impedance (high-Z) state.

Both address decoders and three-state buffers are widely used in microcomputers. The three-state buffers are usually part of MPUs, larger RAMs, ROMs, and peripheral interface adapter ICs.

Self-Test

Supply the missing word or words in each statement.

23. Refer to Fig. 12-9. The _____ in this system selects which RAM will be used.

24. Refer to Fig. 12-9. When not in use, RAMs are isolated from the data bus by _____ .

25. Refer to Fig. 12-10. If the MPU outputs 00001000 on the address bus, RAM _____ (number) will be activated and storage area _____ (decimal number) located in the RAM will be accessed.

26. Refer to Fig. 12-12. If the control input on the three-state buffer is HIGH, output *Y* is _____ (connected to input *A*; in its high-impedance state).

12-8 DATA TRANSMISSION

Most data in digital systems is transmitted directly through wires and PC boards. Many times bits of data must be transmitted from one place to another. Sometimes the data must be transmitted over telephone lines or cables to points far away. If all the data were sent at one time over *parallel* wires, the cost and size of these cables would be too expensive and large. Instead, the data is sent over a single wire in *serial* form and reassembled into parallel data at the receiving end. The devices used for sending and receiving serial data are called *multiplexers* (MUX) and *demultiplexers* (DEMUX).

The basic idea of a MUX and DEMUX is shown in Fig. 12-13. Parallel data from one digital device is changed into *serial* data by the MUX. The serial data is transmitted by a single wire. The serial is reassembled into parallel data at the output by the DEMUX. Notice the control lines that must also connect the MUX and DEMUX. These control lines keep the MUX and DEMUX synchronized. Notice that the 16 input lines are cut down to only a few transmission lines.

The system in Fig. 12-13 works in the following manner. The MUX first connects input 0 to the serial data transmission line. The bit is then transmitted to the DEMUX, which places this bit of data at output 0. The MUX and DEMUX proceed to transfer the data at input 1 to output 1, and so on. The bits are transmitted one bit at a time.

A MUX works much like a single-pole, many-position rotary switch, as shown in Fig. 12-14. Rotary switch 1 shows the action of a MUX. The DEMUX operates like rotary switch 2 in Fig. 12-14. The mechanical control in this diagram makes sure input 5 on SW 1 is delivered to output 5 on SW 2. Notice that the mechanical switches in Fig. 12-14 permit data to travel in either direction. Being made from logic gates, MUXs and DEMUXs permit data to travel only from input to output, as in Fig. 12-13.

You used a MUX before, in Chap. 4. The other name for MUX is *data selector*. DEMUXs are sometimes called *distributors* or *decoders*. The term "distributor" describes the action of SW 2 in Fig. 12-14, as it distributes the serial data first to output 1, then to output 2, then to output 3, and so forth.

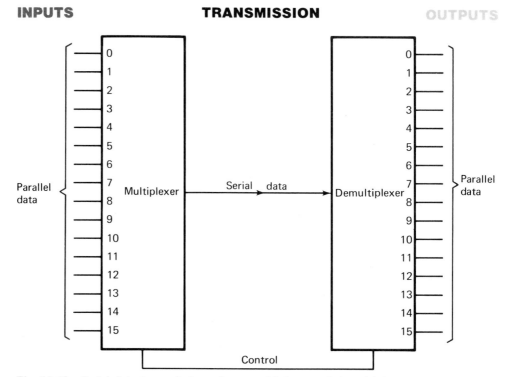

Fig. 12-13   Serial data transmission using a multiplexer and demultiplexer.

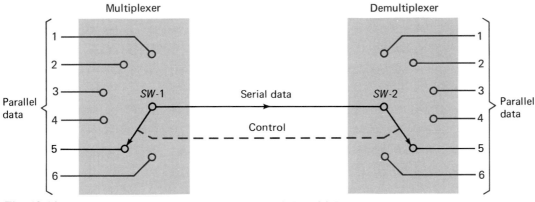

Multiplexer   Demultiplexer

Fig. 12-14   Rotary switches act like multiplexers and demultiplexers.

Figure 12-15 on the next page is a detailed wiring diagram of a transmission system using the MUX/DEMUX arrangement. A word (16 bits long) is entered at the inputs (0 to 15) of the 74150 MUX IC. The 7493 counter starts at binary 0000. This would be shown as 0 on the seven-segment display. With the data select inputs (*D, C, B, A*) of the 74150 MUX at 0000, the data is taken from input 0, which is shown as a logical 0. The logical 0 is transferred to the 74154 DEMUX IC, where it is routed to output 0. Normally the output of the 74154 IC is inverted, as shown by the invert bubbles. A 7404 inverter complements the output back to the original logical 0.

The counter increases to binary 0001. This is shown as a 1 on the decimal readout. This binary 0001 is applied to the data select inputs of both ICs (74150 and 74154). The logical 1 at the input of the 74150 MUX is transferred to the transmission line. The 74154 DEMUX routes the data to output 1. The 7404 inverter complements the output, and the logical 1 appears as a lighted LED, as shown in the diagram. The counter continues to scan each input of the 74150 IC and transfer the contents to the output of the 74154. Notice that the counter must count from binary 0000 to 1111 (16 counts) to transfer just one parallel word from the input to the output of this system. The seven-segment LED readout provides a convenient way of keeping track of which input is being transmitted. If the clock is pulsed very fast, the parallel data can be transmitted quite quickly as serial data to the output.

Notice from Fig. 12-15 that we have saved many pieces of wire by sending the data in *serial* form. This takes somewhat more time, but the rate at which we send data over the transmission line can be very high.

One common example of data transmission is the link between a microcomputer and a peripheral device such as a printer or modem. The computer's interface may send data either in parallel or serial format depending on the design of the printer.

A *parallel interface* transmits 8 bits (1 byte) of data at one time. Figure 12-16 on page 267 shows how the microcomputer's CPU controls a special IC called a *peripheral-interface adapter (PIA)*. The PIA IC communicates with the printer through the *handshaking* line to check if it is ready to receive data. If the printer signals the PIA that it is ready, bytes are transmitted from the CPU to the PIA and then on to the printer's buffer memory. The CPU can send data much faster than the printer can print the information. For this reason, the printer signals the PIA when its buffer memory is full. The PIA then signals the CPU to temporarily stop sending data until there is more room for data in the printer's buffer memory.

Peripheral interface adapters are not standardized. For instance, Motorola calls their unit a 6820 PIA while Intel's name for a similar input/output adapter unit is the 8255 PPI (programmable peripheral interface). The PIAs are general-purpose ICs that can be programmed for either input or output. They have several parallel 8-bit I/O ports.

A serial interface transmits data 1 bit at a time. Very complexed ICs called *UARTs* (*universal asynchronous receiver-transmitters*) are many times used as the interface between the CPU and the data lines (also called *data links*). A UART consists of three sections as shown in Fig. 12-17 on page 267. They are a *receiver*, a *transmitter*, and a *control* block. The receiver converts serial to parallel data. The transmitter section converts parallel data (as from the data bus of the CPU) to serial data. The

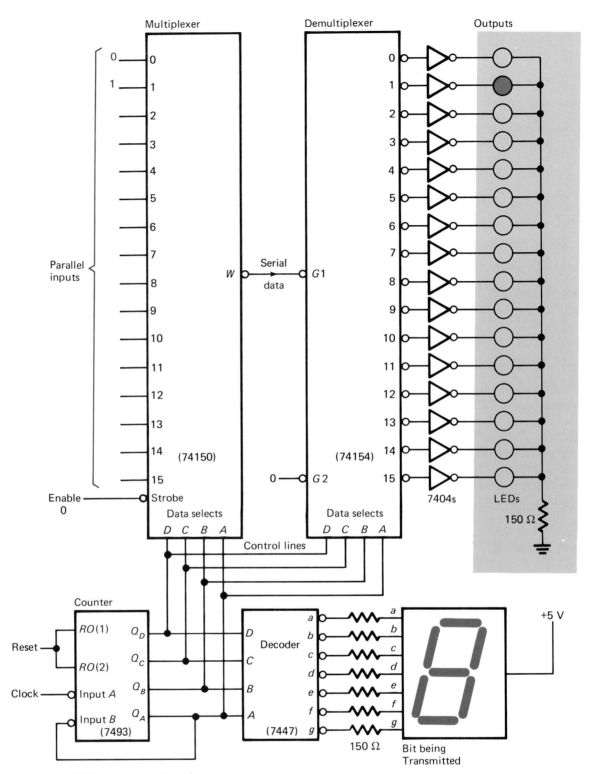

Fig. 12-15   Wiring diagram for a transmission system.

control section manages the UART's functions and handles communications with the CPU and the peripheral device. The UART also encodes and decodes the serial signal including start, stop, and parity bits.

The speed at which serial data is transmit-ted is called the baud rate. The *baud rate* is the number of bits per second being transmit-ted through a data link. The baud rate is *not* the same as the number of characters or words transmitted per second. Common baud rates are 110, 300, 1200, 2400, and 9600.

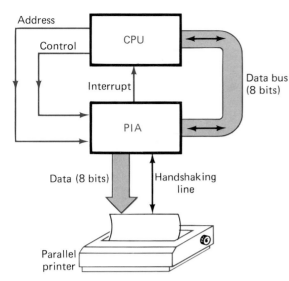

Fig. 12-16   Parallel data transmission from the CPU to a printer using a peripheral-interface adapter (PIA) IC.

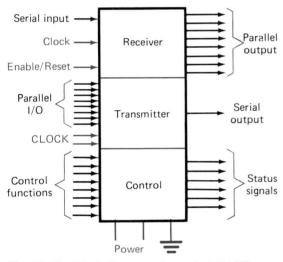

Fig. 12-17   Block diagram of a typical UART.

The signal levels found in data lines are many times defined by standards. Two serial interface standards are the EIA RS-232C standard and the older 20-mA current loop teletype standard.

Two common parallel interfaces are the Centronics standard and the IEEE-488 standard. The Centronics standard is used between many microcomputers and printers. The IEEE-488 interface is used between computers and scientific instrumentation.

Self-Test

Answer the following questions.

27. Refer to Fig. 12-13. A(n) _____ changes parallel data to serial data, whereas a(n) _____ changes serial data to parallel data for transmission.
28. Refer to Fig. 12-15. The 7493 IC is used to sequence the data selects from 0000 through _____ (binary number).
29. Refer to Fig. 12-16. A complex chip called a(n) _____ is used to output parallel data to the printer in many microcomputer systems.
30. An LSI IC used for asynchronous data transmission is called a(n) _____ .
31. A measure of the speed of serial data transmission is called the _____ rate.
32. The EIA RS-232C standard might be used for _____ (parallel, serial) interfacing between a microcomputer and a peripheral device.

12-9 DETECTING ERRORS IN DATA TRANSMISSIONS

Digital equipment, such as a computer, is valuable to people because it is fast and *accurate*. To help make digital devices accurate, special *error detection* methods are used. You can well imagine an error creeping into a system when data is transferred from place to place.

To detect errors we must keep a constant check on the data being transmitted. To check accuracy we generate and transmit an extra *parity bit*. Figure 12-18, on the next page, shows such a system. In this system three parallel bits (*A*, *B*, and *C*) are being transmitted over a long distance. Near the input they are fed into a *parity bit generator* circuit. This circuit generates what is called a parity bit. The parity bit is transmitted with the data, and near the output the results are checked. If an error occurs during transmission, the *error detector* circuit sounds an alarm. If all the parallel data is the same at the output as it was at the input, no alarm sounds.

Table 12-1 will help you understand how the error-detection system works. This table is really a truth table for the parity bit generator in Fig. 12-18. Notice that the inputs are labeled *A*, *B*, and *C* for the three data transmission lines. The output is determined by looking across a horizontal row. We want an *even number of 1s* in each row (zero 1s, or two 1s, or

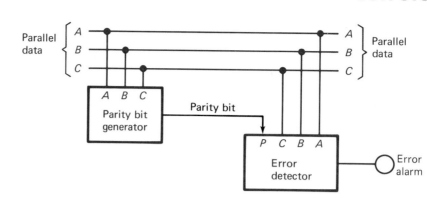

Fig. 12-18 Error-detection system using a parity bit.

Table 12-1 Truth Table for Parity Bit Generator

Inputs			Output
Parallel Data			Parity Bit
C	B	A	P
0	0	0	0
0	0	1	1
0	1	0	1
0	1	1	0
1	0	0	1
1	0	1	0
1	1	0	0
1	1	1	1

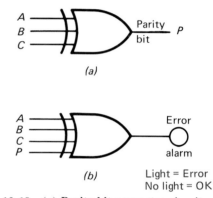

Fig. 12-19 (a) Parity bit generator circuit. (b) Error detector circuit.

four 1s). Notice that row 1 has no 1s. Row 2 has a single 1 plus the parity bit 1. Row 2 now has two 1s. As you look down Table 12-1 you will notice that each horizontal row contains an even number of 1s. Next, the truth table is converted to a logic circuit. The logic circuit for the parity bit generator is drawn in Fig. 12-19(a). You can see that a three-input XOR gate will do the job for generating a parity bit. The three-input XOR gate in Fig. 12-19, then, is the logic circuit you would substitute for the parity bit generator block in Fig. 12-18.

Look at the entire truth table in Table 12-1. We can see that under normal circumstances each horizontal row contains an *even number* of 1s. Were an error to occur, we would then have an *odd number* of 1s appear. A circuit that gives a logical 1 output any time an odd number of 1s appear is shown in Fig. 12-19(b). A four-input XOR gate would detect an odd number of 1s at the inputs and turn on the alarm light. Figure 12-19(b) diagrams the logic circuit that can substitute for the error-detector block in Fig. 12-18.

The use of the parity bit only warns you of an error; the system we used does *not* correct the error. Some codes that are *error correcting*, such as the *Hamming code*, have been developed. The Hamming code uses several extra parity bits when transmitting data.

Self-Test

Supply the missing word or words in each statement.

33. Refer to Table 12-1. This is the truth table for an _____ (even, odd) parity bit generator.
34. Refer to Fig. 12-18. The parity bit generator block could be replaced with a three-input _____ gate, whereas the error detector block could be replaced with a four-input _____ gate.

12-10 ADDER/SUBTRACTOR SYSTEM

In Chap. 10 you worked with adders and subtractors and studied an adder/subtractor system. Figure 12-20 is a block diagram of that system (the complete wiring diagram is shown in Fig. 10-17).

The system diagramed in Fig. 12-20 consists of familiar subsystems. The keyboard is the input device to this system. An encoder (74147 IC) converts the keyboard input to BCD. The load control routes data to either storage unit (register A or register B). Registers A and B hold data at the inputs of the 4-bit adder while the calculation is performed. For subtraction the add/subtract control unit performs the 1s complement and end-around carry procedure. The sum from the 7483 IC adder is applied to the decoder (7447). The decoder translates from BCD to the seven-segment code. The seven-segment display reads out the sum or difference in decimal. Displays A and B show in BCD the contents at the outputs of the two 74194 registers.

Self-Test

Supply the missing word in each statement.

35. Refer to Fig. 12-20. The _____ translates the decimal input from the keyboard to BCD.
36. Refer to Fig. 12-20. Registers A and B hold data at the inputs of the _____ while a calculation is performed.

12-11 THE DIGITAL CLOCK

We introduced a digital electronic clock in Chap. 8 and noted that various *counters* are the heart of a digital clock system. Figure 12-21(*a*) on the next page is a simple block diagram of a digital clock system. Many clocks use the power-line frequency of 60 Hz as their input. This frequency is divided into seconds, minutes, and hours by the *frequency divider* section of the clock. The one-per-second, one-per-minute, and one-per-hour pulses are then counted and stored in the *count accumulator* section of the clock. The stored contents of the count accumulators (seconds, minutes, hours)

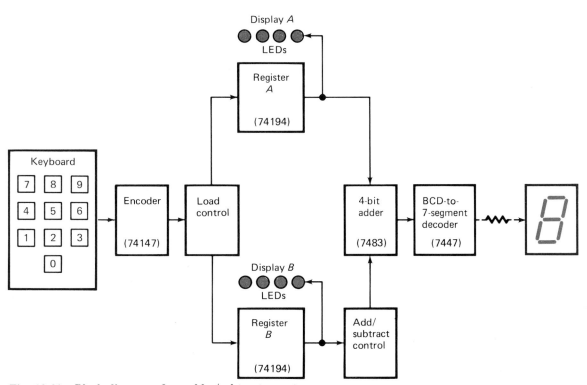

Fig. 12-20   Block diagram of an adder/subtractor system.

INPUT

OUTPUTS

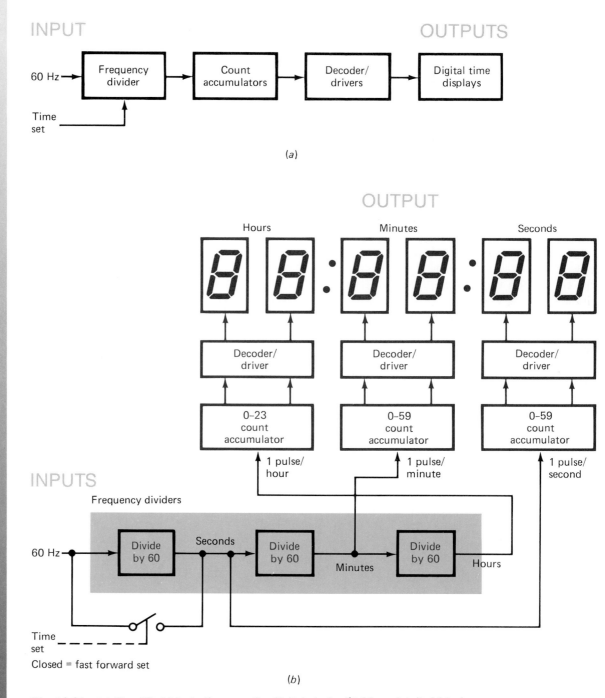

60 Hz → | Frequency divider | → | Count accumulators | → | Decoder/drivers | → | Digital time displays |

Time set

(a)

OUTPUT

Hours   Minutes   Seconds

Decoder/driver   Decoder/driver   Decoder/driver

0–23 count accumulator   0–59 count accumulator   0–59 count accumulator

1 pulse/hour   1 pulse/minute   1 pulse/second

INPUTS

Frequency dividers

60 Hz → Divide by 60 → Seconds → Divide by 60 → Minutes → Divide by 60 → Hours

Time set

Closed = fast forward set

(b)

Fig. 12-21 (*a*) **Simplified block diagram of a digital clock.** (*b*) **More detailed block diagram of a digital clock.**

are then *decoded,* and the correct time is shown on the output *time displays.* The digital clock has the typical elements of a system. The input is the 60-Hz alternating current. The processing takes place in the frequency divider, count accumulator, and decoder sections. Storage takes place in the count accumulators. The control section is illustrated by the *time-set* control, as shown in Fig. 12-21(*a*). The output section is the digital time display.

It was mentioned that all systems consist of

logic gates, flip-flops, and subsystems. The diagram in Fig. 12-21(*b*) shows how subsystems are organized to display time in hours, minutes, and seconds. This is a more detailed diagram of a digital clock. The input is still a 60-Hz signal. The 60 Hz may be from the low-voltage secondary coil of a transformer. The 60 Hz is divided by 60 by the first frequency divider. The output of the first divide-by-60 circuit is 1 pulse per second. The 1 pulse per second is fed into an *up counter* that counts upward from 00 through 59 and then resets to 00. The seconds counters are then decoded and displayed on the two 7-segment LED displays at the upper right, Fig. 12-21(*b*).

Consider the middle frequency-divider circuit in Fig. 12-21(*b*). The input to this divide-by-60 circuit is 1 pulse per second; the output is 1 pulse per minute. The 1-pulse-per-minute output is transferred into the 0 to 59 minutes counter. This up counter keeps track of the number of minutes from 00 through 59 and then resets to 00. The output of the minutes count accumulator is decoded and displayed on the two 7-segment LEDs at the top center, Fig. 12-21(*b*).

Now for the divide-by-60 circuit on the right in Fig. 12-21(*b*). The input to this frequency divider is 1 pulse per minute. The output of this circuit is 1 pulse per hour. The 1-pulse-per-hour output is transferred to the hours counter on the left. This hours count accumulator keeps track of the number of hours from 0 to 23. The output of the hours count accumulator is decoded and transferred to the two 7-segment LED displays at the upper left, Fig. 12-21(*b*). You probably have noticed already that this is a 24-h digital clock. It easily could be converted to a 12-h clock by changing the 0 to 23 count accumulator to a 0 to 11 counter.

For setting the time a time-set control has been added to the digital clock in Fig. 12-21(*b*). When the switch is closed (a logic gate may be used), the display counts forward at a fast rate. This enables you to set the time quickly. The switch bypasses the first divide-by-60 frequency divider so that the clock moves forward at 60 times its normal rate. An even faster *fast-forward* set could be used by bypassing both the first and the second divide-by-60 circuits. The latter technique is common in digital clocks.

What is inside the divide-by-60 frequency dividers in Fig. 12-21(*b*)? In Chap. 8 we spoke of a counter being used to divide frequency. Figure 12-22(*a*) is a block diagram of how a divide-by-60 frequency divider might be organized. Notice that a divide-by-6 counter is feeding a divide-by-10 counter. The entire unit divides the incoming frequency by 60. In this example, the 60-Hz input is reduced to 1 Hz at the output.

A detailed wiring diagram for a divide-by-60 counter circuit is drawn in Fig. 12-22(*b*). The three JK flip-flops and NAND gate form the divide-by-6 counter while the 74192 decade counter performs as a divide-by-10 unit. If 60 Hz enters at the left, the frequency will be reduced to 1 Hz at output Q_D of the 74192 counter.

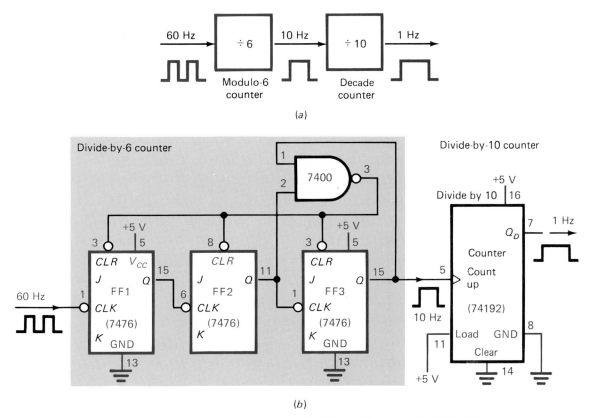

Fig. 12-22 **Divide-by-60 counter.** (*a*) **Block diagram.** (*b*) **Wiring diagram using TTL ICs.**

The seconds and minutes count accumulators in Fig. 12-21(*b*) are also counters. The 0 to 59 is a decade counter cascaded with a 0 to 5 counter. The decade counter is coupled with the 1s place of the displays. The mod-6 counter is coupled with the 10s place of the displays. In a like manner, the hours count accumulator is a decade counter cascaded with a 0 to 2 counter. The decade counter is coupled to the 1s place in the hours display. The mod-3 counter is coupled with the 10s place of the hours display.

In many practical digital clocks the output may be in hours and minutes only. Most digital clocks are based upon one of many inexpensive ICs. Large-scale-integrated *clock chips* have all the frequency dividers, count accumulators, and decoders built into a single IC. For only a few extra dollars, clock chips have other features, such as 12- or 24-h outputs, calendar features, alarm controls, and radio controls.

An added feature you will use when you construct a digital timepiece is shown in Fig. 12-23(*a*). A *wave-shaping circuit* has been added to the block diagram of our digital clock. The IC counters that make up the frequency-divider circuit do not work well with a sine-wave input. The sine wave [shown at the left in Fig. 12-23(*a*)] has a slow *rise time* that does not trigger the counter properly. The sine-wave input must be converted to a square wave. The wave-shaping circuit changes the sine wave to a square wave. The square wave will now properly trigger the frequency-divider circuit.

Commercial LSI clock chips have wave-shaping circuitry built into the IC. In the lab-

oratory you may use a Schmitt trigger inverter IC to square up the sine waves as you did in Chap. 7. A simple wave-shaping circuit is shown in Fig. 12-23(*b*). This circuit uses the TTL 7414 Schmitt trigger inverter IC to convert the sine wave to a square wave. The circuit in Fig. 12-23(*b*) also contains a *start/stop control*. When the control input is HIGH, the square wave from the Schmitt trigger inverter passes through the AND gate. When the control input goes LOW, the square-wave signal is inhibited and does not pass through the AND gate. The counter is stopped.

You will want to get some practical knowledge of how counters are used in dividing frequency. Remember that the counter subsystem is used for two jobs in the digital timepiece: first to divide frequency and second to count upward and keep track of the number of pulses at its input.

Self-Test

Answer the following questions.

37. Refer to Fig. 12-21(*a*). Counters are used in the _____ and _____ sections of a digital clock.
38. Refer to Fig. 12-23. When operating a clock with a sine wave, a(n) _____ circuit is added to the clock.
39. Refer to Fig. 12-23(*b*). What is the purpose of the Schmitt trigger inverter in this circuit?
40. Refer to Fig. 12-23(*b*). What is the purpose of the AND gate in this circuit?

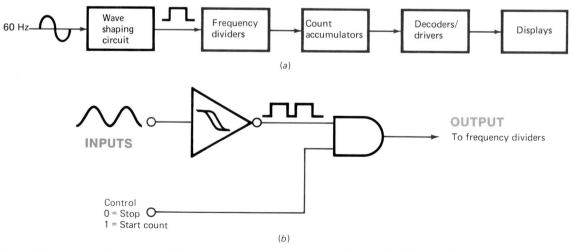

Fig. 12-23 Wave-shaping. (*a*) Adding a wave-shaping circuit to the input of the digital clock system. (*b*) Schmitt trigger inverter used as a wave shaper.

12-12 THE LSI DIGITAL CLOCK

The LSI clock chip forms the heart of modern digital timepieces. These digital clock chips are made as monolithic MOS ICs. Many times, the MOS LSI chip, or die, is mounted in an 18-, 24-, 28-, or 40-pin DIP IC. Other times, the MOS LSI chip is mounted directly on the PC board of a clock module. The tiny silicon die is sealed under an epoxy coating. Examples of the two packaging methods are shown in Fig. 12-24. An MOS LSI clock IC packaged in a 24-pin DIP is illustrated in Fig. 12-24(*a*). Pin 1 of the DIP IC is identified in the normal manner (pin 1 is immediately ccw from the notch). A National Semiconductor clock module is sketched in Fig. 12-24(*b*). The back is a PC board with 22 edge connectors. The numbering of the edge connectors is shown. A four-digit LED display is premounted on the board with all connections complete. Some clock modules have some discrete components and a DIP clock IC mounted on the board. The clock module in Fig. 12-24(*b*) has the tiny silicon chip, or die, mounted on the PC board. It is sealed with a protective epoxy coating.

A block diagram of National Semiconductor's MM5314 MOS LSI clock IC is shown in Fig. 12-25(*a*). The pin diagram is shown in Fig. 12-25(*b*). Refer to Fig. 12-25(*a*) and (*b*) on the next page for the following functional description of the MM5314 digital clock IC.

50- OR 60-HZ INPUT (PIN 16)

Alternating current or rectified ac is applied to this input. The *wave-shaping circuit* squares up the waveform. The shaper circuit drives a chain of counters which perform the time-keeping job.

50- OR 60-HZ SELECT INPUT (PIN 11)

This input programs the *prescale counter* to divide by either 50 or 60 to obtain a 1-Hz, or 1-pulse-per-second time base. The counter is programmed for 60-Hz operation by connecting this input to V_{DD} (GND). If the 50/60-Hz select input pin is left unconnected, the clock is programmed for 50-Hz operation.

TIME-SETTING INPUTS (PINS 13, 14, AND 15)

Slow- and fast-setting inputs as well as a hold input are provided on this clock IC. These inputs are enabled when they are connected to V_{DD} (GND). Typically, a normally open push-button switch is connected from these pins to V_{DD}. The three gates in the counter chain are used for setting the time. For *slow set*, the prescale counter is bypassed. For *fast set*, the prescale counter and seconds counter are bypassed. The *hold* input inhibits any signal from passing through gate *A* to the prescale counter. This stops the counters, and time does not advance on the output display.

12- OR 24-H SELECT INPUT (PIN 10)

This input is used to program the hours counter to divide by either 12 or 24. The 12-h display format is selected by connecting this input to V_{DD} (GND). Leaving pin 10 unconnected selects the 24-h format.

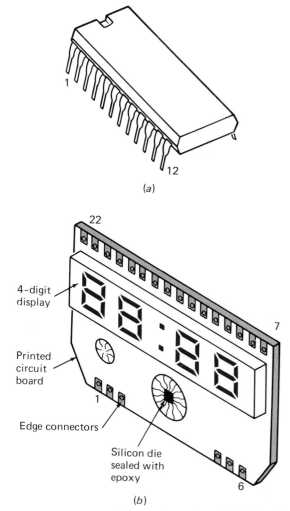

4-digit display

Printed circuit board

Edge connectors

Silicon die sealed with epoxy

(*b*)

Fig. 12-24 (*a*) **An LSI clock chip in a 24-pin dual in-line package.** (*b*) **A typical clock module containing a MOS/LSI die.**

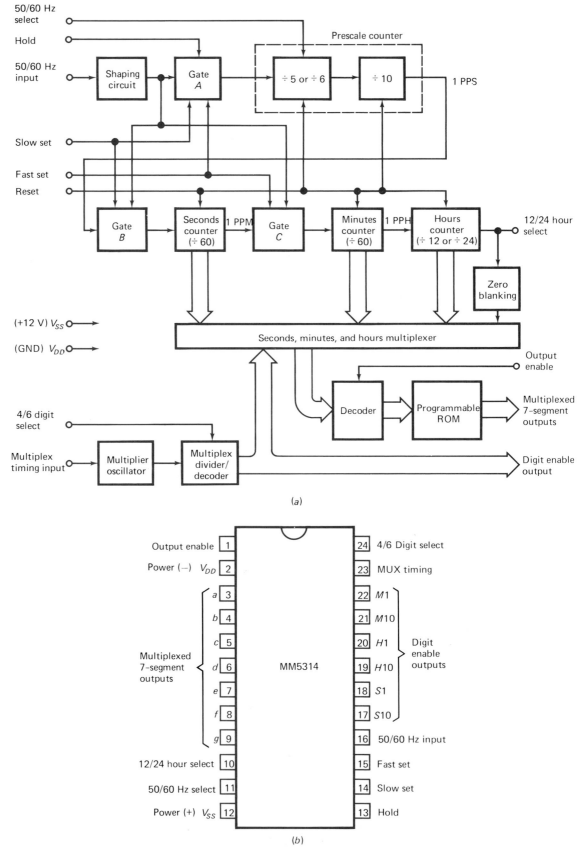

Fig. 12-25 (*a*) **Functional block diagram of the MM5314 MOS/LSI clock chip.** (*Courtesy of National Semiconductor Corporation*) (*b*) **Pin diagram for the MM5314 digital clock IC.** (*Courtesy of National Semiconductor Corporation*)

OUTPUT MUX OPERATION (PINS 3 TO 9 AND 17 TO 22)

The seconds, minutes, and hours counters continuously reflect the time of day. Outputs from each counter are *multiplexed* to provide digit-by-digit sequential access to the time data. In other words, only one display digit is turned on for a very short time, then the second, then the third, and so forth. By multiplexing the displays instead of using 48 leads to the 6 displays (8 pins each × 6 = 48), only 13 output pins are required. These 13 outputs are the multiplexed seven-segment outputs (pins 3 through 9) and the digit enable outputs (pins 17 through 22).

The MUX is addressed by a *multiplex divider/decoder*, which is driven by a *multiplex oscillator*. The oscillator uses external timing components (resistor and capacitor) to set the frequency of the multiplexing function. The four/six-digit select input controls if the MUX turns on all six or just four displays in sequence. The *zero-blanking* circuit suppresses the 0 that would otherwise sometimes appear in the tens-of-hours display. The MUX addresses also become the display digit enable outputs (pins 17 to 22). The MUX outputs are applied to a decoder which is used to address a PROM. The ROM generates the final seven-segment output code. The displays are enabled in sequence from the unit seconds through the tens-of-hours display.

MULTIPLEX TIMING INPUT (PIN 23)

Adding a resistor and capacitor to the MM5314 clock IC forms a *relaxation oscillator*. The external resistor and capacitor are connected to the MUX timing input as shown in Fig. 12-26. Typical timing resistor and capacitor values might be 470 kΩ and 0.01 μF.

FOUR/SIX-DIGIT SELECT INPUT (PIN 24)

The four/six-digit select input controls the MUX. With no input connection, the clock outputs data for a four-digit display. Applying V_{DD} (GND) to this pin provides a six-digit display.

OUTPUT ENABLE INPUT (PIN 1)

With this pin unconnected, the seven-segment outputs are enabled. Switching V_{DD} (GND) to this input inhibits these outputs.

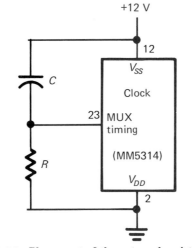

Fig. 12-26 Placement of the external resistor and capacitor used to set the frequency of the multiplex oscillator in the MM5314 clock IC.

POWER INPUTS (PINS 2 AND 12)

A dc 11- to 19-V nonregulated power supply operates the clock IC. The positive of the power supply connects to the V_{SS} (pin 12), while the negative connects to V_{DD} (pin 2).

Self-Test

Supply the missing word in each statement.

41. Digital clock LSI chips are made using _____ (bipolar, MOS) technology.
42. Refer to Fig. 12-25. If pin 16 of the MM5314 is at GND, the clock IC is programmed for _____ -Hz operation.
43. Refer to Fig. 12-25. If the slow set input to the MM5314 IC is grounded, the _____ counter is bypassed.
44. Refer to Fig. 12-25. The MM5314 MOS/LSI clock chip requires a(n) _____ -V nonregulated power supply.

12-13 A PRACTICAL LSI DIGITAL CLOCK SYSTEM

A six-digit clock using the MM5314 IC is sketched in Fig. 12-27(*a*) on the next page. This student-built unit uses six common-anode seven-segment LED displays. Also notice the many extra parts used in the clock system. A block diagram of this system is shown in Fig. 12-27(*b*). The National Semiconductor MM5314 clock chip is used in this system. The 60 Hz is divided down to seconds, minutes, and hours by

Multiplex divider/decoder

Multiplex oscillator

Zero-blanking circuit

PROM

MM5314 clock IC

Relaxation oscillator

Practical LSI digital clock system

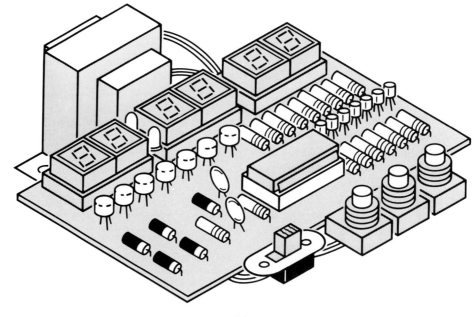

(a)

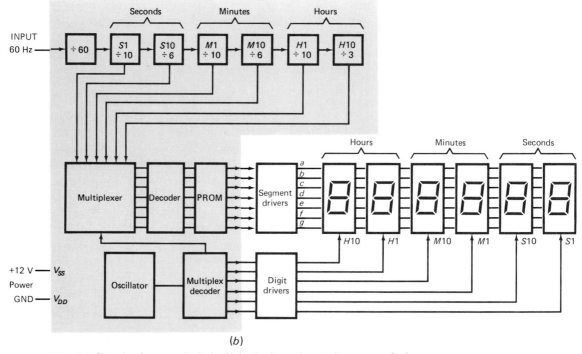

(b)

Fig. 12-27 (a) **Sketch of a practical six-digit clock project.** (*Courtesy of Electronic Kits International, Inc.*) (b) **Block diagram of the practical six-digit clock project using the MM5314 clock chip.**

counters across the top in Fig. 12-27(b). These feed the MUX. The *oscillator* at the lower left produces a frequency of about 1 kHz.

Outside the MM5314 clock chip are six 7-segment common-anode LED displays. Be-cause of the higher currents used by the LED displays, *segment drivers* are used to sink the current from the cathodes of the displays. The *digit drivers* furnish an adequate amount of cur-rent to the anodes of the selected digit.

To help explain how the MUX works, suppose the time is 12:34:56. This information is held in the counters in the clock chip. The multiplex decoder first selects the $S1$ display. The MUX takes data from the $S1$ counter and places it on the decoder PROM. Segment lines c, d, e, f, and g are *activated to all the displays*. The multiplex decoder activates *only* the $S1$ line of the digit driver. The decimal number 6 flashes on for an instant, as shown in Fig. 12-28(*a*) on the next page. The segments c, d, e, f, and g have been activated on all the displays, but only the right $S1$ unit has had its common-anode lead activated, or connected to V_{SS}. Therefore only the $S1$ display lights up.

Second, the clock IC's multiplex decoder selects the $S10$ display. The MUX finds that the $S10$ counter holds 5. The decoder-PROM-segment driver activates segments a, c, d, f, and g. Next, the common anode of the $S10$ display is activated, or connected to V_{SS}. A decimal 5 flashes on the $S10$ display. This is shown in Fig. 12-28(*b*).

One at a time, each display is activated by the multiplex decoder and digit driver. At the same time the MUX-decoder-PROM activates the proper segments. This is based on the present contents of the counters. Look over Fig. 12-28. This is one cycle through the six displays. This whole sequence (*a* through *f*) happens more than 100 times each second. This multiplexing, or scanning, occurs at a fast rate, and so the eye does not notice a flickering in the displays.

A schematic diagram of the digital clock system using the MM5314 IC is shown in Fig. 12-29 on page 279. The step-down 12-V transformer ($T1$) with the bridge rectifier ($D1$–4) and filter capacitor ($C1$) form the dc power supply section of the clock. Alternating current voltage is taken off the transformer and coupled to the 50/60-Hz input (pin 16) of the clock chip through resistor $R3$. Capacitor $C3$ and resistor $R4$ determine the frequency of the multiplex oscillator. Placing a much larger value capacitor (perhaps 1 to 5 µF) across $C3$ slows down the multiplexing process to a point where you can see each display light in sequence.

The fast-set, slow-set, and hold normally open push-button switches (S_2, S_3, and S_4) are located at the lower left in Fig. 12-29. The action (fast set, slow set, or hold) is taken when these pins are connected through the switch to V_{DD}.

The *segment drivers* are the seven NPN transistors (Q_7 to Q_{13}) on the right of the IC in Fig. 12-29. These transistors sink the current from the displays when activated. The *digit drivers* are the six PNP transistors (Q_1 to Q_6) at the upper left in Fig. 12-29. These transistors connect *only one display anode at a time* to V_{SS}. The digit drivers scan the six displays at a frequency of about 500 to 1500 Hz. This activates each display about 100 to 200 times each second.

The two LEDs (D_6 and D_7 in Fig. 12-29) are activated 100 to 200 times each second and appear lit continuously. These LEDs form the colon between the hours and minutes displays on the completed clock. This colon may be seen in Fig. 12-27(*a*). Resistor $R3$, capacitor $C2$, and diode $D5$ in Fig. 12-29 form an *RC* filter network. This network is used to remove possible line voltage transients that could either cause the clock to gain time or damage the IC.

The 12/24-h select input (pin 10) on the MM5314 clock chip in Fig. 12-29 is connected to V_{DD}. This selects the 12-h format. The 50/60-Hz select input (pin 11) is connected to V_{DD}. This programs the IC for 60-Hz operation. The four/six-digit select input (pin 24) is connected to V_{DD}. This programs the multiplex decoder to provide a six-digit display.

Self-Test

Answer the following questions.

45. Refer to Fig. 12-28. This is an example of the MM5314 clock chip _____ (counting, decoding, multiplexing) six LED decimal displays.
46. Refer to Fig. 12-29. The six PNP transistors function as _____ in this digital clock circuit.
47. Refer to Fig. 12-29. The seven NPN transistors function as _____ in this digital clock circuit.
48. Refer to Fig. 12-29. With the 12/24-h select input grounded, the clock is programmed as a(n) _____ -h clock.
49. Refer to Fig. 12-29. Which two components (outside the clock chip) are responsible for determining the frequency of the multiplexer?
50. Refer to Fig. 12-29. The 4/6 digital select (pin 24) input to the MM5314 clock chip is connected to _____ (V_{DD}, V_{SS}), which selects the 6-digit mode.

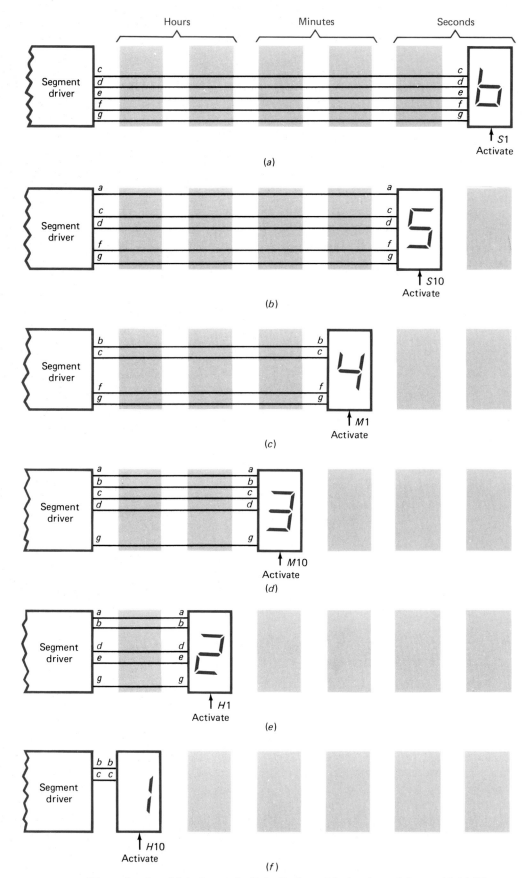

Hours   Minutes   Seconds

Segment driver

*S*1
Activate

(*a*)

Segment driver

*S*10
Activate

(*b*)

Segment driver

*M*1
Activate

(*c*)

Segment driver

*M*10
Activate

(*d*)

Segment driver

*H*1
Activate

(*e*)

Segment driver

*H*10
Activate

(*f*)

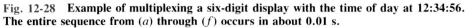

Fig. 12-28 Example of multiplexing a six-digit display with the time of day at 12:34:56. The entire sequence from (*a*) through (*f*) occurs in about 0.01 s.

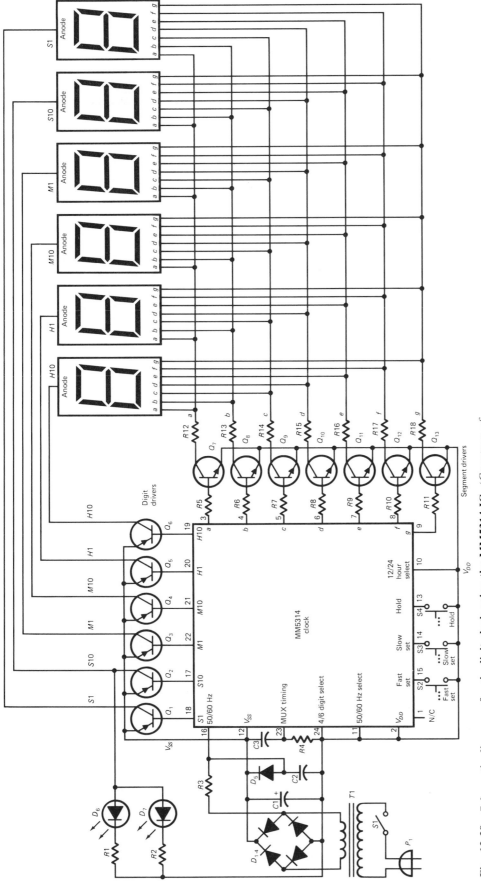

Fig. 12-29 Schematic diagram of a six-digit clock using the MM5314 IC. *(Courtesy of Electronic Kits International, Inc.)*

12-14 THE FREQUENCY COUNTER

An instrument used by technicians and engineers is the *frequency counter*. A digital frequency counter shows in decimal numbers the frequency in a circuit. Counters can measure from low frequencies of a few cycles per second (hertz, Hz) up to very high frequencies of thousands of megahertz (MHz). Like a digital clock, the frequency counter uses decade counters.

As a review, the block diagram for a digital clock is shown in Fig. 12-30(*a*). The known frequency is divided properly by the counters in the clock. The counter outputs are decoded and displayed in the time display. Figure 12-30(*b*) shows a block diagram of a frequency counter. Notice that the frequency counter circuit is fed an *unknown* frequency instead of the known frequency in a digital clock. The counter circuit in the frequency counter in Fig. 12-30(*b*) also contains a *start/stop control*.

The frequency counter has been redrawn in Fig. 12-31(*a*). Notice that an AND gate has been added to the circuit. The AND gate controls the input to the decade counters. When the start/stop control is at logical 1, the unknown frequency pulses pass through the AND gate and on to the decade counters. The

counters count upward until the start/stop control returns to logical 0. The 0 turns off the control gate and stops the pulses from getting to the counters.

Figure 12-31(*b*) is a more exact timing diagram of what happens in the frequency counter. Line *A* shows the start/stop control at logical 0 on the left and then going to 1 for *exactly 1 s*. The start/stop control then returns to logical 0. Line *B* diagrams a continuous string of pulses from the unknown frequency input. The unknown frequency and the start/stop control are ANDed together as we saw in Fig. 12-31(*a*). Line *C* in Fig. 12-31(*b*) shows only the pulses that are allowed through the AND gate. These pulses trigger the up counters. Line *D* shows the count observed on the displays. Notice that the displays start cleared to 00. The displays then count upward to 11 during the 1 s. The unknown frequency in line *B* in Fig. 12-31(*b*) is shown as 11 Hz (11 pulses/s).

A somewhat higher frequency is fed into the frequency counter in Fig. 12-31(*c*). Again line *A* shows the start/stop control beginning at 0. It is then switched to logical 1 for *exactly 1 s*. It is then returned to logical 0. Line *B* in Fig. 12-31(*c*) shows a string of higher-frequency pulses. This is the unknown frequency being measured by this digital frequency counter. Line *C* shows the pulses that trigger the decade counters during the 1-s count-up period. The decade counters sequence upward to 19, as shown in line *D*. The unknown frequency in Fig. 12-31(*c*) is then 19 Hz.

If the unknown frequency were 870 Hz, the counter would count from 000 to 870 during the 1-s count period. The 870 would be displayed for a time, and then the counters would be reset to 000 and the frequency counted again. This *reset-count-display sequence* is repeated over and over.

Notice that the start/stop control pulse (count pulse) must be *very accurate*. Figure 12-32 on page 282 shows how a count pulse can be generated by using an accurate known frequency, such as the 60 Hz from the power line. The 60-Hz sine wave is converted to a square wave by the wave-shaping circuit. The 60-Hz square wave triggers a counter that divides the frequency by 60. The output is a pulse *exactly 1 s* in length. This *count pulse* turns on the control circuit when it goes high and permits the unknown frequency to trigger the counters. The unknown frequency is applied to the counters for exactly 1 s.

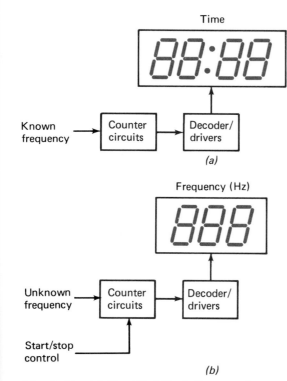

Fig. 12-30 (*a*) Simplified block diagram of a digital clock. (*b*) Simplified block diagram of a digital frequency counter.

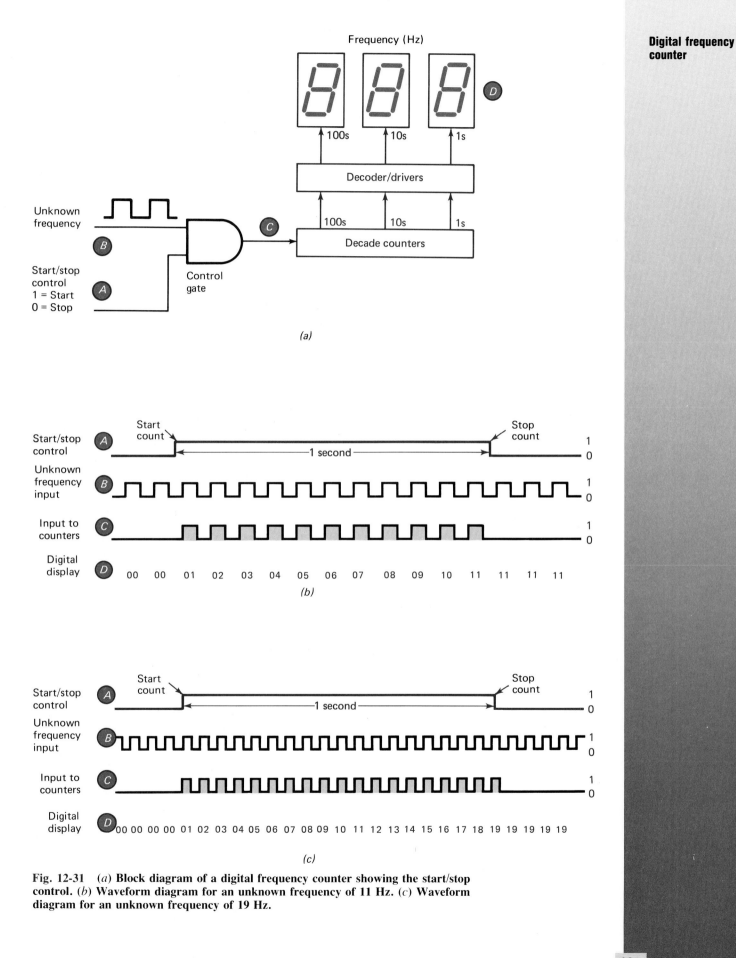

Fig. 12-31 (*a*) **Block diagram of a digital frequency counter showing the start/stop control.** (*b*) **Waveform diagram for an unknown frequency of 11 Hz.** (*c*) **Waveform diagram for an unknown frequency of 19 Hz.**

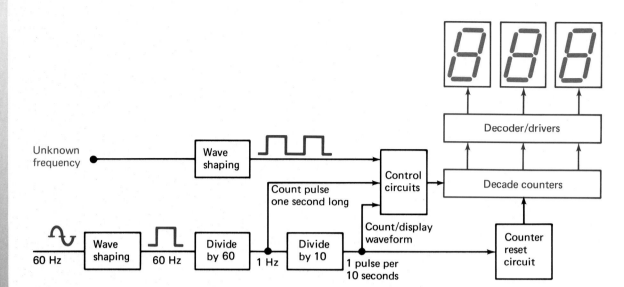

Unknown frequency

Wave shaping

Count pulse one second long

Control circuits

Decoder/drivers

Decade counters

60 Hz

Wave shaping

60 Hz

Divide by 60

1 Hz

Divide by 10

Count/display waveform

1 pulse per 10 seconds

Counter reset circuit

Fig. 12-32   More detailed block diagram of a digital frequency counter.

Remember that the frequency counter goes through the reset-count-display sequence. So far we have shown only the count part of this sequence. The *counter reset* circuit is a group of gates that reset, or clear, the decade counters to 000 at the correct time—just before the count starts. Next, the 1-s count pulse permits the counters to count upward. The count pulse ends, and the unknown frequency is *displayed* on the seven-segment displays. In this circuit, the frequency reads in hertz. It is convenient to leave this display on the LEDs for a time. To do this, the divide-by-10 counter sends a pulse to the control circuit, which turns off the count sequence for 9 s. Events then happen like this. *Reset* the counters to 000. *Count* upward for 1 s. *Display* the unknown frequency for 9 s with no counts. Repeat the reset-count-display procedure every 10 s.

The frequency counter in Fig. 12-32 measures frequencies from 1 to 999 Hz. Notice the extensive use of counters in the divide-by-60, divide-by-10, and three decade counter circuits—hence the name frequency counter. The digital frequency counter actually counts the pulses in a given amount of time.

One limitation of the counter diagramed in Fig. 12-32 is its top frequency; the top frequency that can be measured is 999 Hz. There are two ways to increase the top frequency of our counter. The first method is to add one or more counter-decoder-display units. We could extend the range of the frequency counter in Fig. 12-32 to a top limit of 9999 Hz by adding a single counter-decoder-display unit.

The second method of increasing the frequency range is to count by 10s instead of 1s. This idea is illustrated in Fig. 12-33. A divide-by-6 counter replaces the divide-by-60 unit in our former circuit. This makes the *count pulse* only 0.01 s long. The count pulse permits only one-tenth as many pulses through the control as with the 1-s pulse. This is the same as counting by 10s. Only three LED displays are used. The 1s display in Fig. 12-33 is only to show that a 0 must be added to the right of the three LED displays. The range of this frequency counter is from 10 to 9990 Hz.

In the circuit in Fig. 12-33, the decade counters count upward for 0.1 s. The display is held on the LEDs for 0.9 s. The counters are then reset to 000. The count-display-reset procedure is then repeated. The circuit in Fig. 12-33 has one other new feature: during the count time the displays are blanked out. They are then turned on again when the unknown frequency is on the display. The sequence for this frequency counter is then reset, count (with displays blank), and, finally, the longer display period. This sequence repeats itself once every second while the instrument is being used.

The frequency counter diagramed in Fig. 12-33 is similar to one you can assemble in the laboratory from gates, flip-flops, and subsystems. It is suggested that you set up this complicated digital system because practical experience will teach you the details of the frequency counter system.

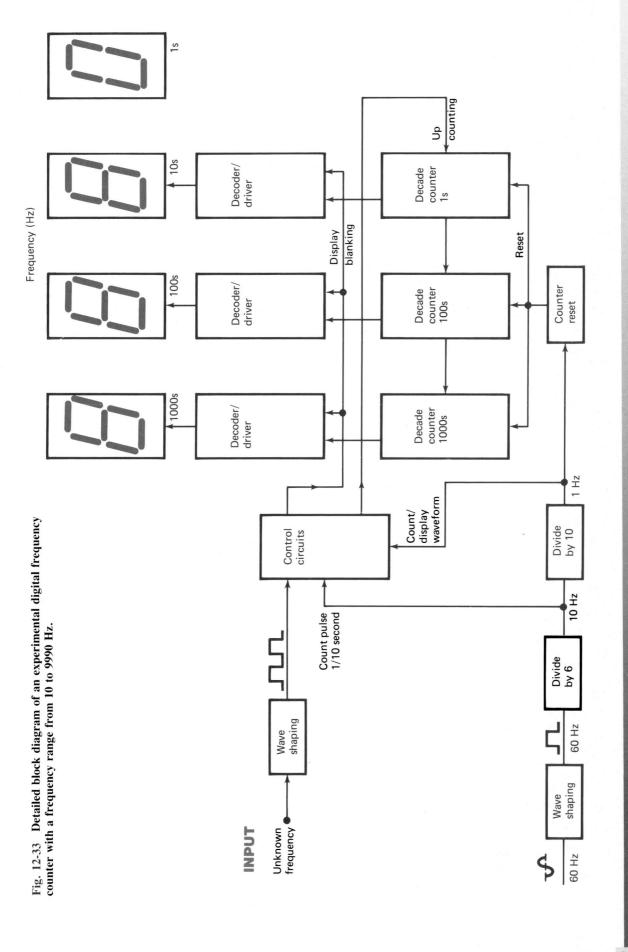

Fig. 12-33 Detailed block diagram of an experimental digital frequency counter with a frequency range from 10 to 9990 Hz.

Self-Test

Supply the missing word in each statement.

51. Refer to Fig. 12-31. This digital frequency counter counts the number of pulses passing through the AND gate in _____ s.
52. Refer to Fig. 12-31. If the start/stop control to the AND gate is LOW, the signal at output C is a _____ (HIGH, LOW, square wave).
53. Refer to Fig. 12-32. The wave-shaping blocks might be implemented using TTL _____ (XOR gates, Schmitt trigger inverters).
54. Refer to Fig. 12-32. The divide-by-60 block might be implemented using _____ (counters, shift registers).
55. Refer to Fig. 12-33. The count pulse is _____ s long in this frequency counter.
56. Refer to Fig. 12-33. The unknown input waveform is conditioned by a _____ circuit before entering the control circuitry of the frequency counter.
57. Refer to Fig. 12-33. The decade counters serve the dual purpose of counting upward and _____ the count for display.

12-15 AN EXPERIMENTAL FREQUENCY COUNTER

This section is based upon a frequency counter you can construct in the laboratory. Figure 12-34 is a detailed wiring diagram of that frequency counter. This instrument was purposely designed using only components you have already used earlier in the book. This experimental frequency counter is not quite as accurate or stable as commercial units. Its maximum frequency is also limited to 9990 Hz, and its inputs are somewhat primitive.

The purposes for including the experimental frequency counter are as follows:

1. To show how SSI and MSI chips may be used to build digital subsystems and systems.

2. To demonstrate the concepts involved in the design and operation of a frequency counter.

Figure 12-33 is a block diagram of the frequency counter. Most components in the wiring diagram are in the same general position as in the block diagram.

At the lower left in Fig. 12-34, a 60-Hz sine wave is being shaped into a square wave. The 60-Hz signal may come from the secondary of a low-voltage power transformer. The *wave shaping* is done by the 7414 Schmitt trigger inverter. This is the same unit we used earlier in Chap. 7 and in the digital clock to square up a sine wave. Remember that the divide-by-6 counter needs a square-wave input to operate properly.

To the right of the lower 7414 inverter is a *divide-by-6 counter*. Three flip-flops (FF1, FF2, and FF3) and a NAND gate are wired to form the mod-6 counter. The frequency going into the divide-by-6 counter is 60 Hz; the frequency coming out of the counter (at Q of FF3) is 10 Hz. The 10 Hz is fed into the 7493 IC wired as a decade or divide-by-10 counter.

Figure 12-34 shows that the four outputs from the 7493 counter are NORed together (OR gate and inverter). The four-input NOR gate generates a 1-Hz signal. This 1-Hz signal is called the *count/display waveform*. The count/display waveform is HIGH for exactly 0.1 s and low for 0.9 s. The count/display waveform is fed back into the 7400 control gate. When the counter/display waveform is HIGH for 0.1 s, the unknown frequency passes through the NAND gate on to the clock input of the 10s counter. When the count/display waveform is LOW for 0.9 s, the unknown frequency is blocked from passing through the NAND control gate. It is during the 0.9 s that you may read the frequency off the seven-segment LED displays.

The frequency counter goes through a *reset-count-display* sequence. The *reset* pulse is generated by the five-input AND gate near the lower right in Fig. 12-34. It clears the 10s, 100s, and 1000s counters to zero. The reset (or counter clear pulse) is a very short positive pulse that occurs just before the counting occurs.

Next in the reset-count-display sequence is the *count* or *sampling time*. When the count/display waveform goes HIGH, the control gate is enabled and the unknown frequency passes through the NAND gate to the clock input of the 10s counter. Each pulse during this *sampling time* increments the 10s counter. When the 10s counter goes from 9 to 10, it carries the 1 to the 100s counter. After 0.1 s the count/display waveform goes LOW. This is the end of the sampling time. You will notice that the unknown frequency that was sampled causes the frequency to increase by 10s.

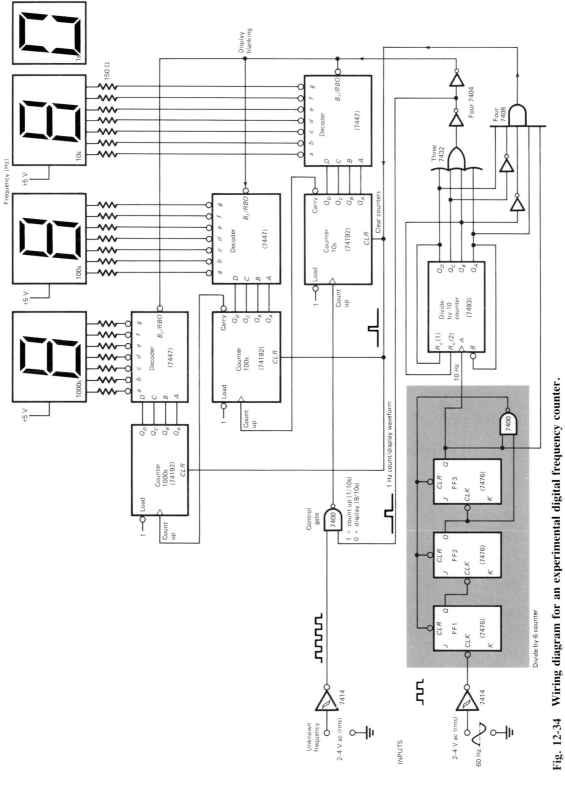

Fig. 12-34 Wiring diagram for an experimental digital frequency counter.

Last in the reset-count-display sequence is the *display time*. When the count/display waveform goes LOW the control gate is disabled. It is during this time that a stable frequency display may be read from the LEDs. Notice that an extra 1s display has been added in Fig. 12-34 to remind you that a 0 must be added to the right of the three active displays for a readout in hertz.

To make the displays look better, *display blanking* occurs during the count time of the reset-count-display sequence. The displays light normally with a stable readout during the display time. The display blanking waveform is a 0.1 s negative pulse generated by a 7404 inverter off the count/display waveform line. It causes the three displays to blank out for 0.1 s during the count time. The blanking does cause the displays to flicker. This problem could be cured by the use of latches to hold data on the inputs of the decoders.

For the most part commercial frequency counters operate just like the one in Fig. 12-34. Commercial counters usually have more displays and read out in kilohertz and megahertz. The experimental frequency counter needs an input signal of about 3 to 8 V to make it operate. Commercial counters usually have an amplifier circuit before the first wave-shaping circuit to amplify weaker signals to the proper level. Overvoltage protection is also provided with a zener diode. To get rid of the blinking of the display, commercial counters usually use a slightly different method of storing and displaying the contents of the counters. We used the power-line frequency of 60 Hz as our known frequency. Commercial frequency counters usually use an accurate high-frequency crystal oscillator to generate their known frequency.

Some of the important specifications of commercial frequency counters are the *frequency range, input sensitivity, input impedance, input protection, accuracy, gate intervals,* and *display time*. For instance, the frequency counter pictured in Fig. 1-7(*c*) has a frequency range from a low of 5 Hz to a high of 500 MHz. Its input sensitivity ranges from about 25 to 250 mV depending which range is selected. It has selectable 1-MΩ or 50-Ω input impedances. It uses a 3.579545-MHz crystal oscillator with an accuracy of ±4 ppm (parts per million). The gate intervals (count time) are switch selectable at 0.1 and 1.0 s.

Self-Test

Supply the missing word(s) or number in each statement.

58. Refer to Fig. 12-34. FF1, FF2, FF3, and the NAND gate form a(n) _____ counter.
59. Refer to Fig. 12-34. The count time for this frequency counter is _____ s, while the display time is _____ s.
60. Refer to Fig. 12-34. The sampling time of the frequency counter is also called the _____ (count, display) time.
61. Refer to Fig. 12-34. The five-input AND gate generates a counter clear or _____ pulse. This is a _____ (negative, positive) pulse.
62. Refer to Fig. 12-34. The display blanking pulse is generated during the _____ (count, display) time. This is a _____ (negative, positive) pulse.
63. Refer to Fig. 12-34. The 7414 IC is called a hex _____ _____ inverter. The 7414 inverters are used for _____ shaping in this circuit.
64. Refer to Fig. 12-34. Each pulse from the unknown frequency that reaches the counter increases the frequency reading by _____ (1, 10, 100) Hz.

12-16 LCD TIMER WITH ALARM

Most microwave ovens and kitchen stoves feature at least one timer with an alarm. Older appliances used mechanical timers, but modern microwaves and ranges feature electronic timers using digital circuitry. The concept of a timer system is sketched in Fig. 12-35(*a*). In this system, the keypad is the input and both the digital display and alarm buzzer are the output devices. The processing and storage of data occur within the digital circuits block in Fig. 12-35(*a*).

A somewhat more detailed block diagram of a digital timer is shown in Fig. 12-35(*b*). The digital circuits block has been subdivided into four blocks. They are the *time-base clock*, the *self-stopping down counter*, the *latch/decoder/driver*, and the *magnitude comparator*. The *input controls* block presets the time held in the down counters. The time base is an astable multivibrator which generates a known frequency. In this case, the signal is a 1-Hz square wave. The accuracy of the entire timer depends on the accuracy of the time-base clock. Activating the *start* input control causes the

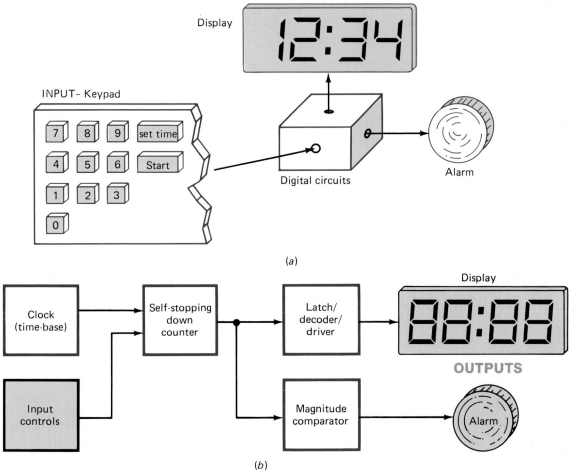

Display

INPUT– Keypad

7	8	9	set time
4	5	6	Start
1	2	3
0

Digital circuits

Alarm

(a)

Clock (time-base) → Self-stopping down counter → Latch/decoder/driver → Display

Input controls

Magnitude comparator → Alarm

OUTPUTS

(b)

Fig. 12-35   Digital timer system. (*a*) **Concept sketch of timer with alarm.** (*b*) **Simple block diagram of timer with alarm.**

down counter to decrement. Each lower number is latched and decoded by the next block. This block also drives the display.

The illustrations in Fig. 12-35 might be preliminary sketches used by a designer in visualizing a timer system. The design might next decide on what type of input, output, and processing technologies to use to implement the system.

A somewhat more detailed block diagram of a digital electronic timer system is drawn in Fig. 12-36 on page 288. The designer decided to use a two-digit liquid-crystal display along with low-power CMOS ICs. This system was designed with your lab trainers in mind, so the inputs are logic switches to simplify the input section. The designer decided on seconds as the time interval. Notice that each block roughly corresponds to an MSI digital IC or input/output device. A wiring diagram could then be developed from the detailed block diagram in Fig. 12-36.

The block diagram in Fig. 12-36 on the next page represents an experimental LCD timer with alarm that you might construct in the laboratory. The timer is operated as follows:

1. Set the load/start control to 0 (load mode).
2. Load the 1s counter by setting a BCD number using the top four switches.
3. Load the 10s counter by setting a BCD number using the bottom four switches.
4. A two-digit number should now be displayed on the LCD.
5. Move the load/start control to 1 (start down count mode).

The timer will start counting downward in seconds. The LCD shows the time remaining before the alarm sounds. When both counters reach zero, the LCD reads 00 and the alarm will sound. The final step is to disconnect the power to the circuit to turn off the alarm.

A wiring diagram for the experimental LCD timer circuit is detailed in Fig. 12-37

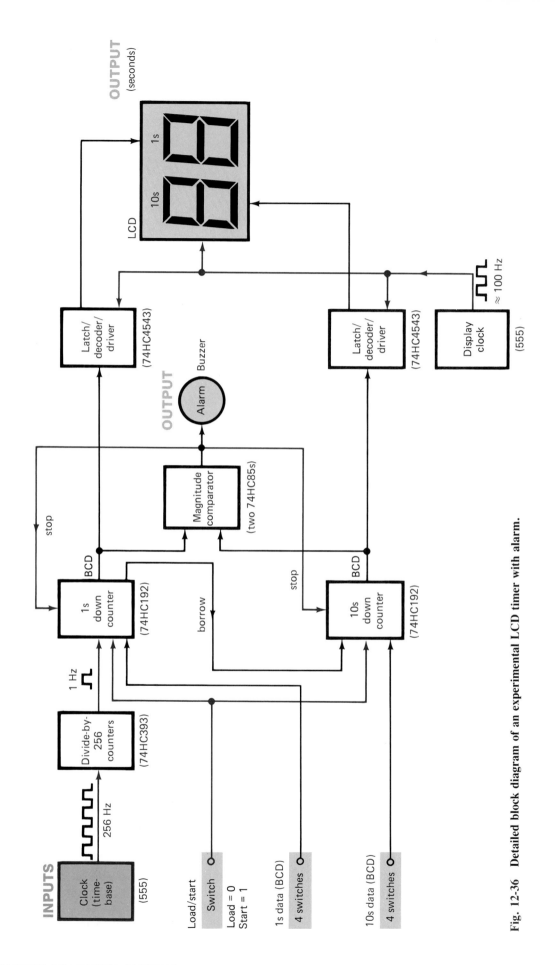

Fig. 12-36   Detailed block diagram of an experimental LCD timer with alarm.

(see pages 290 and 291). Notice that each IC is placed in the same relative position on the wiring diagram as in the block diagram in Fig. 12-36.

Detailed operation of the LCD timer circuit in Figs. 12-36 and 12-37 follows.

TIME BASE

The *time-base clock* is a 555 timer IC wired as a free-running MV. It is designed to generate a 256-Hz square wave. The time-base clock in this experimental timer is not very accurate or stable. It can be calibrated by adjusting the value of resistor R_1. The nominal value of R_1 should be about 20 Ω.

The second part of the time base is the *divide-by-256-counter* block. The function of this block is to output a 1-Hz signal. The divide-by-256-counter block is actually two 4-bit counters wired together. Figure 12-38 on page 292 shows the two 4-bit units wired as divide-by-16 counters. Note that the 1A and 2A inputs are clock inputs and only the Q_D outputs are used. The first divide-by-16 counter divides the frequency from 256 to 16 Hz (256/16 = 16 Hz). The second counter divides the frequency down to the required 1-Hz output (16/16 = 1 Hz).

SELF-STOPPING DOWN COUNTERS

The two 74HC192 decade counters are the 75HCXXX series equivalent to the 74192 TTL IC detailed in Chap. 8, Section 8-7. When the load inputs to the 74HC192 counters are activated by a LOW, data at the data inputs (A, B, C, D) is immediately transferred into the counter's flip-flops. It then appears at the outputs of the counter (Q_A, Q_B, Q_C, Q_D). The data loaded should be in BCD (binary-coded decimal) form. When the load/start control goes HIGH, the 1-Hz signal activates the count down input of the 1s counter. The count decreases by 1 on each L-to-H transition of the clock pulse. The *borrow out* output of the 1s down counter goes from L-to-H when the 1s counter goes from 0 to 9. This decrements the 10s counter. The down counters are actually wired as a self-stopping down counter because of the *counter stop line* fed back to the CLR input of both 74HC192 counters. When this line goes HIGH, both counters stop at 0000.

8-BIT MAGNITUDE COMPARATOR

The 74HC85 4-bit comparators are shown cascaded in Fig. 12-37 to form an 8-bit-magnitude comparator. Their purpose in this circuit is to detect when the outputs of the counters reach 0000 0000_{BCD}. When both counters reach zero, the output of the 8-bit magnitude comparator ($A = B_{out}$) goes HIGH. This serves two purposes. First, it stops both 74HC192 counters at 0000. Second, the HIGH at the output of the comparator turns transistor Q_1 on. This allows current to flow up through the transistor, sounding the buzzer. The diode across the buzzer suppresses transient voltages that may be generated by the buzzer.

DECODER/DRIVER

The two 74HC4543 ICs used in the timer circuit serve three purposes. The functions of the 74HC4543 IC are summarized in Fig. 12-39. The latch enable (*LE*) input is permanently tied HIGH in the timer circuit (Fig. 12-37), which disables the latches. The BCD data flows through the latch to the BCD-to-seven-segment decoder. The decoder translates the BCD input to seven-segment code. Finally, the driver circuitry in the 74HC4543 chip energizes the correct segments on the LCD.

The *display clock* shown at the lower right in Fig. 12-37 generates a 100-Hz square wave. This is sent to the common (back plane) connection on the LCD and the Ph inputs of the 74HC4543 ICs. The LCD driver in the 74HC4543 chip sends inverted or 180° out-of-phase signals to the LCD segments that are to be activated. Segments that are not activated receive an in-phase square-wave signal from the LCD driver section of the 74HC4543 IC.

Self-Test

Answer the following questions.

65. Refer to Fig. 12-37. The accuracy of the entire timer depends on the frequency generated by the _____ clock.
66. Refer to Fig. 12-37. The initial numbers to be loaded into the counters in the timer must be entered in _____ (BCD, binary, decimal) form.
67. Refer to Fig. 12-37. What two components are considered output devices in this timer circuit?
68. Refer to Fig. 12-37. When the count reaches zero, the output ($A = B_{out}$) of the 8-bit magnitude comparator goes _____ (HIGH, LOW). This causes the counter stop line to go _____

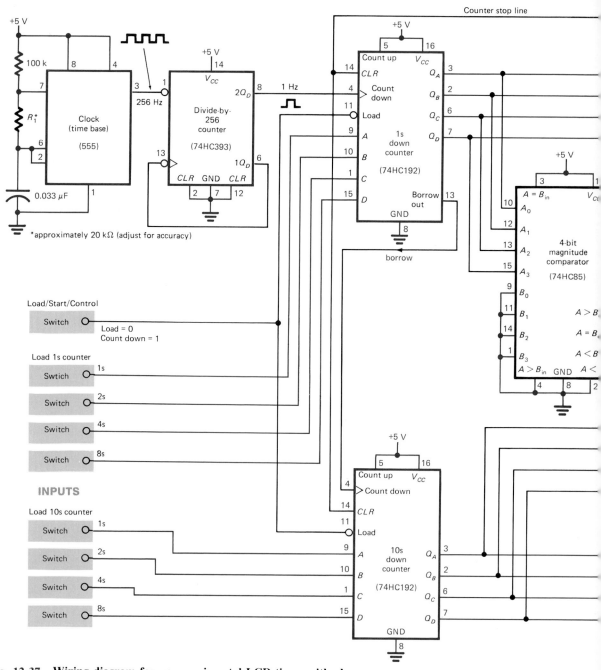

Fig. 12-37 Wiring diagram for an experimental LCD timer with alarm.

(HIGH, LOW), stopping the counters. This also turns _____ (off, on) the transistor causing it to conduct electricity and sound the buzzer.

69. Refer to Fig. 12-37. The driver section of the 74HC4543 IC sends an _____ (in-phase, out-of-phase) square-wave signal to the LCD segments that are to be activated.

70. Refer to Fig. 12-37. The display clock sends a 100-Hz square-wave signal to the _____ inputs of both 74HC4543 ICs

and the _____ connection on the LCD.

71. The timer circuit in Fig. 12-37 is calibrated in (minutes, seconds, tenths of a second).

12-17 ELECTRONIC GAMES

Electronics has been a popular hobby for more than a half century. A favorite task for most electronic hobbyists, young and old, is to construct electronic games. Electronic games and

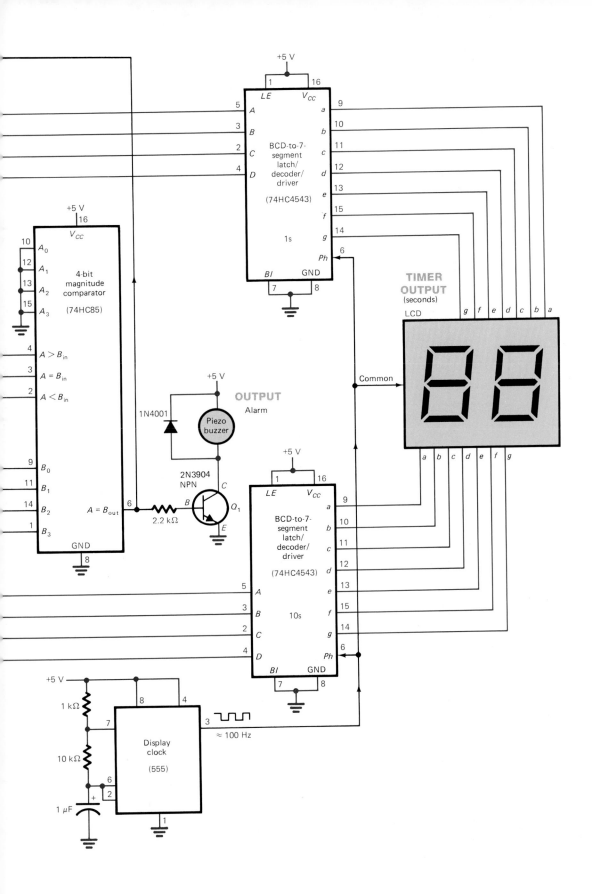

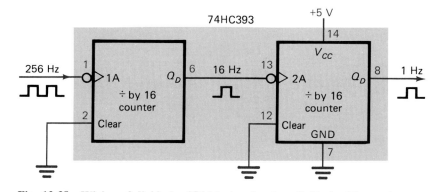

Fig. 12-38 Wiring of divide-by-256 block using two divide-by-16 counters.

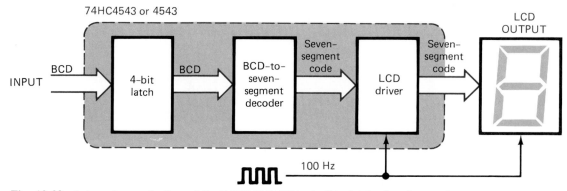

Fig. 12-39 Internal organization of the 74HC4543 IC including latch, decoder, and driver sections.

toys are also very popular with students studying electronics in high schools, technical schools, and colleges.

Electronic games may be classified as simple self-contained, computer, arcade, or TV games. The simple self-contained type are the games and toys most often constructed by students and hobbyists. Several simple electronic games using SSI and MSI digital ICs will be surveyed in this section.

A block diagram of a simple *digital dice game* is sketched in Fig. 12-40. When the push button is pressed, a signal from the clock is sent to the counter. The counter is wired to have a counting sequence of 1, 2, 3, 4, 5, 6, 1, 2, 3, etc. The binary output from the counter

is translated to seven-segment code by the decoder block. The decoder block also contains a seven-segment LED display driver. The output device in this circuit is an LED display. When the push-button switch is opened, the counter *stops at a random number* from 1 through 6. This simulates the roll of a single die. The binary number stored in the counter is decoded and shown as a decimal number on the display. This circuit could be doubled to simulate the rolling of a pair of dice.

A wiring diagram for the digital dice game is detailed in Fig. 12-41. Pressing the input switch causes the counter to sequence through the binary numbers 001, 010, 011, 100, 101, 110, 001, 010, 011, etc. When the switch is opened the

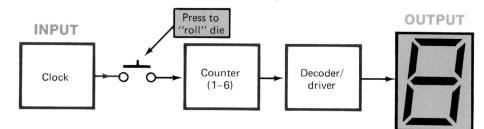

Fig. 12-40 Simple block diagram of a digital dice game.

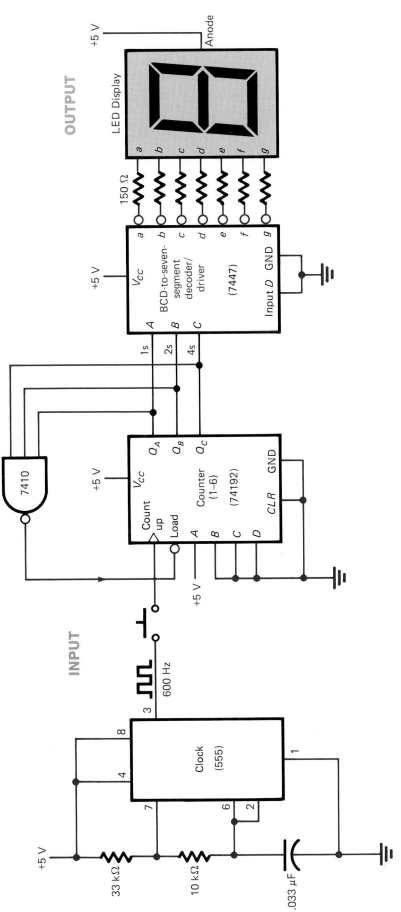

Fig. 12-41 Wiring diagram for an experimental digital dice game using TTL ICs.

last binary count is stored in the flip-flops of the 74192 counter. It is decoded by the 7447 IC and lights the seven-segment LED display.

The 555 timer IC is wired as an astable MV in Fig. 12-41. It generates a 600-Hz square-wave signal.

The 74192 IC is wired as a mod-6 (1 to 6) up counter. The three-input NAND gate is activated when the count reaches binary 111. The LOW signal from the NAND gate loads the next number in the count sequence, which is binary 001. It should be noted that the three outputs of the counter (Q_A, Q_B, Q_C) all go HIGH for only an extremely short time (less than a microsecond) while the counter is being loaded with 0001. Therefore, the temporary count of binary 111 never appears as a 7 on the LED display.

The 7447 BCD-to-seven-segment decoder chip translates the binary inputs (A, B, C) to seven-segment code. The 7447 IC drives the LED segments with active LOW outputs (a to g). The seven 150-Ω resistors serve to limit the current flow through the LEDs to a safe level. Note that the seven-segment LED display used in Fig. 12-41 is a common-anode type.

The digital dice game featured in Fig. 12-41 used TTL ICs with an unrealistic seven-segment display. A more realistic dice simulation is implemented in the circuit shown in Figs. 12-42 and 12-43.

A block diagram for a second digital dice game is sketched in Fig. 12-42(a). This unit uses individual LEDs for the output device.

Depressing the push button in Fig. 12-42(a) causes the clock to generate a square-wave signal. This signal causes the down counter to cycle through the count sequence 6, 5, 4, 3, 2, 1, 6, 5, 4, etc. The logic block lights the proper LEDs to represent various decimal counts. The LED pattern for each possible decimal output is diagramed in Fig. 12-42(b).

The wiring diagram for the second digital dice game is detailed in Fig. 12-43. The circuit features the use of 4000 series CMOS ICs and a 12-V dc power supply. The push-button switch on the left is the input device, while the LEDs ($D1$ to $D7$) on the right form the output. Physically, the LEDs are arranged as shown at the lower right in Fig. 12-43.

When the "roll dice" switch is closed, the two NAND Schmitt trigger gates at the left in Fig. 12-43 produce a 100-Hz square-wave signal. The two NAND gates and associated resistors and capacitors are wired to form an astable MV. The 100-Hz signal is fed into the clock input of the *4029 presettable binary/ decade up/down counter.* In this circuit, the 4029 IC is wired as a down counter whose outputs produce binary 110, 101, 100, 011, 010, 001, 110, 101, 100, etc.

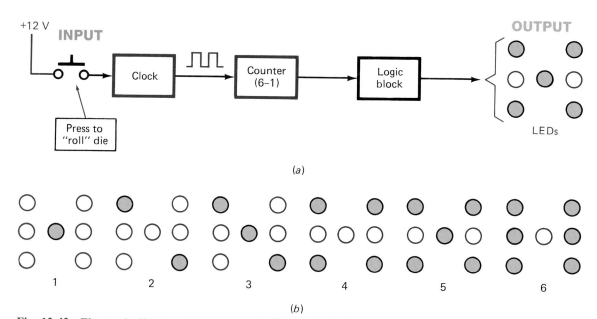

(a)

(b)

Fig. 12-42 Electronic dice simulation game. (*a*) **Simple block diagram.** (*b*) **LED patterns used for representing dice rolls of 1 through 6.**

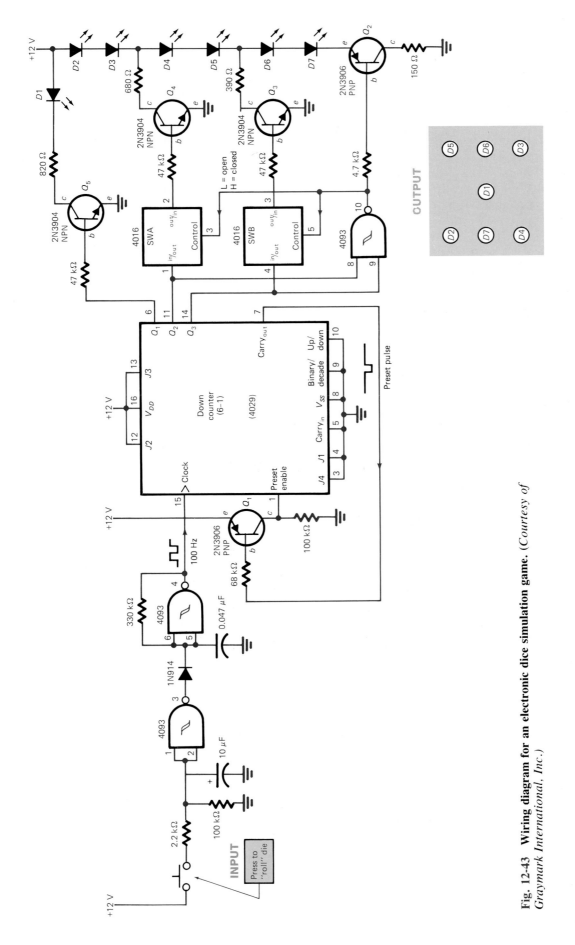

Fig. 12-43 Wiring diagram for an electronic dice simulation game. (*Courtesy of Graymark International, Inc.*)

Consider the situation when the down counter in Fig. 12-43 reaches binary 001. On the next LOW-to-HIGH transition of the clock pulse, the *carry out* output (pin 7) of the 4029 IC drops LOW. This signal is fed back and turns on transistor Q_1. This causes the *preset enable* input of the 4029 counter to go HIGH. When the preset enable input goes HIGH, the data at inputs J4, J3, J2, and J1 (the "jam inputs") is asynchronously loaded into the counter's flip-flops. In this example, binary 0110 is loaded on the preset pulse. Once the flip-flops have been loaded, the carry out pin goes back HIGH and transistor Q_1 turns off.

The final sections of the dice game at the right in Fig. 12-43 contain many components. The table in Fig. 12-44 below will help explain the complexities of the *logic* and *output* sections of this dice game.

The input section of the table in Fig. 12-44 indicates the logic levels present at the output of the 4029 counter. The top line of the table shows a binary 110 (HHL) stored in the 4s, 2s, and 1s flip-flops of the counter. The middle column of the table lists only the components that are activated to light the proper LEDs.

Consider the first line of the table in Fig. 12-44. The output of the NAND gate goes LOW, which turns on the PNP transistor Q_2.

The transistor conducts and all six LEDs (D2 to D7) on the right in Fig. 12-43 light. This would simulate a dice roll of 6.

Consider line 2 of the table in Fig. 12-44. The binary data equals 101 (HLH). The HIGH on the 1s line (pin 6) turns on transistor Q_5. Transistor Q_5 conducts and lights LED D1. The output of the NAND gate is HIGH, which causes both bilateral switches to be in the closed condition (low impedance from in/out to out/in). The bilateral switches pass the logic level from the 2s and 4s lines to the base of transistors Q_4 and Q_3. Transistor Q_3 is turned on by the HIGH and conducts. Light-emitting diodes D2, D3, D4, and D5 light. A decimal 5 is represented when these five LEDs (D1 to D5) are lit.

You may look over the remaining lines of the table in Fig. 12-44 to determine the operation of the logic and output sections of this CMOS digital dice game.

The 4016 IC used in Fig. 12-43 is described by the manufacturer as a *quad bilateral switch*. It is an electronically operated SPST switch. A HIGH at the control input of the 4016 bilateral switch causes it to be in the "closed" or ON" position. In the "closed" position the internal resistance from in/out to out/in terminals is quite low (400 Ω typical). A LOW at the control input of the bilateral switch causes it to be in the open or OFF position. The 4016 IC acts

INPUTS			ACTIVE COMPONENTS	OUTPUT	
4s (pin 14)	2s (pin 11)	1s (pin 6)		LEDs LIT	DECIMAL
H	H	L	NAND output LOW transistor Q_2 turned on	D2, D3, D4, D5, D6, D7	6
H	L	H	transistor Q_5 turned on bilateral switch SWB closed transistor Q_3 turned on	D1 D2, D3, D4, D5	5
H	L	L	bilateral switch SWB closed transistor Q_3 turned on	D2, D3, D4, D5	4
L	H	H	transistor Q_5 turned on bilateral switch SWA closed transistor Q_4 turned on	D1 D2, D3	3
L	H	L	bilateral switch SWA closed transistor Q_4 turned on	D2, D3	2
L	L	H	transistor Q_5 turned on	D1	1

Fig. 12-44 Explaining the logic and output sections of the electronic dice simulation game.

like an open switch when the control is LOW. Unlike a gate, a bilateral switch can pass data in either direction. It can pass either dc or ac signals. A bilateral switch is also referred to as a *transmission gate*.

Self-Test

Answer the following questions.

72. Refer to Fig. 12-41. The 555 timer is wired as a(n) _____ multivibrator in this digital system.
73. Refer to Fig. 12-41. When the 74192 IC increases its count from 110 to 111, the output of the NAND gate goes _____ (HIGH, LOW) immediately loading _____ (binary) into the counter.
74. Refer to Fig. 12-41. List the possible digits that can appear on the seven-segment LED display when the input switch is released.
75. Refer to Fig. 12-43. Two gates packaged in the _____ (4016, 4029, 4093) IC are wired as a free-running MV in this digital dice game circuit.
76. Refer to Fig. 12-43. List the binary counting sequence of the 4029 IC in this circuit.
77. Refer to Fig. 12-43. When the output of the counter is binary 001, only $D1$ lights because only transistor _____ is turned on and conducts.
78. Refer to Fig. 12-43. When the output of the counter is binary 010, LED(s) _____ light because the bilateral switches are _____ (closed, open) and transistor _____ is turned on grounding the cathode of light-emitting diode of D3.
79. A(n) _____ is an IC device available in CMOS that acts much like a SPST switch that can conduct either ac or dc signals.

SUMMARY

1. An assembly of digital subsystems connected correctly forms a digital system.
2. Digital systems have six common elements: input, transmission, storage, processing, control, and output.
3. Manufacturers produce ICs that are classified as small-, medium-, large-, and very large-scale integrations.
4. A calculator is a complex digital system often based upon a single LSI IC. A typical modern calculator might use a CMOS IC, LCD, and be battery- and/or solar-powered.
5. The computer is one of the most complex digital systems. It is unique because of its vast size, high speed, and stored program.
6. The microcomputer is slower and less expensive than its larger counterparts. The microcomputer is a microprocessor-based digital system.
7. The microcomputer makes extensive use of both ROM and RAM for internal storage. Both floppy and hard disks are widely used for secondary bulk storage. Microcomputers support many peripheral input and output devices.
8. Microprocessing unit instructions are composed of the operation and operand parts.

The MPU follows a fetch-decode-execute sequence when running a program.
9. Combinational logic gates can be used for microcomputer address decoding.
10. Three-state devices, such as buffers, must be used when several memories and microprocessors transfer information over a common data bus.
11. Multiplexers and demultiplexers can be used for data transmission. More complex UARTs may also be used for serial data transmission.
12. Data transmission can be either serial or parallel. Various interface ICs are available for sending and receiving parallel and serial data.
13. Errors occurring during data transmission can be detected using parity bits.
14. A digital clock and a digital frequency counter are two closely related digital systems. Both make extensive use of counters.
15. Many LSI digital clock chips are available. Most clock ICs need other components to produce a working digital clock.
16. Multiplexing is a commonly used method of driving seven-segment LED displays.
17. All digital systems are basically constructed from AND gates, OR gates, and inverters.

18. A frequency counter is an instrument that accurately counts input pulses in a given time interval and displays it in digital form. It constantly cycles through a reset-count-display sequence.

19. Block diagrams communicate the organization of a digital system. The most detailed block diagrams break the system down to the chip level.

20. Electronic games are popular construction projects. Many are simulations of older games like throwing dice.

CHAPTER REVIEW QUESTIONS

Answer the following questions.

12-1. List at least five common devices that are considered digital systems.

12-2. List at least four devices you have used that are considered digital subsystems.

12-3. List the six elements found in most digital systems.

12-4. What do the following letters stand for when referring to ICs?

    a. IC   **e.** VLSI

    b. SSI   **f.** ROM

    c. MSI   **g.** RAM

    d. LSI   **h.** PROM

12-5. The term "chip" usually is taken to mean a(n) _____ (IC, sliver of plastic) in digital electronics.

12-6. What do the following letters stand for when referring to organizations?

    a. IEEE

    b. JEDEC

12-7. Inexpensive pocket calculators are usually based on an _____ (LSI, MSI).

12-8. The organization of the circuits within a calculator is called the _____ (architecture, dimensions) of the IC.

12-9. A calculator is *timed* by the _____ (internal clock, pressing of keys on the keyboard).

12-10. A simple calculator contains a rather large _____ (RAM, ROM).

12-11. _____ (Calculators, Computers, Both calculators and computers) contain a controller or control circuitry.

12-12. A _____ (computer, digital wristwatch) is usually based upon a single LSI IC.

12-13. Draw a diagram of the organization of the five main parts of a computer. Show the flow of *program* information and *data* through the system.

12-14. The CPU of a computer contains what three sections?

12-15. The _____ (ALU, MUX) section of a computer performs calculations and logic functions.

12-16. The most complex digital system is a _____ (computer, digital multimeter).

12-17. An IC called a _____ is the CPU of a microcomputer.

12-18. Refer to Fig. 12-5. The parts of a microcomputer system are connected by control lines, an _____ bus, and a two-way _____ _____ .

12-19. The input-store-print operation shown in Fig. 12-7 required three instructions which use _____ bytes of program memory.

12-20. A gating circuit called a(n) _____ decoder is used to select one of many memory devices for sending or retrieving data via the data bus.

12-21. Classify these microcomputer peripheral devices as input, output, storage, or input/output units.
  a. CRT monitor
  b. Floppy disk drive
  c. Keyboard
  d. Mouse
  e. Modem
  f. Laser printer
  g. Hard disk drive
  h. Plotter

12-22. Microcomputer memory addresses are commonly listed in _____ (Gray code, hexadecimal).

12-23. What do the following letters stand for when referring to a microcomputer system?
  a. CPU
  b. PIA
  c. PPI
  d. UART

12-24. The baud rate is the number of _____ per second being transmitted serially through a data link.

12-25. The IEEE-488 standard is a common _____ (parallel, serial) interface standard for the data link between a computer and scientific instrumentation.

12-26. Draw the logic symbol and truth table for a three-state buffer.

12-27. What do the following letters stand for?
  a. MUX
  b. DEMUX

12-28. A MUX/DEMUX system converts parallel input data to _____ (asynchronous, serial) data for transmission.

12-29. A MUX/DEMUX system operates somewhat like two _____ (rotary, three-way) switches.

12-30. Errors in transmission can be detected by using a _____ (parity, 16-word) bit.

12-31. An _____ (AND, XOR) gate can detect an odd number of 1s at its input.

12-32. The _____ (Gray, Hamming) code is an error-correcting code.

12-33. The adder/subtractor in Fig. 12-20 contains which of the elements of the system listed in Fig. 12-1?

12-34. A digital clock system is closely related to a _____ (computer, frequency counter) system.

12-35. A digital clock makes extensive use of _____ (counter, shift register) subsystems.

12-36. A known frequency is the main input to a digital _____ (clock, frequency counter) system.

12-37. Counters are used for counting upward and _____ (shifting data, storing data) in the digital clock system.

12-38. Most LSI digital clock chips are manufactured using _____ (bipolar, MOS) technology.

12-39. The National Semiconductor MM5314 clock chip _____ (directly drives, multiplexes) the output displays.

12-40. The multiplex oscillator's frequency in Fig. 12-25(a) is set by _____ (connecting an external capacitor and resistor to the correct IC pins; the factory and cannot be changed).

12-41. In practice, the segment driver block shown in Fig. 12-27(b) may consist of _____ (a VLSI chip; seven transistors with associated resistors).

12-42. The multiplexed displays in Fig. 12-29 are _____ (all turned on and off together to save power; turned on and off one at a time in rapid succession).

12-43. The *known frequency* entering the MM5314 clock chip in Fig. 12-29 is _____ Hz. This frequency comes from the _____ (oscillator, transformer).

12-44. Counters are used for counting upward and _____ (counting downward, dividing frequency) in a digital frequency counter.

12-45. The three J-K flip-flops (FF1, FF2, FF3) and the NAND gate in Fig. 12-34 function as a _____ (down counter, frequency divider).

12-46. The 7408 AND gate in Fig. 12-34 serves to _____ (clear, inhibit) the counters.

12-47. The frequency counter in Fig. 12-34 counts from a low of _____ Hz to a high of _____ Hz.

12-48. What IC(s) are being used as wave-shaping circuits in the frequency counter in Fig. 12-34?

12-49. Refer to Fig. 12-34. The unknown frequency is allowed to pass through the control gate for 0.1 s when the count/display waveform goes _____ (HIGH, LOW).

12-50. Refer to Fig. 12-34. The displays are blanked during the _____ portion of the count/display waveform.

12-51. Refer to Fig.12-37. List two ICs that form the time base clock section of the LCD timer.

12-52. Refer to Fig. 12-37. List the IC(s) that detect when the count of the timer reaches 00.

12-53. Refer to Fig. 12-37. When the count on the timer reaches 00, the output of the magnitude comparator goes _____ (HIGH, LOW). This turns on transistor Q_1, sounding the alarm, and activates the _____ _____ line.

12-54. Refer to Fig. 12-37. When the LCD reads 88, the signals on all of the lines from the 74HC4543 drivers to the displays are _____ (in phase, 180° out of phase) with the signal at the output of the display clock.

12-55. Refer to Fig. 12-37. The accuracy of the entire timer depends on the accuracy of the _____ clock.

12-56. Refer to Fig. 12-35(*b*). A commercial timer would probably use a(n) _____ -controlled oscillator (astable MV) for the time-base clock to assure maximum accuracy.

12-57. Refer to Fig. 12-37. The 74HC4543 ICs have the _____ (decoder, driver, latch) section of the chip disabled in this circuit.

12-58. Refer to Fig. 12-37. The divide-by-256 block consists of _____ (two, four, 256) divide-by-16 counters.

12-59. Refer to Fig. 12-41. When the push button is _____ (closed, opened), the display will stop and indicate a random number from 1 to _____ (number) simulating the roll of a single die.

12-60. Refer to Fig. 12-41. This circuit uses _____ (CMOS, TTL) ICs.

12-61. Refer to Fig. 12-41. If a "1" shows on the seven-segment LED display, then outputs _____ (letters) of the 7447 IC are activated with a _____ (HIGH, LOW).

12-62. Refer to Fig. 12-41. When the 74192 IC tries to count upward from 110 to 111, the NAND gate is activated driving the load input _____ (HIGH, LOW). This immediately loads _____ (binary number) into the counter's flip-flops.

12-63. Refer to Fig. 12-43. Two _____ trigger NAND gates and associated resistors and capacitors form the clock section in this digital dice game.

12-64. Refer to Fig. 12-43. Grounding pin 10 of the 4029 IC converts this unit to a(n) _____ (down, up) counter.

12-65. Refer to Fig. 12-43. When the preset pulse goes _____ (HIGH, LOW), transistor Q_1 is turned on and the preset enable input to the 4029 counter is activated with a _____ (HIGH, LOW).

12-66. Refer to Fig. 12-43. When the counter's outputs are 110 (HHL), LEDs _____ light. This is caused by the output of the NAND gate going _____ (HIGH, LOW) and transistor Q_2 being turned _____ (on, off).

12-67. Refer to Fig. 12-43. When the counter's outputs are 011 (LHH), LEDs _____ light. This is caused by transistor Q_5 being turned on. Also bilateral switch SWA is _____ (closed, open) and transistor _____ (Q_3, Q_4) is turned on.

12-68. A bilateral switch is also called a _____ gate.

Answers to Self-Tests

1. control
2. input
3. 12 to 100
4. 1000
5. LSI
6. output
7. input
8. 1. power supply problems
   2. keyboard problems
9. peripheral
10. size, stored program, speed
11. Program information
12. data
13. programs
14. address
15. modem
16. CRT monitor or TV
17. mouse
18. operation, operand
19. 100, 101
20. MPU (microprocessor)
21. decode-execute
22. sequential
23. address decoder
24. three-state buffers or tristate buffers
25. 0, 8
26. in its high-impedance state
27. multiplexer, demultiplexer
28. 1111
29. PIA (peripheral-interface adapter)
30. UART (universal asynchronous receiver-transmitter)
31. baud
32. serial
33. even
34. XOR, XOR
35. encoder
36. 4-bit adder
37. frequency divider, count accumulator
38. wave-shaping
39. wave-shaping
40. control gate (start/stop control)
41. MOS
42. 60
43. prescale
44. 11 to 19
45. multiplexing
46. digit drivers
47. segment drivers
48. 12
49. $C3$ and $R4$
50. V_{DD}
51. 1.0
52. LOW
53. Schmitt trigger inverters
54. counters
55. 0.1
56. wave-shaping
57. storing
58. divide-by-6
59. 0.1, 0.9
60. count
61. reset, positive
62. count, negative
63. Schmitt trigger, wave
64. 10
65. time-base
66. BCD
67. piezo buzzer, LCD (liquid-crystal display)
68. HIGH, HIGH, on
69. out-of-phase
70. common (back plane)
71. seconds
72. LOW
73. 0001
74. 1, 2, 3, 4, 5, 6
75. 4093
76. 110, 101, 100, 011, 010, 001
77. Q_5
78. $D2$ and $D3$, closed, Q_4
79. bilateral switch or transmission gate

CHAPTER 13

Connecting with Analog Devices

To this point in our study, most data entering or leaving a digital system has been digital information. Many digital systems, however, have analog *inputs that vary continuously between two voltage levels. In this chapter we shall discuss the interfacing of analog devices to digital systems.*

Most real-world information is analog. For instance, it was mentioned in Chap. 1 that time, speed, weight, pressure, light intensity, and position measurements are all *analog in nature*.

The digital system in Fig. 13-1 has an analog input. The voltage varies continuously from 0 to 3 V. The *encoder* is a special device that converts the analog signal to digital information. The encoder is called an *analog-to-digital converter* or, for short, an *A/D converter*. The A/D converter, then, converts analog information to digital data.

The digital system diagramed in Fig. 13-1 also has a *decoder*. This decoder is a special type: it converts the digital information from the digital processing unit to an analog output. For instance, the analog output may be a continuous voltage change from 0 to 3 V. We call this decoder a *digital-to-analog converter* or, for short, a *D/A converter*. The D/A converter, then, decodes digital information to analog form.

The entire system in Fig. 13-1 might be called a *hybrid* system because it contains both digital and analog devices. The encoders and decoders that convert from analog to digital and digital to analog are called *interface devices* by engineers and technicians. The word "interface" is generally used when referring to a device or circuit that converts from one mode of operation to another. In this case we are converting between analog and digital data.

Note that the input block in Fig. 13-1 refers to an analog voltage ranging from 0 to 3 V. This voltage could be produced by a transducer. A *transducer* is defined as a device that converts one form of energy to another. For instance, a photocell could be used as an input transducer to give a voltage proportional to light intensity. In this example, light energy is being converted into electrical energy by the photocell. Other transducers might include microphones, speakers, strain gauges, photoresistive cells, temperature sensors, potentiometers, and r/min pickup coils.

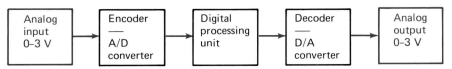

Fig. 13-1 A digital system with analog input and analog output.

13-1 D/A CONVERSION

Refer to the D/A converter in Fig. 13-1. Let us suppose we want to convert the binary from the processing unit to a 0- to 3-V output. As with any decoder, we must first set up a truth table of all the possible situations. Table 13-1 shows four inputs (D, C, B, A) into the D/A converter. The inputs are in binary form. Each 1 is about +3 to +5 V. Each 0 is about 0 V. The outputs are shown as voltages in the rightmost column in Table 13-1. According to the table, if binary 0000 appears at the input of the D/A converter, the output is 0 V. If binary 0001 is the input, the output is 0.2 V. If binary 0010 appears at the input, then the output is 0.4 V. Notice that for each row you progress downward in Table 13-1, the analog output increases by 0.2 V.

Table 13-1   Truth Table for D/A Converter

	Digital Input				Analog Output
	D	C	B	A	Volts
Row 1	0	0	0	0	0
Row 2	0	0	0	1	0.2
Row 3	0	0	1	0	0.4
Row 4	0	0	1	1	0.6
Row 5	0	1	0	0	0.8
Row 6	0	1	0	1	1.0
Row 7	0	1	1	0	1.2
Row 8	0	1	1	1	1.4
Row 9	1	0	0	0	1.6
Row 10	1	0	0	1	1.8
Row 11	1	0	1	0	2.0
Row 12	1	0	1	1	2.2
Row 13	1	1	0	0	2.4
Row 14	1	1	0	1	2.6
Row 15	1	1	1	0	2.8
Row 16	1	1	1	1	3.0

A block diagram of a D/A converter is shown in Fig. 13-2. The digital inputs (D, C, B, A) are at the left. The decoder consists of two sections: the *resistor network* and the *summing amplifier*. The output is shown as a voltage reading on the voltmeter at the right.

The resistor network in Fig. 13-2 must take into account that a 1 at input B is worth twice as much as a 1 at input A. Also, a 1 at input C is worth four times as much as 1 at input A. Several arrangements of resistors are used to do this job. These circuits are called *resistive ladder networks*.

The summing amplifier in Fig. 13-2 takes the output voltage from the resistor network and amplifies it the proper amount to get the voltages shown in the rightmost column of Table 13-1. The summing amplifier typically uses an IC unit called an *operational amplifier*. An operational amplifier is often simply called an *op amp*. The summing amplifier is also called a *scaling amplifier*.

The special decoder called a D/A converter consists of two parts: a group of resistors forming a resistive ladder network and an op amp used as the summing amplifier.

Self-Test

Supply the missing word in each statement.

1. A special encoder that converts from analog to digital information is called a(n) _____ .

2. A special decoder that converts from digital to analog information is called a(n) _____ .

3. A D/A converter consists of a _____ network and a(n) _____ amplifier.

4. The name "op amp" stands for _____ .

13-2 OPERATIONAL AMPLIFIERS

The special amplifiers called *op amps* are characterized by high input impedance, low output impedance, and a variable voltage gain that can be set with external resistors. The symbol for an op amp is shown in Fig. 13-3(*a*). The op amp shown has two inputs. The top input is labeled an *inverting input*. The inverting input is shown by the minus sign ($-$) on the symbol. The other input is labeled a *noninverting input*. The noninverting input is shown by the plus sign ($+$) on the symbol. The output of the amplifier is also shown on the right of the symbol.

The operational amplifier is almost never used alone. Typically, the two resistors shown in Fig. 13-3(*b*) are added to the op amp to set the voltage gain of the amplifier. Resistor R_{in} is called the input resistor. Resistor R_f is called

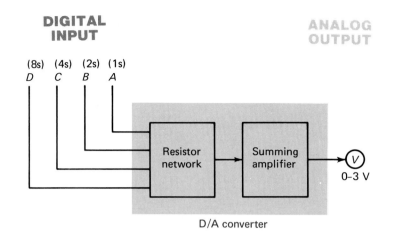

Fig. 13-2   Block diagram of a D/A converter.

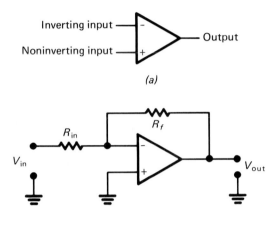

(a)

(b)

Fig. 13-3   Operational amplifier. (*a*) **Symbol.**
(*b*) **With input and feedback resistors setting gain.**

the feedback resistor. The *voltage gain* of this
amplifier is found by using the simple formula

$$A_v \text{ (voltage gain)} = \frac{R_f}{R_{\text{in}}}$$

Suppose the values of the resistors connected to the op amp are $R_f = 10$ kΩ and
$R_{\text{in}} = 10$ kΩ. Using our voltage gain formula,
we find that

$$A_v = \frac{R_f}{R_{\text{in}}} = \frac{10,000}{10,000} = 1$$

The gain of the amplifier is 1. In our example,
if the input voltage at V_{in} in Fig. 13-3(*b*) is
5 V, the output voltage at V_o is 5 V. The inverting input is being used, and so if the input
voltage is $+5$ V, then the output voltage is
-5 V. The voltage gain of the op amp can also
be calculated using the formula

$$A_v = \frac{V_o}{V_{\text{in}}}$$

The voltage gain for the circuit above is then

$$A_v = \frac{V_o}{V_{\text{in}}} = \frac{5}{5} = 1$$

The voltage gain is again found to be 1.
   Suppose the input and feedback resistors are
1 kΩ and 10 kΩ, as shown in Fig. 13-4 on the
next page. What is the voltage gain for this circuit? The voltage gain is calculated as

$$A_v = \frac{R_f}{R_{\text{in}}} = \frac{10,000}{1000} = 10$$

The voltage gain is 10. If the input voltage is
$+0.5$ V, then the voltage at the output is how
many volts? If the gain is 10, then the input
voltage of 0.5 V times 10 equals 5 V. The output voltage at V_o is -5 V, as measured on the
voltmeter in Fig. 13-4.
   You have seen how the voltage gain of an
op amp can be changed by changing the ratio
between the input and feedback resistors. You
should know how to set the gain of an opera-
tional amplifier by using different values for R_{in}
and R_f.
   In summary, the op amp is part of a D/A converter; it is used as a summing amplifier in the
converter. The gain of the op amp is easily
set by the ratio of the input and feedback
resistors.

Self-Test

Answer the following questions.

5. Refer to Fig. 13-3(*b*). The resistor labeled
   R_f in this op amp circuit is called the
   _____ resistor.

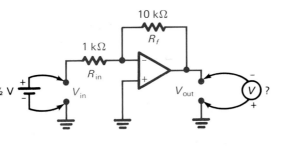

Fig. 13-4 Amplifier circuit using an op amp.

6. Refer to Fig. 13-3(b). The resistor labeled R_{in} in this op amp circuit is called the _____ resistor.
7. What is the voltage gain (A_v) of an op amp such as the one shown in Fig. 13-3(b) if $R_{in} = 1\ k\Omega$ and $R_f = 20\ k\Omega$?
8. What is the output voltage (V_o) from the op amp in question 7 if the input voltage is +0.2 V?

13-3 A BASIC D/A CONVERTER

A simple D/A converter is shown in Fig. 13-5. The D/A converter is made in two sections. The resistor network on the left is made up of resistors R_1, R_2, R_3, and R_4. The summing amplifier on the right consists of an op amp and a feedback resistor. The input (V_{in}) is 3 V applied to switches D, C, B, and A. The output voltage (V_o) is measured on a voltmeter. Notice that the op amp requires a rather unusual dual

power supply: a +10-V supply and a −10-V supply.

With all switches at GND (0 V), as shown in Fig. 13-5, the input voltage at point A is 0 V and the output voltage is 0 V. This corresponds to row 1, Table 13-1. Suppose we move switch A to the logical 1 position in Fig. 13-5. The input voltage of 3 V is applied to the op amp. We next calculate the gain of the amplifier. The gain is dependent upon the feedback resistor (R_f), which is 10 kΩ), and the input resistor (R_{in}), which is the value of R_1, or 150 kΩ. Using the gain formula, we have

$$A_v = \frac{R_f}{R_{in}} = \frac{10,000}{150,000} = 0.066$$

To calculate the output voltage, we multiply the gain by the input voltage as shown here:

$$V_o = A_v \times V_{in} = 0.066 \times 3 = 0.2\ V$$

The output voltage is 0.2 V when the input is binary 0001. This satisfies the requirements of row 2, Table 13-1.

Let us now apply binary 0010 to the D/A converter in Fig. 13-5. Switch B is moved to the logical 1 position, applying 3 V to the op amp. The gain is

$$A_v = \frac{R_f}{R_{in}} = \frac{10,000}{75,000} = 0.133$$

Multiplying the gain by the input voltage gives us 0.4 V. The 0.4 V is the output voltage. This satisfies row 3, Table 13-1.

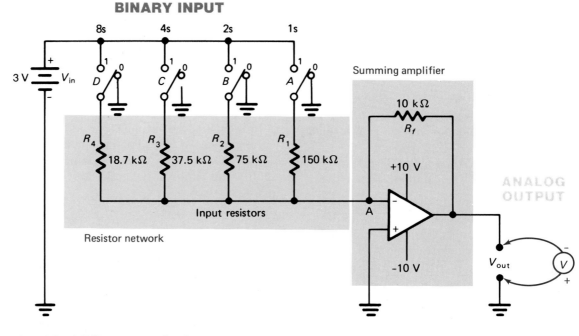

Fig. 13-5 A D/A converter circuit.

Notice that for each binary count in Table 13-1 the output voltage of the D/A converter increases by 0.2 V. This increase occurs because of the increased voltage gain of the op amp as we switch in different resistors (R_1, R_2, R_3, R_4). If only resistor R_4 from Fig. 13-5 were connected by activating switch D, the gain would be

$$A_v = \frac{R_f}{R_{in}} = \frac{10,000}{18,700} = 0.535$$

The gain multiplied by the input voltage of 3 V gives 1.6 V at the output of the op amp. This is what is required by row 9, Table 13-1.

When all switches are activated (at logical 1) in Fig. 13-5, the op amp puts out the full 3 V because the gain of the amplifier has increased to 1.

Any input voltage up to the limits of the operational amplifier power supply (± 10 V) may be used. More binary places may be added by adding switches. If a 16s place value switch is added in Fig. 13-5, it needs a resistor with half the value of resistor R_4. Its value would then have to be 9350 Ω. The value of the feedback resistor would also be changed to about 5 kΩ. The input would then contain a 5-bit binary number; the output would still be an analog output varying from 0 to 3 V.

The basic D/A converter shown in Fig. 13-5 does have two disadvantages: it takes a large range of resistor values and it has low accuracy.

Self-Test

Answer the following questions.

9. Calculate the voltage gain of the op amp in Fig. 13-5 when only switch C (the 4s switch) is at logical 1.
10. Using the voltage gain from question 9, calculate the output voltage of the D/A converter in Fig. 13-5 when only switch C is at logical 1.
11. List two disadvantages of the basic D/A converter shown in Fig. 13-5.

13-4 LADDER-TYPE D/A CONVERTERS

Digital-to-analog converters consist of a resistor network and a summing amplifier. Figure 13-6 on the next page diagrams a type of resistor network that provides the proper weighting for the binary inputs. This resistor network is sometimes called the *R-2R ladder* network. The advantage of this arrangement of resistors is that only two values of resistors are used. Resistors R_1, R_2, R_3, R_4, and R_5 are 20 kΩ each. Resistors R_6, R_7, R_8, and R_f are each 10 kΩ. Notice that all the horizontal resistors on the "ladder" are exactly twice the value of the vertical resistors, hence the title R-2R ladder network.

The summing amplifier in Fig. 13-6 is the same one used in the last section. Again notice the use of the dual power supply on the op amp.

The operation of this D/A converter is similar to the basic one in the last section. Table 13-2 details the operation of this D/A converter. Notice that we are using an input voltage of 3.75 V on this converter. Each binary count increases the analog output by 0.25 V, as shown in the rightmost column of Table 13-2. Remember that each 0 on the input side of the table means 0 V applied to that input. Each 1 on the input side of the table means 3.75 V applied to that input. The input voltage of 3.75 V is used because this is very close to the output of TTL counters and other ICs you may have used. The inputs (D, C, B, A) in Fig. 13-6, then, could be connected directly to the outputs of a TTL IC and operate according to Table 13-2. In actual practice, however, the

Table 13-2 Truth Table for D/A Converter

Binary Input				Analog Output
8s	4s	2s	1s	
D	C	B	A	Volts
0	0	0	0	0
0	0	0	1	0.25
0	0	1	0	0.50
0	0	1	1	0.75
0	1	0	0	1.00
0	1	0	1	1.25
0	1	1	0	1.50
0	1	1	1	1.75
1	0	0	0	2.00
1	0	0	1	2.25
1	0	1	0	2.50
1	0	1	1	2.75
1	1	0	0	3.00
1	1	0	1	3.25
1	1	1	0	3.50
1	1	1	1	3.75

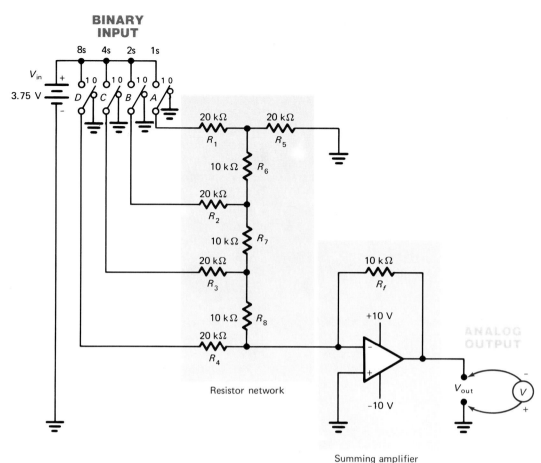

BINARY INPUT

Resistor network

Summing amplifier

Fig. 13-6 A D/A converter circuit using an R-2R ladder resistor network.

outputs of a TTL IC are not accurate enough; they have to be put through a level translator to get a very precise voltage output.

More binary places (16s, 32s, 64s, and so on) can be added to the D/A converter in Fig. 13-6. Follow the pattern of resistor values shown in this diagram when adding place values.

Two types of special decoders called digital-to-analog converters have been covered. The R-2R ladder-type D/A converter has some advantages over the more basic unit. The heart of the D/A converter consists of the resistor network and the summing amplifier.

Self-Test

Answer the following questions.

12. The digital-to-analog converter in Fig. 13-6 is a(n) _____ -type D/A converter.

13. Refer to Fig. 13-6. The gain of the op amp is greatest when all input switches are at logical _____ (0, 1).

14. Refer to Fig. 13-6 and Table 13-2. The gain of the op amp is *least* when switch _____ (A, B, C, D) is the only switch at logical 1.

13-5 AN A/D CONVERTER

An analog-to-digital converter is a special type of encoder. A basic block diagram of an A/D converter is shown in Fig. 13-7. The input is a single variable voltage. The voltage in this case varies from 0 to 3 V. The output of the A/D converter is in binary. The A/D converter translates the analog voltage at the input into a 4-bit binary word. As with other encoders, it is well to define exactly the expected inputs and outputs. The truth table in Table 13-3 shows how the A/D converter should work. Row 1 shows 0 V being applied to the input of the A/D converter. The output is binary 0000. Row 2 shows a 0.2-V input. The output is binary 0001. Notice that each increase of 0.2 V increases the binary count by 1. Finally, row 16 shows that when the maximum 3 V is ap-

A/D converter

Voltage
comparator

AND gate

BCD counter

D/A converter

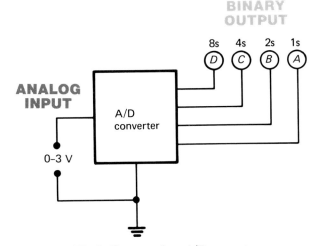

Fig. 13-7 Block diagram of an A/D converter.

plied to the input, the output reads binary 1111. Notice that the truth table in Table 13-3 is just the reverse of the D/A converter truth table in Table 13-1; the inputs and outputs have just been reversed.

Table 13-3 Truth Table for A/D Converter

	Analog Input	Binary Output			
		8s	4s	2s	1s
	Volts	D	C	B	A
Row 1	0	0	0	0	0
Row 2	0.2	0	0	0	1
Row 3	0.4	0	0	1	0
Row 4	0.6	0	0	1	1
Row 5	0.8	0	1	0	0
Row 6	1.0	0	1	0	1
Row 7	1.2	0	1	1	0
Row 8	1.4	0	1	1	1
Row 9	1.6	1	0	0	0
Row 10	1.8	1	0	0	1
Row 11	2.0	1	0	1	0
Row 12	2.2	1	0	1	1
Row 13	2.4	1	1	0	0
Row 14	2.6	1	1	0	1
Row 15	2.8	1	1	1	0
Row 16	3.0	1	1	1	1

The truth table for the A/D converter looks quite simple. The electronic circuits that perform the task detailed in the truth table are somewhat more complicated. One type of A/D converter is diagrammed in Fig. 13-8 on page 310. The A/D converter contains a *voltage comparator,* an AND gate, a BCD counter, and a D/A converter. All the sections of the A/D converter except the comparator are familiar to you.

The analog voltage is applied at the left of Fig. 13-8 on the next page. The comparator checks the voltage coming from the D/A converter. If the analog input voltage at *A* is *greater than* the voltage at input *B* of the comparator, the clock is allowed to *increase* the count of the BCD counter. The count on the counter increases until the feedback voltage from the D/A converter becomes greater than the analog input voltage. At this point the comparator stops the counter from advancing to a higher count. Suppose the input analog voltage is 2 V. According to Table 13-3, the binary counter increases to 1010 before it is stopped. The counter is reset to binary 0000, and the counter starts counting again.

Now for more detail on the A/D converter in Fig. 13-8. Let us assume that there is a logical 1 at point *X* at the output of the comparator. Also assume that the BCD counter is at binary 0000. Assume, too, that 0.55 V is applied to the analog input. The 1 at point *X* enables the AND gate, and the first pulse from the clock appears at the CLK input of the BCD counter. The counter advances its count to 0001. The 0001 is displayed on the lights in the upper right of Fig. 13-8. The 0001 is also fed back to the D/A converter.

According to Table 13-1, a binary 0001 produces 0.2 V at the output of the D/A converter. The 0.2 V is fed back to the *B* input of the com-

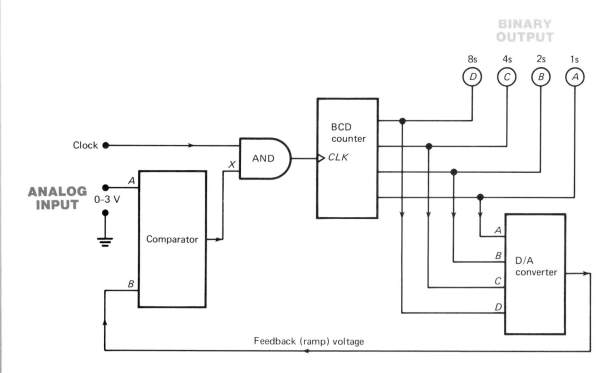

BINARY
OUTPUT

Fig. 13-8 **Block diagram of a counter-ramp-type A/D converter.**

parator. The comparator checks its inputs. The
A input is higher (0.55 V as opposed to 0.2 V),
and so that comparator puts out a logical 1. The
1 enables the AND gate, which lets the next
clock pulse through to the counter. The
counter advances its count by 1. The count is
now 0010. The 0010 is fed back to the D/A converter.

According to Table 13-1, a 0010 input produces a 0.4-V output. The 0.4 V is fed back to
the B input of the comparator. The comparator again checks the B input against the A input; the A input is still larger (0.55 V as
opposed to 0.4 V). The comparator puts out a
logical 1. The AND gate is enabled, letting the
next clock pulse reach the counter. The
counter increases its count to binary 0011. The
0011 is fed back to the D/A converter.

According to Table 13-1, a 0011 input produces a 0.6 V output. The 0.6 V is fed back to
the B input of the comparator. The comparator
checks input A against input B; for the first time
the B input is larger than the A input. The comparator puts out a logical 0. The logical 0 disables
the AND gate. No more clock pulses can reach
the counter. The counter stops at binary 0011.
Binary 0011 must then equal 0.55 V. A look at
row 4, Table 13-3, shows that 0.6 V gives the
readout of binary 0011. Our A/D converter has
worked according to the truth table.

If the input analog voltage were 1.2 V, the
binary output would be 0110, according to

Table 13-3. The counter would have to count
from binary 0000 to 0110 before being stopped
by the comparator. If the input analog voltage
were 2.8 V, the binary output would be 1110.
The counter would have to count from binary
0000 to 1110 before being stopped by the comparator. Notice that it does take some time for
the conversion of the analog voltage to a binary readout. However, in most cases the
clock runs fast enough so that this time lag is
not a problem.

You now should appreciate why we studied
the D/A converter before the A/D converter.
This *counter-ramp A/D converter* is fairly complex and needs a D/A converter to operate. The
term "ramp" in the name for this converter refers to the gradually increasing voltage from the
D/A converter that is fed back to the comparator. If you drew a graph of the voltage being
fed to input B of the comparator, it would appear as a *ramp* or a *sawtooth waveform*.

Self-Test

*Supply the missing word or words in each
statement.*

15. An A/D converter will translate a(n)
_____ input voltage into a(n)
_____ output.

16. Refer to Table 13-3. If the analog input voltage is 1 V, the binary output will be _____ .

17. Refer to Fig. 13-8. When point *B* is less than *A*, the output of the comparator at point *X* is _____ (HIGH, LOW). This causes the clock pulses to _____ (be blocked by, pass through) the AND gate.

18. The unit diagramed in Fig. 13-8 is a(n) _____ -type A/D converter.

13-6 COMPARATORS

In the last section we used a *comparator*. We found that a comparator compares two voltages and tells us which is the larger of the two. Figure 13-9 is a basic block diagram of a comparator. If the voltage at input *A* is larger than at input *B*, the comparator gives a logical 1 output. If the voltage at input *B* is larger than at input *A*, the output is a logical 0. This is written *A > B* = 1 and *B > A* = 0 in Fig. 13-9.

INPUTS   OUTPUT

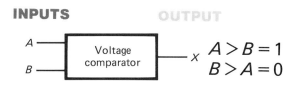

$$A > B = 1$$
$$B > A = 0$$

Fig. 13-9 Block diagram of a voltage comparator.

The heart of a comparator is an op amp. Figure 13-10(*a*) on the next page shows a comparator circuit. Notice that input *A* has 1.5 V applied and input *B* has 0 V applied. The output voltmeter reads about 3.5 V, or a logical 1.

Figure 13-10(*b*) shows that the input *B* voltage has been increased to 2 V. Input *A* is still at 1.5 V. Input *B* is larger than input *A*. The output of the comparator circuit is about 0 V (actually the voltage is about −0.6 V), or a logical 0.

The comparator in the A/D converter in Fig. 13-8 works exactly like this unit. The zener diode in the comparator in Fig. 13-10 is there to clamp the output voltage at about +3.5 and 0 V. Without the zener diode the output voltages would be about +9 and −9 V. The +3.5 and 0 V are more compatible with the TTL ICs you may have used.

Self-Test

Answer the following questions.

19. The comparator block shown in Fig. 13-8 compares two _____ (binary numbers, decimal numbers, dc voltages).

20. A voltage comparator circuit can be constructed using a(n) _____ IC, several resistors, and a zener diode.

21. Refer to Fig. 13-10 on the next page. When input *B* increases and becomes higher than input *A*, the output of the op amp will change from _____ (HIGH, LOW) to _____ (HIGH, LOW).

13-7 AN ELEMENTARY DIGITAL VOLTMETER

One use for an A/D converter is in a *digital voltmeter*. You have already used all the subsystems needed to make an elementary digital voltmeter system. A block diagram of a simple digital voltmeter is shown in Fig. 13-11 on the next page. The A/D converter converts the analog voltage to binary form. The binary is sent to the decoder, where it is converted to a seven-segment code. The seven-segment readout indicates the voltage in decimal numbers. With 7 V applied to the input of the A/D converter, the unit puts out binary 0111, as shown. The decoder activates lines *a* to *c* of the seven-segment display; segments *a* to *c* light on the display. The display reads as a decimal 7. Note that the A/D converter is also a decoder; it decodes from an analog input to a binary output.

A wiring diagram of a digital voltmeter is shown in Fig. 13-12 on page 313. Notice the voltage comparator, the AND gate, the counter, the decoder, the seven-segment display, and the D/A converter. Several power supplies are needed to set up this circuit. A dual ±10-V supply (or two individual 10-V supplies) is used for the 741 op amps. A 5-V supply is used for the 7408, 7493, and 7447 TTL ICs and the seven-segment LED display. A 0-to 10-V variable dc power supply is also needed for the analog input voltage.

Let us assume a 2-V input to the analog input of the digital voltmeter in Fig. 13-12. Reset the counter to 0000. The comparator checks inputs *A* and *B*; *A* is larger (*A* = 2 V, *B* = 0 V). The comparator output is a logical 1. This 1 enables the AND gate. The pulse from the clock passes through the AND gate. The pulse causes the counter to advance one count. The

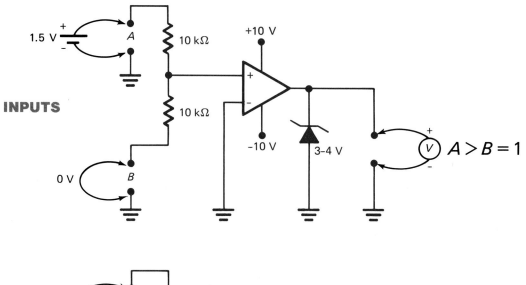

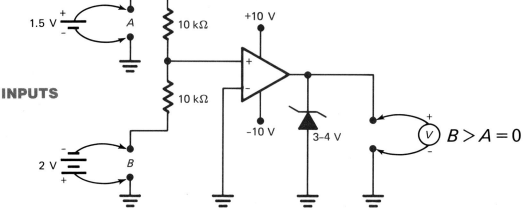

Fig. 13-10 Voltage comparator circuit. (*a*) With greater voltage at input *A*.
(*b*) With greater voltage at input *B*.

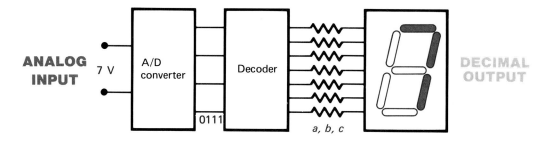

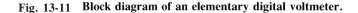

Fig. 13-11 Block diagram of an elementary digital voltmeter.

count is now 0001. The 0001 is applied to the decoder. The decoder enables lines *b* and *c* of the seven-segment display; segments *b* and *c* light on the display, giving a decimal readout of 1. The 0001 is also applied to the D/A converter. About 3.2 V from the counter is applied through the 150-kΩ resistor to the input of the op amp. The voltage gain of the op amp is

$$A_v = \frac{R_f}{R_{in}} = \frac{47,000}{150,000} = 0.31$$

The gain is 0.31. The voltage gain times the input voltage equals the output voltage:

$$V_o = A_v \times V_{in} = 0.31 \times 3.2 = 1 \text{ V}$$

The output voltage of the D/A converter is −1 V. The 1 V is fed back to the comparator.

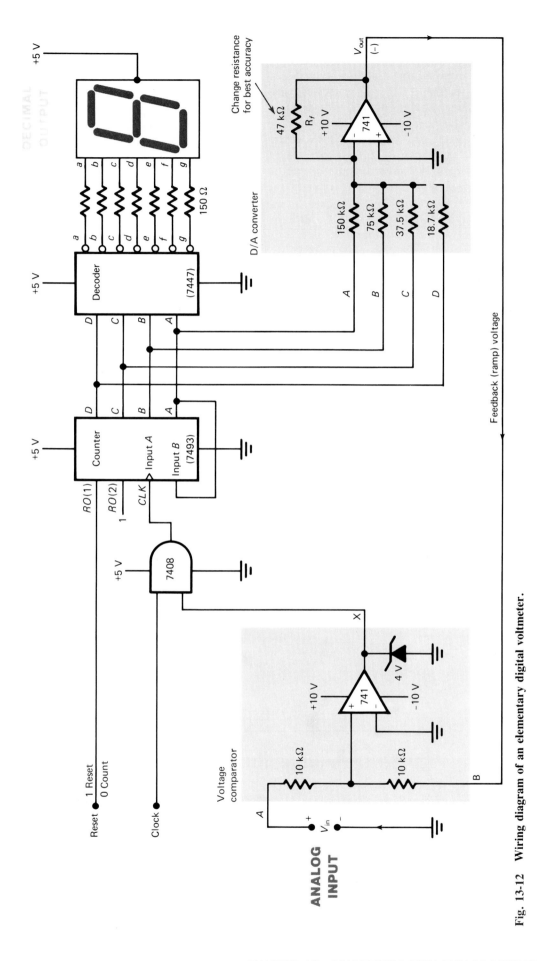

Fig. 13-12 Wiring diagram of an elementary digital voltmeter.

Now, with 2 V still applied to the input, the comparator checks A against B; input A is larger. The comparator applies a logical 1 to the AND gate. The AND gate passes the second clock pulse to the counter. The counter advances to 0010. The 0010 is decoded and reads out as a decimal 2 on the seven-segment display. The 0010 also is applied to the D/A converter. The D/A converter puts out about 2 V, which is fed back to the B input of the comparator.

The display now reads 2. The 2 V is still applied to input A of the comparator. The comparator checks A against B; B is just slightly larger. Output X of the comparator goes to logical 0. The AND gate is disabled. No clock pulses reach the counter. The count has stopped at 2 on the display. This is the voltage applied at the analog input.

The digital voltmeter in Fig. 13-12 is an experimental circuit. The circuit is included because it demonstrates the fundamentals of how a digital voltmeter works. It shows how SSI and MSI ICs can be used to build more complex functions. It is a simple example of a *hybrid* electronic system containing both digital and analog devices.

Modern digital voltmeters and DMMs are based on LSI ICs. These specialized A/D converters are available from many manufacturers. Large-scale-integrated digital voltmeter chips include all the active devices on a single CMOS IC. Included are A/D converters, seven-segment decoders, display drivers, and a clock.

Self-Test

Answer the following questions.

22. One application for an A/D converter is in a(n) _____ .
23. Refer to Fig. 13-12. The 7493 IC is wired as a mod- _____ counter in this circuit.
24. Refer to Fig. 13-12. The elementary digital voltmeter is considered a _____ (digital, hybrid) system because it contains both digital and analog ICs.
25. Refer to Fig. 13-12. With the counter *reset* to 0000, the feedback (ramp) voltage will be about _____ V.
26. Refer to Fig. 13-12. If the analog input voltage is 3.5 V and the counter is reset, how many clock pulses reach the 7493 IC before the counter stops?
27. Refer to Fig. 13-12. If the analog input voltage is 4.6 V, the display will read _____ V after the reset/count sequence.

13-8 OTHER A/D CONVERTERS

In Sec. 13-5 we studied the counter-ramp A/D converter. Several other types of A/D converters are also used; in this section we shall discuss two other types of converters.

A *ramp A/D converter* is shown in Fig. 13-13. This A/D converter works very much like the counter-ramp A/D converter in Fig. 13-8. The *ramp-generator* at the left in Fig. 13-13 is the

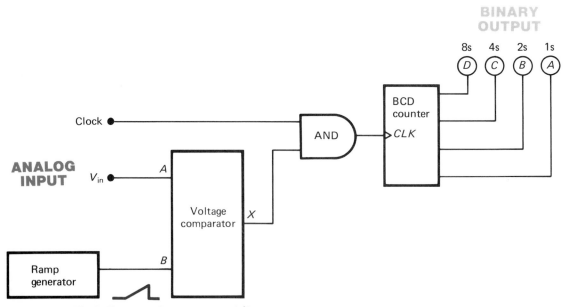

Fig. 13-13 Block diagram of a ramp-type A/D converter.

only new subsystem. The ramp generator produces a sawtooth waveform, which looks like a triangle-shaped wave in Fig. 13-14(a).

Suppose 3 V is applied to the analog voltage input of the A/D converter in Fig. 13-13. This situation is diagramed in Fig. 13-14(a). The ramp voltage starts to increase but is still lower than input A of the comparator. The comparator output is at a logical 1. This 1 enables the AND gate so that a clock pulse can pass through. In Fig. 13-14(a) the diagram shows three clock pulses getting through the AND gate before the ramp voltage gets larger than the input voltage. At point Y in Fig. 13-14(a) the comparator output goes to a logical 0. The AND gate is disabled. The counter stops counting at binary 0011. The binary 0011 means 3 V is applied at the input.

Figure 13-14(b) gives another example. The input voltage to the ramp-type A/D converter is 6 V in this situation. The ramp voltage begins to increase from left to right. The comparator output is at a logical 1 because input A is larger than the ramp generator voltage at input B. The counter continues to advance. At point Z on the ramp voltage the ramp genera-

tor voltage is larger than V_{in}. At this point the comparator output goes to a logical 0. This 0 disables the AND gate. The clock pulses no longer reach the counter. The counter is stopped at binary 0110. The binary 0110 stands for the 6-V analog input.

The difficulty with ramp-type A/D converters is the long time it takes to count up to higher voltages. For instance, if the binary output were eight binary places, the counter might have to count up to 255. To eliminate this slow conversion time we use a different type of A/D converter. A converter that cuts down on conversion time is a *successive-approximation A/D converter*.

A block diagram of a successive-approximation A/D converter is shown in Fig. 13-15 on the next page . The converter consists of a voltage comparator, a D/A converter, and a new logic block. The new logic block is called the successive-approximation logic section.

Suppose we apply 7 V to the analog input. The successive-approximation A/D converter first makes a "guess" at the analog input voltage. This guess is made by setting the MSB to 1. This is shown in block 1, Fig. 13-16. This

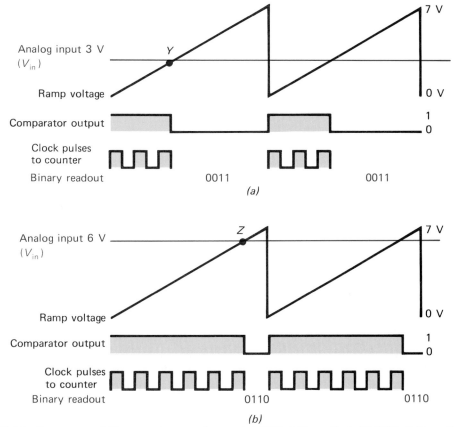

Fig. 13-14   Ramp-type A/D converter waveforms. (*a*) **With 3 V applied.** (*b*) **With 6 V applied.**

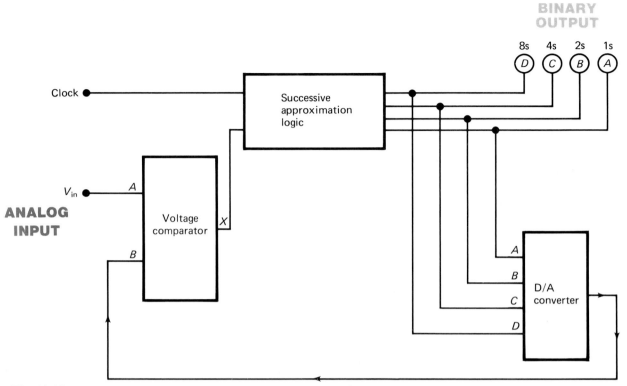

Fig. 13-15 Block diagram of a successive-approximation-type A/D converter.

job is performed by the successive-approximation logic unit. The result (1000) is fed back to the comparator through the D/A converter. The comparator answers the question in block 2, Fig. 13-16: is 1000 high or low compared with the input voltage? In this case the answer is "high." The successive-approximation logic then performs the task in block 3. The 8s place is cleared to 0, and the 4s place is set to 1. The result (0100) is sent back to the comparator unit through the D/A converter. The comparator next answers the question in block 4: is 0100 high or low compared with the input voltage? The answer is "low." The successive-approximation logic then performs the task in block 5. The 2s place is set to 1. The result (0110) is sent back to the comparator. The comparator answers the question in block 6: is 0110 high or low compared with the input voltage? The answer is "low." The successive-approximation logic then performs the task in block 7. The 1s place is set to 1. The final result is binary 0111. This stands for the 7 V applied at the input of the A/D converter.

Notice in Fig. 13-16 that the items in the blocks are performed by the successive-approximation logic unit. The questions are answered by the comparator. Also notice that the task performed by the successive-approxi-

mation logic depends upon whether the answer to the previous question is "low" or "high" (see blocks 3 and 5).

The advantage of the successive-approximation A/D converter is that it takes fewer guesses to get the answer. The *digitizing* process is thus faster. The successive-approximation A/D converter is very widely used.

Self-Test

Answer the following questions.

28. List three types of A/D converter circuits.
29. The counter-ramp A/D converter uses a D/A converter to generate the ramp voltage fed back to the comparator, whereas the ramp type uses a _____ to do this job.
30. The successive-approximation A/D converter is _____ (faster, slower) than ramp-type units.
31. Refer to Fig. 13-13. The ramp generator produces a _____ (sawtooth, square) waveform.
32. Refer to Fig. 13-13. If the input voltage (V_{in}) is 2 V and the ramp voltage is 0 V, the output of the voltage comparator is

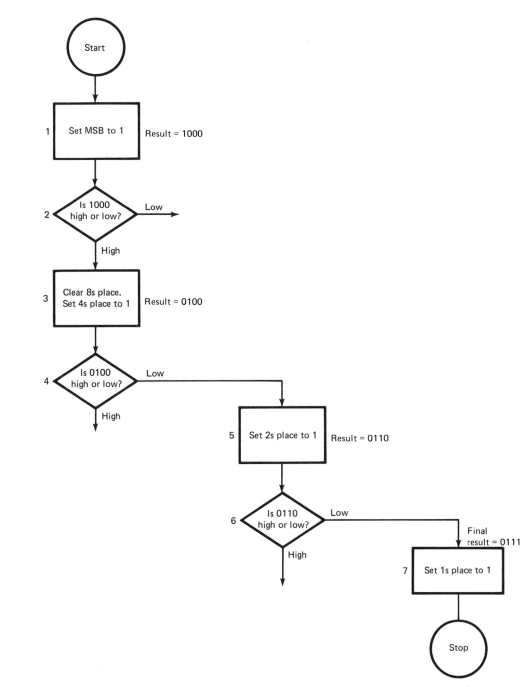

Fig. 13-16   Flowcharting the operation of the successive-approximation-type A/D converter.

_____ (HIGH, LOW) and the AND gate allows clock pulses to pass to the counter.

33. Refer to Fig. 13-15. When starting a new conversion, the _____ first sets the MSB to 1 and the voltage comparator checks to see if the input voltage (V_{in}) is higher than the feedback voltage at B.

13-9 A/D CONVERTER SPECIFICATIONS

Manufacturers produce a wide variety of A/D converters. One recent publication lists over 300 different A/D converters fabricated by many manufacturers.

Some of the more common specifications of A/D converters are detailed on the next page.

TYPE OF OUTPUT

Generally, A/D converters are classified as having either binary or decimal outputs. Analog-to-digital converters with decimal outputs are commonly used as digital voltmeters and used in digital panel meters and DMMs. Analog-to-digital converters with binary outputs have from 4 to 16 outputs. Analog-to-digital converters with binary outputs are common input devices to microprocessor-based systems. The latter are sometimes referred to as *μP-type A/D converters*.

RESOLUTION

The *resolution* of an A/D converter is given as the *number of bits* at the output for a binary-type unit. For decimal-output A/D converters (used in DMMs), the resolution is given as the *number of digits* in the readout (such as 3½ or 4½). Typical A/D converters with binary outputs have resolutions of 4, 6, 8, 10, 12, 14, and 16 bits. The slight errors that occur due to the use of discrete binary steps to represent a continuous analog voltage are called *quantizing errors*.

A 16-bit A/D converter is more "accurate" than a 4-bit unit because it divides the input or reference voltage into smaller discrete steps. For instance, each step in a 4-bit A/D converter would be one-fifteenth ($2^4 - 1 = 15$) of the input voltage. This would be a resolution of 6.7 percent ($1/15 \times 100 = 6.7$ percent). However, an 8-bit A/D converter would have finer increments. An 8-bit unit would have 255 ($2^8 - 1 = 255$) discrete steps. This would equal a resolution of 0.39 percent ($1/255 \times 100 = 0.39$ percent). The 8-bit unit has better resolution or "accuracy" than the 4-bit A/D converter.

ACCURACY

The resolution of an A/D converter can be thought of as the inherent "digital" error due to the discrete steps available at the output of the IC. Another source of error in an A/D converter might be the analog component, such as the comparator. Other errors might be introduced by the resistor network. The overall precision of an A/D converter is called the *accuracy* of the A/D converter IC.

The accuracy of an A/D converter IC with binary outputs ranges from ±½ LSB to ±2 LSB. The accuracy of an A/D converter IC with decimal output might range from 0.01 to 0.05 percent.

CONVERSION TIME

The *conversion time* is another important specification of an A/D converter. It is the time it takes for the IC to convert the analog input voltage to binary (or decimal) data the outputs. Typical conversion times range from 0.05 to 100,000 μs for A/D converter ICs with binary outputs. Conversion times for A/D converters with decimal outputs are somewhat longer and might typically be 200 to 400 ms.

OTHER SPECIFICATIONS

Four other common characteristics given for A/D converters are the power supply voltage, output logic levels, input voltage, and maximum power dissipation. Power supply voltages are commonly +5 V. However, some A/D converter ICs operate on voltages from +5 to +15 V. The output logic levels are either TTL, CMOS, or tristate. The input voltage range is commonly 5 V. Maximum power dissipation for an A/D converter IC might be in a range from about 15 to 3000 mW.

Self-Test

Supply the missing word or number in each statement.

34. An A/D converter with binary outputs is sometimes referred to as a _____ (meter, μP) -type unit.
35. The _____ of an A/D converter is given as the number of bits at the output of a binary-type unit.
36. An 8-bit A/D converter has greater resolution than a _____ (4, 12) -bit chip.
37. A typical conversion time for an A/D converter might be about _____ (110 μs, 1 s).
38. A typical A/D converter might have a maximum power dissipation of about _____ (850 mW, 10 μW).
39. The conversion time for meter-type A/D converters is _____ (longer, shorter) than for μP-type units.

13-10 AN A/D CONVERTER IC

A commercial A/D converter IC will be featured in this section. Figure 13-17(*a*) shows the pin diagram for an ADC0804 8-bit A/D converter IC. The table in Fig. 13-17(*b*) lists the name and function of each pin on the ADC0804 IC.

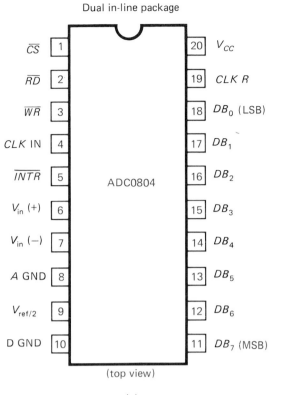

Dual in-line package

$\overline{CS}$	1	20	V_{CC}
$\overline{RD}$	2	19	$CLK\ R$
$\overline{WR}$	3	18	DB_0 (LSB)
CLK IN	4	17	DB_1
$\overline{INTR}$	5	16	DB_2
V_{in} (+)	6	15	DB_3
V_{in} (−)	7	14	DB_4
A GND	8	13	DB_5
$V_{ref/2}$	9	12	DB_6
D GND	10	11	DB_7 (MSB)

ADC0804

(top view)

(a)

ADC0804 A/D converter IC

Pin No.	Symbol	Input/Output or Power	Description
1	$\overline{CS}$	Input	Chip select line from µP-control
2	$\overline{RD}$	Input	Read line from µP-control
3	$\overline{WR}$	Input	Write line from µP-control
4	CLK IN	Input	Clock
5	$\overline{INTR}$	Output	Interrupt line goes to µP interrupt input
6	V_{in} (+)	Input	Analog voltage (positive input)
7	V_{in} (−)	Input	Analog voltage (negative input)
8	A GND	Power	Analog ground
9	$V_{ref/2}$	Input	Alternate voltage reference (+)
10	D GND	Power	Digital ground
11	DB_7	Output	MSB data output
12	DB_6	Output	Data output
13	DB_5	Output	Data output
14	DB_4	Output	Data output
15	DB_3	Output	Data output
16	DB_2	Output	Data output
17	DB_1	Output	Data output
18	DB_0	Output	LSB data output
19	CLK R	Input	Connect external resistor for clock
20	V_{CC} (or ref)	Power	Positive of 5-V power supply and primary reference voltage

(b)

Fig. 13-17   ADC0804 A/D converter IC. (*a*) **Pin diagram.** (*b*) **Pin labels and functions.**
(*Courtesy of National Semiconductor Corporation*)

The ADC0804 A/D converter was designed to interface directly with the 8080, 8085, or Z80 microprocessors. Some pin labels on the ADC0804 IC correspond to pins on popular microprocessors. For instance, the ADC0804 uses $\overline{RD}$, $\overline{WR}$, and $\overline{INTR}$ as pin labels which correspond to the *RD*, *WR*, and *INTR* pins on the 8085 microprocessor. The ADC0804 can also be interfaced with other popular 8-bit microprocessors such as the 6800 and 6502. The $\overline{CS}$ control input to the ADC0804 A/D converter receives its signal (chip select) from the microprocessor address-decoding circuitry.

The ADC0804 is a CMOS 8-bit successive-approximation A/D converter. It has three-state outputs so that it can interface directly with a microprocessor-based system data bus. The ADC0804 has binary outputs and features a short conversion time of only 100 μs. Its inputs and outputs are both MOS- and TTL-compatible. It has an on-chip clock generator. The on-chip generator does need two external components (resistor or capacitor) to operate. The ADC0804 IC operates on a standard +5-V dc power supply and can encode input analog voltages ranging from 0 to 5 V.

The ADC0804 A/D converter IC can be tested using the circuit shown in Fig. 13-18. The function of the circuit is to encode the difference in voltage between $V_{in}(+)$ and $V_{in}(-)$ compared to the reference voltage (5.12 V in this example) to a corresponding binary value. For instance, the resolution of the ADC0804 IC is 8 bits or 0.39 percent. This means that for each 0.02 V (5.12 V × 0.39 percent = 0.02 V) increase in voltage at the analog inputs the binary count increases by 1.

The "start switch" in Fig. 13-18 is first closed and then opened to start this free-running A/D converter. It is "free-running" because it continuously converts the analog input to digital outputs. The start switch should be left open once the A/D converter is operating. The $\overline{WR}$ input can be thought of as a clock input with the interrupt output ($\overline{INTR}$) pulsing the $\overline{WR}$ input at the end of each analog-to-digital conversion. A L-to-H transition of the signal at the $\overline{WR}$ input starts the A/D conversion process. When the conversion is finished, the binary display is updated and the $\overline{INTR}$ output emits a negative pulse. The negative interrupt pulse is fed back to clock the $\overline{WR}$ input and it initiates another A/D conversion. The circuit in Fig. 13-18 will perform about 5,000 to 10,000 conversions per second. The conversion rate of the ADC0804 is high because it uses the successive-approximation technique in the conversion process.

The resistor (R_1) and capacitor (C_1) connected to the *CLK R* and *CLK IN* inputs to the ADC0804 IC in Fig. 13-18 cause the internal clock to operate. The data outputs (DB7-DB0) drive the LED binary displays. The data outputs are active HIGH three-state outputs.

What is the binary output in Fig. 13-18 if the analog input voltage is 1.0 V? Recall that each 0.02 V equals a single binary count. Dividing 1.0 V by 0.02 V equals 50 in decimal. Converting decimal 50 to binary equals 00110010_2. The output indicators will show binary 00110010 (LLHHLLHL).

Self-Test

Supply the missing word in each statement.

40. The ADC0804 A/D converter is manufactured using _____ (CMOS, TTL) technology.
41. The ADC0804 IC is a _____ (meter, microprocessor) -type A/D converter.
42. The ADC0804 is an A/D converter with a resolution of _____ .
43. The ADC0804 IC's inputs and outputs meet both MOS and _____ voltage-level specifications.
44. The conversion time for the ADC0804 IC is about _____ (100 μs, 400 ms).
45. Refer to Fig. 13-18. Components R_1 and C_1 are used by the ADC0804 IC's internal _____ (clock, comparator).
46. Refer to Fig. 13-18. If the analog input voltage is 2.0 V, the binary output is _____ .
47. Refer to Fig. 13-18. An _____ (H-to-L, L-to-H) signal at the $\overline{WR}$ input to the ADC0804 IC starts a new A/D conversion.
48. Refer to Fig. 13-18. What output terminal of the ADC0804 IC produces a negative pulse immediately after each A/D conversion?

13-11 DIGITAL LIGHT METER

The A/D converter is the electronic device used to encode analog voltages to digital form. These analog voltages are often generated by transducers. For instance, light intensity may be converted to a variable resistance using a photocell.

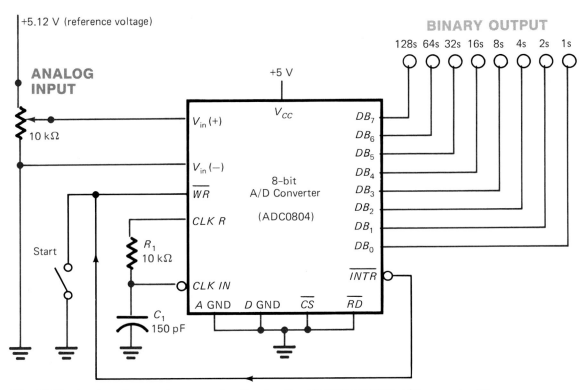

Fig. 13-18 Wiring diagram for a test circuit using the ADC0804 CMOS A/D converter IC.

A schematic diagram for a basic *digital light meter* is drawn in Fig. 13-19 on the next page. The ADC0804 IC is wired as a free-running A/D converter as in the last section. The push-button switch is pressed only once to start the A/D converter. The analog input voltage is being measured across resistor R_2. The photocell (R_3) is the light sensor or transducer in this circuit. As the light intensity increases, the resistance of the photocell (R_3) decreases. Decreasing the resistance of R_3 causes an increase in current through series resistances R_2 and R_3. The increased current through R_2 causes a proportional increase in the voltage drop across the resistor. The voltage drop across R_2 is the analog input voltage to the A/D converter. An increase in the analog input voltage causes an increase in the reading at the binary outputs.

The *cadmium sulfide photocell* used in Fig. 13-19 is a variable resistor. As the intensity of the light striking the photocell increases, its resistance decreases. The Radio Shack 276-116A photocell specified in Fig. 13-19 has a maximum resistance of about 500 kΩ and a minimum of about 100 Ω. The 276-116A cadmium sulfide photocell is most sensitive in the green-to-yellow portion of the light spectrum. The photocell is also referred to as a *photoresistor* or a *photoresistive cell*.

Other photocells may be substituted for the 276-116A unit specified in Fig. 13-19. If the substitute photocell has different resistance specifications, you will have to change the value of resistor R_2 in the light meter circuit to properly scale the binary output.

A second digital light meter circuit is drawn in Fig. 13-20 on page 323. This light meter indicates the relative brightness of the light striking the photocell in decimal (0 to 9). The new light meter is similar to the circuit in Fig. 13-19. The new light meter has a clock added to the circuit. The clock consists of a 555 timer IC, two resistors, and a capacitor wired as an astable MV. The clock generates a TTL output with a frequency of about 1 Hz. This means the analog input voltage is only converted into digital form one time per second. The very low conversion rate keeps the output from ''jittering'' between two readings on the seven-segment LED display.

The 7447A IC decodes the four MSBs (DB7, DB6, DB5, DB4) from the output of the ADC0804 A/D converter. The 7447A IC also drives the segments on the seven-segment LED display. The seven 150-Ω resistors between the 7447A IC and seven-segment LED display limit the current through an ''on'' segment to a safe level.

As in the previous circuit (Fig. 13-19), the output of the new light meter may have to be

INPUTS

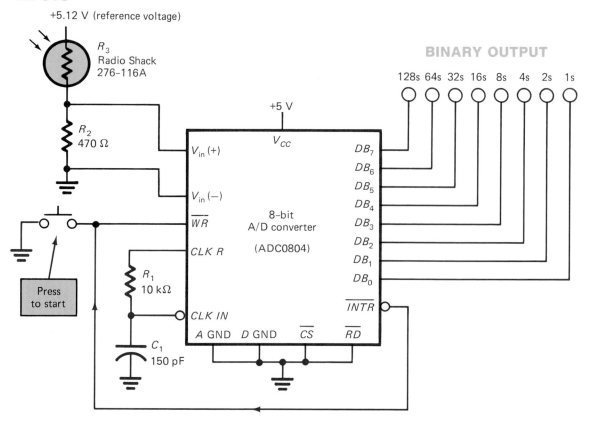

Fig. 13-19 Wiring diagram for a digital light meter circuit using binary outputs.

scaled so that low light reads 0 and high light intensity reads 9 on the seven-segment LED display. The value of resistor R_2 can be changed to scale the output. If R_2 is substituted with a lower-value resistor, the decimal output will read lower for the same light intensity. However, if the resistance value of R_2 is increased, the output will read higher.

Self-Test

Supply the missing word or number in each statement.

49. Refer to Fig. 13-19. As the light intensity striking the surface of the photocell increases, the binary value at the output of the light meter circuit _____ (decreases, increases).
50. Refer to Fig. 13-19. As the light intensity striking the surface of the photocell increases, the resistance of the photocell _____ (decreases, increases).
51. Refer to Fig. 13-20. If current through series resistances R_2 and R_3 increases, the analog input voltage to the A/D converter _____ (decreases, increases).
52. Refer to Fig. 13-20. The conversion rate of the ADC0804 IC in this digital light meter circuit is about _____ (1, 400) A/D conversion(s) per second.
53. Refer to Fig. 13-20. Substituting R_2 with a resistor of a lower ohmic value would cause the output display to read _____ (higher, lower) for the same light intensity.
54. Refer to Fig. 13-20. The part labeled R_3 in the light meter circuit is a _____ (transducer, transformer) that converts light intensity into a variable resistance.
55. Refer to Fig. 13-20. The component labeled R_3 is a cadmium _____ .

13-12 DIGITAL VOLTMETER

At least one manufacturer of chips groups A/D converters as being either microprocessor type or display type. The display-type A/D converters are used in constructing digital voltmeters, digital thermometers, and digital multimeters.

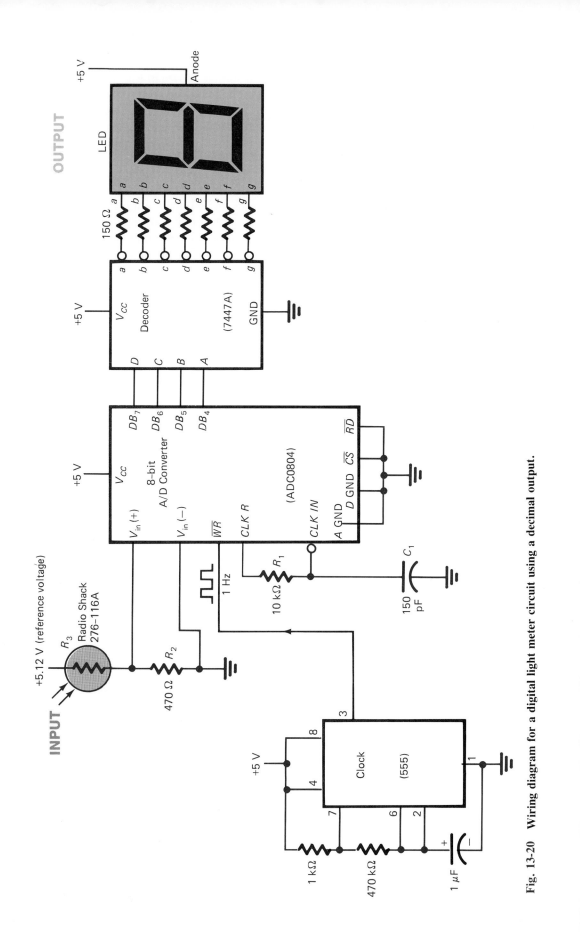

Fig. 13-20 Wiring diagram for a digital light meter circuit using a decimal output.

One display-type A/D converter will be featured in this section. Pin configurations for the Intersil ICL7106 3½-digit LCD single-chip A/D converter are detailed in Fig. 13-21. This complex CMOS IC is packaged in both a DIP and the newer *surface-mount package*. The top view of both IC package styles is shown in Fig. 13-21. Notice the location of pin 1 on the surface-mount package in Fig. 13-21(*b*). Pin 1 is immediately counterclockwise from the dot at the upper left corner of the package. Note that the pin numbering is *not the same* on the DIP and the surface-mount package.

The ICL7106 A/D converter needs only 10 external passive components plus an LCD display to make an accurate 3½-digit panel meter. A schematic diagram of such a panel meter using a 3½-digit LCD is drawn in Fig. 13-22(*a*). The panel meter detailed in Fig. 13-22(*a*) is available as an *evaluation kit* (ICL7106 EV/KIT) from Intersil, Inc. A sketch of a completed evaluation kit is shown in Fig. 13-22(*b*). The circuit shown in Fig. 13-22(*a*) will measure voltage from 0 to 200.0 mV.

The limited voltage range (0 to 0.2 V) of the panel meter in Fig. 13-22 can be extended. The easiest method of extending the range of the voltmeter is illustrated in Fig. 13-23 on page 326. The input voltage (0 to 19.99 V) is divided by 100 by the series resistors R_1 and R_2. The panel meter has a potentiometer used to adjust the reference voltage for accurate calibration even if resistors R_1 and R_2 are not extremely accurate. External components can be added to the panel meter in Fig. 13-22 for measuring current, ac voltage, and resistance.

The ICL7106 A/D converter is manufactured using CMOS technology. It typically consumes less than 10 mW of power and operates on a single 9-V battery. The ICL7106 features a built-in clock, voltage reference, decoders, and direct display drivers for 3½-digit seven-segment LCDs.

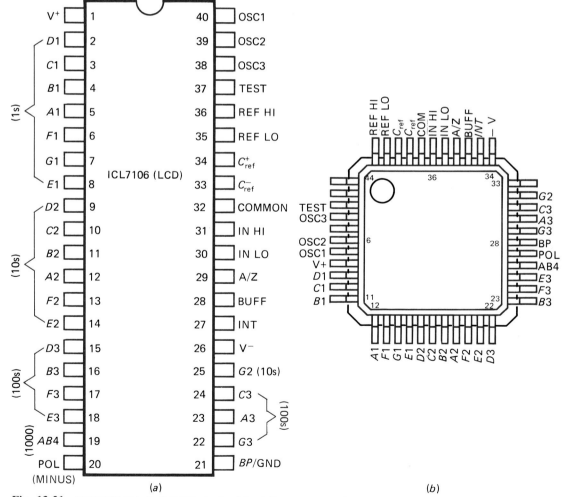

Fig. 13-21 ICL7106 3½-digit LCD single-chip A/D converter IC. (*a*) Pin diagram—dual in-line package. (*b*) Pin diagram—surface-mount package. (*Courtesy of Intersil, Inc.*)

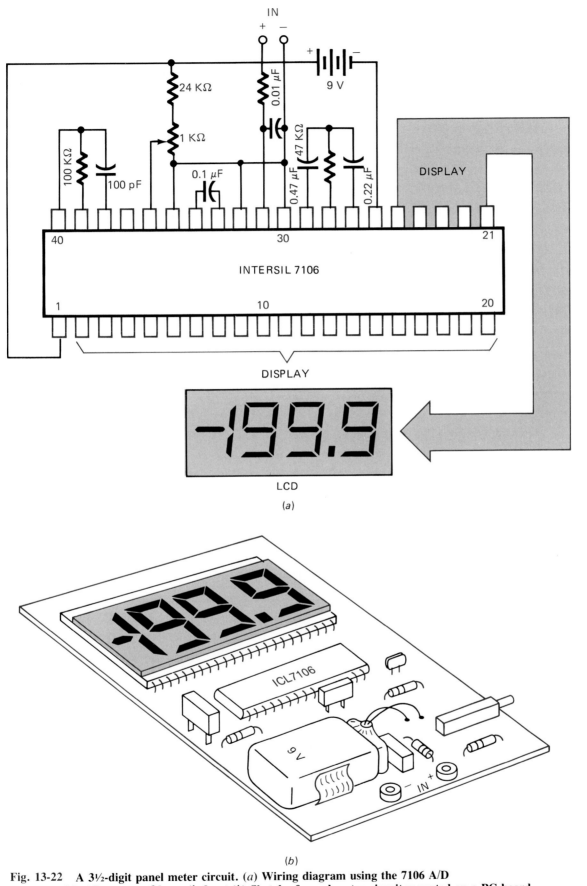

Fig. 13-22   A 3½-digit panel meter circuit. (*a*) **Wiring diagram using the 7106 A/D
converter IC.** *(Courtesy of Intersil, Inc.)* (*b*) **Sketch of panel meter circuit mounted on a PC board.**

The A/D converter is very accurate and has an auto zero feature and high input impedance.

Another version of the ICL7106 IC is the ICL7107 3½-digit LED single-chip A/D converter. It performs the same as the ICL7106 unit except it requires a +5 V/−5 V power supply. The ICL7107 will directly drive a 3½-digit LED display. These chips can also be used to design digital thermometers.

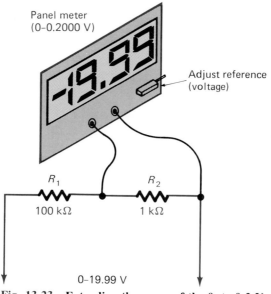

Fig. 13-23 **Extending the range of the 0- to 0.2-V panel meter using external resistors.**

Self-Test

Supply the missing word(s) in each statement.

56. Besides the dual-in-line packaging of chips, a newer _____ package is becoming more popular.
57. The ICL7106 IC is a _____ (display, microprocessor) -type A/D converter.
58. The ICL7106 IC contains internal decoding and drivers for a 3½-digit _____ (LED display, LCD).
59. The ICL7106 IC operates on a _____ (9 V, +5 V/−5 V) power supply.
60. Refer to Fig. 13-23. If the voltage range is to be extended to measure 2.000 V and if $R_1 = 100$ kΩ, then resistor R_2 would be _____ kΩ.

SUMMARY

1. Special interface encoders and decoders are used between analog and digital devices. These are called D/A converters and A/D converters.
2. A D/A converter consists of a resistor network and a summing amplifier.
3. Operational amplifiers are used in D/A converters and comparators. Gain can be easily set with external resistors on the op amp.
4. Several different resistor networks are used for weighting the binary input to a D/A converter.
5. Common A/D converters are the counter-ramp, ramp-generator, and successive-approximation types.
6. A comparator compares two voltages and determines which is larger. An operational amplifier is the heart of the comparator.
7. Common specifications used for A/D converters include such characteristics as type of output, resolution, accuracy, conversion time, power supply voltage, output logic levels, input voltage, and power dissipation.
8. The ADC0804 IC is a CMOS 8-bit A/D converter. It features fast conversion times, microprocessor compatibility, three-state outputs, TTL level inputs and outputs, and an on-chip clock.
9. A photocell can be used as a transducer to drive an A/D converter in a digital light meter circuit.
10. An A/D converter is at the heart of a digital voltmeter. Most commercial digital voltmeters and DMMs use complex meter-type A/D converter LSI ICs.

CHAPTER REVIEW QUESTIONS

Answer the following questions.

13-1. An A/D converter is a special type of _____ (decoder, encoder).

13-2. A D/A converter is a(n) _____ (decoder, encoder).

13-3. The _____ (A/D, D/A) converter digitizes information.

13-4. The _____ (A/D, D/A) converter translates from binary to an analog voltage.

13-5. A D/A converter consists of a _____ network and a summing _____ .

13-6. The term "operational amplifier" is frequently shortened to _____ .

13-7. The voltage gain of the operational amplifier in Fig. 13-3(*b*) is determined by dividing the value of _____ (R_f, R_{in}) by the value of _____ (R_f, R_{in}).

13-8. Draw a symbol for an operational amplifier. Label the inverting input with a minus sign and the noninverting input with a plus sign. Label the output. Label the +10-V and −10-V power supply connections.

13-9. Refer to Fig. 13-4. What is the gain (A_v) of the op amp in this diagram if $R_{in} = 1$ kΩ and $R_f = 100$ kΩ?

13-10. Refer to Fig. 13-4. With the input voltage at +½V, the output voltage is _____ (+, −) 5 V. This is because we are using the _____ (inverting, noninverting) input of the op amp.

13-11. Refer to Fig. 13-5. What is the voltage gain of the op amp in this circuit with only switch *A* at logical 1?

13-12. Refer to Fig. 13-5. What is the combined resistance of parallel resistors R_1 and R_2 if both switches *A* and *B* are at logical 1?

13-13. Refer to Fig. 13-5. What is the gain (A_v) of the op amp with switches *A* and *B* at logical 1? (Use the resistance value from question 12.)

13-14. Refer to Fig. 13-5. What is the output voltage when binary 0011 is applied to the D/A converter? (Use the A_v from question 13.)

13-15. The arrangement of resistors in Fig. 13-6 is called the _____ ladder network.

13-16. Compare Tables 13-1 and 13-2. The difference between the two is the _____ (binary inputs, scaling of the analog output).

13-17. A high, or logical 1, from a TTL device is about _____ (0, 3.75, 5.5) V.

13-18. The letters "TTL" stand for _____ .

13-19. The _____ (A/D, D/A) converter is the more complicated electronic system.

13-20. The counter-ramp A/D converter consists of an _____ (AND, OR) gate, a _____ (comparator, resistor network), a _____ (counter, shift register), and a D/A _____ .

13-21. Refer to Fig. 13-8. If point *X* is at a logical _____ (0, 1), the counter advances one count as a pulse comes from the clock.

13-22. Refer to Fig. 13-8. If input *B* of the comparator has a higher voltage than input *A*, the AND gate is _____ (disabled, enabled).

13-23. Refer to Fig. 13-9. If input *A* of the comparator equals 5 V and input *B* equals 2 V, then output *X* is a logical _____ (0, 1). Output *X* is about _____ (0, 4) V.

13-24. The primary component in a comparator is a(n) _____ (counter, op amp).

13-25. Refer to Fig. 13-12. This digital voltmeter uses a _____ (counter-ramp, successive-approximation) A/D converter.

13-26. A ramp A/D converter consists of an _____ (AND, OR) gate, a (counter, register), a ramp _____ , and a _____ (comparator, D/A converter).

13-27. The _____ (ramp, successive-approximation) A/D converter is faster at digitizing information.

13-28. Devices such as microphones, speakers, strain gauges, photocells, temperature sensors, and potentiometers convert one form of energy to another and are generally called _____ .

13-29. An A/D converter with binary outputs might be classified as a _____ (meter, microprocessor) -type unit.

13-30. Refer to Fig. 13-18. What is the resolution of the ADC0804 A/D converter?

13-31. A(n) _____ (8, 16) -bit A/D converter has a lower quantization error and is considered more "accurate."

13-32. Conversion times are somewhat longer for _____ (meter, microprocessor) -type A/D converters.

13-33. The ADC0804 (Fig. 13-17) has _____ (binary, decimal) outputs.

13-34. The A/D converter wired in Fig. 13-18 performs about _____ (3, 5 to 10,000) A/D conversions per second.

13-35. Refer to Fig. 13-18. If the analog input voltage is 3.0 V, the binary output is _____ .

13-36. Refer to Fig. 13-20. Decreasing the light intensity striking R_3 causes the resistance of the photocell to _____ (decrease, increase).

13-37. Refer to Fig. 13-20. Decreasing the light intensity striking the photocell causes the decimal output to _____ (decrease, increase).

13-38. The ADC0804 IC in Fig. 13-20 performs about _____ (1, 10,000) analog-to-digital conversion(s) per second.

13-39. Refer to Fig. 13-20. If current through series resistances R_2 and R_3 decreases, the analog input voltage to the A/D converter _____ (decreases, increases).

13-40. The ICL7106 IC is a _____ (meter, microprocessor) -type A/D converter.

13-41. Refer to Fig. 13-22. This digital panel meter can measure from 0 to _____ dc volts.

13-42. Refer to Fig. 13-23. The value of resistor R_1 would have to be increased to _____ Ω to extend the range of the panel meter to 200 V dc.

13-43. The ICL7106 A/D converter IC features a built-in clock, voltage reference, decoders, and direct display drivers for 3½-digit seven-segment _____ (LCDs, LED displays).

Answers to Self-Tests

1. analog-to-digital converter (A/D converter)
2. digital-to-analog converter (D/A converter)
3. resistor, summing (scaling)
4. operational amplifier
5. feedback
6. input
7. $A_v = 20$
8. $V_o = -4$ V
9. $A_v = 0.266$
10. $V_o = -0.8$ V
11. 1. low accuracy
    2. a large range of resistor values needed
12. ladder (R-2R ladder)
13. 1
14. A
15. analog, digital (binary)
16. 0101
17. HIGH, pass through
18. counter-ramp
19. dc voltages
20. op amp
21. HIGH, LOW
22. digital voltmeter
23. 10 (decade counter)
24. hybrid
25. 0
26. four
27. 5
28. 1. counter-ramp
    2. ramp
    3. successive-approximation
29. ramp generator

30. faster
31. sawtooth
32. HIGH
33. successive-approximation logic
34. μP (microprocessor)
35. resolution
36. 4
37. 110 μs
38. 850 mW
39. longer
40. CMOS
41. microprocessor
42. 8 bits (0.39 percent)
43. TTL
44. 100 μs
45. clock

46. 01100100_2
47. L-to-H
48. INTR
49. increases
50. decreases
51. increases
52. 1
53. lower
54. transducer
55. sulfide photocell
56. surface mount
57. display
58. LCD (liquid-crystal display)
59. 9 V
60. 10 kΩ

Index